Amplifications

Amplifications

Poetic Migration, Auditory Memory

PAUL CARTER

BLOOMSBURY ACADEMIC
NEW YORK · LONDON · OXFORD · NEW DELHI · SYDNEY

BLOOMSBURY ACADEMIC
Bloomsbury Publishing Inc
1385 Broadway, New York, NY 10018, USA
50 Bedford Square, London, WC1B 3DP, UK

BLOOMSBURY, BLOOMSBURY ACADEMIC and the Diana logo are trademarks
of Bloomsbury Publishing Plc

First published in the United States of America 2019

Copyright © Paul Carter, 2019

Cover design by Louise Dugdale
Cover image © Paul Carter

All rights reserved. No part of this publication may be reproduced or transmitted in any form or by any means, electronic or mechanical, including photocopying, recording, or any information storage or retrieval system, without prior permission in writing from the publishers.

Bloomsbury Publishing Inc does not have any control over, or responsibility for, any third-party websites referred to or in this book. All internet addresses given in this book were correct at the time of going to press. The author and publisher regret any inconvenience caused if addresses have changed or sites have ceased to exist, but can accept no responsibility for any such changes.

A catalog record for this book is available from the Library of Congress.

ISBN: HB: 978-1-5013-4447-3
PB: 978-1-5013-4448-0
ePDF: 978-1-5013-4450-3
eBook: 978-1-5013-4449-7

Typeset by Deanta Global Publishing Services, Chennai, India

To find out more about our authors and books visit www.bloomsbury.com and sign up for our newsletters.

For
Andrew McLennan

CONTENTS

Prolude 1

1 Charms 7

2 Returns 41

3 Rattles 81

4 Sirens 103

5 Echoes 135

6 Recordings 167

7 Voices 197

8 Callings 225

Collect 241

Notes 249
Index 291

Prolude

Voice 3: Every single word that's ever been spoken.
Voice 4: Who is it?
Voice 1: Every single sound that's ever been made since the world came into being.
Voice 2: Speaking. Who is it?
Voice 1: Is still here …[1]

When I arrived in Australia in the early 1980s, I had the sensation of entering a world of echoes. I had the feeling that there existed another place behind the one in which I stood and walked. The current physical arrangements seemed without foundations; normal appearances were defensively clear-cut as if repressing unease; there was a lurking puzzlement about the meaning of things, as if they existed in a mirror state, equal and opposite to the ideas they represented. There was a shortage of poetic associations. Human utterances lacked emotional resonance; either my own voice fell on deaf ears or those who addressed me seemed to borrow personae not their own. A shadow fell between physical presence and metaphysical projection because of a misfit between the history of the place and its spirits and the cultural theatre imposed on it in colonization. Although the sun was bright, I tunnelled through an underworld, like the deaf and blind feeling my way forward in the dark, discovering obstacles where others saw clear ways; the folk accounts of origins reported in what seemed to be a thin-reeded English existed like the wind in the grass ahead of storm. A heightening of the senses detected tiny fissures in reality, discrepancies in ordinary communication – distorted echoes or resonant amplifications, which tended to transmute the sign into the enigma of communication itself. I became aware of the phenomenon of echoic mimicry, a desire to ground human relations differently and to extend this solicitation to the environment.

In *The Road to Botany Bay* I treated these sensations visually and kinetically. I projected my disorientation backwards to reimagine Australian colonization. I described a history whose content was spatial rather than temporal because the names, definitions and associations of phenomena had yet to be fixed. The mismatch between the appearance of the country, the character of the people and the cultural templates laid over them was a history in itself – perhaps *the* distinctive history of colonization. It was spatial, as in this prehistory of places the work to be done was precisely one of navigation. To be all at sea in this way induces vertigo, but is liberating: when the ground slips away from under one's feet, there is the chance to reground oneself (and perhaps colonial footings) differently. The title of that book referred to a supposed escape route that convicts took from the Sydney Cove settlement: it described a return to the geographical site of Australia's colonization motivated by the desire to begin again, and differently. Approaching the 'desired port', First Fleet officer and annalist Watkin Tench had exclaimed, 'Ithaca itself was scarcely more longed for by Ulysses, than Botany Bay by the adventurers who had traversed so many thousands of miles to take possession of it.'[2] A short while later his comparison had acquired an ironic inflection. When Governor Phillip departed in December 1791, David Collins reported, 'The cove and the settlement were now resuming that dull uniformity of uninteresting circumstances which had generally prevailed.'[3] This is historical time and space as an 'immense tautology',[4] where the colonist is a sleepwalker or *hypnon*.[5] Now, Tench wrote, all was quiet but also *stupid*, and he longed for nightfall when he could retreat into dreams of transport to *another* country.[6] I recalled Emily Dickinson's resumé of imperial dreams: 'Finding is the first Act / The second, loss, / Third, Expedition for / The "Golden Fleece" / Fourth, no Discovery – Fifth, no Crew – / Finally, no Golden Fleece – / Jason – sham – too.'[7]

If space and time were intractable, communication, it seemed to me, could be changed. As I finished *The Road to Botany Bay*, I made my first foray into writing for radio. The focus of 'Memory as Desire' was an imaginary encounter between two figures I had studied while researching the book; but it located their meeting within a soundscape of birds and transposed the environmental perplexity that I had discussed into the auditory realm. Charlotte Sturt, wife of the explorer Charles Sturt, asks Aboriginal Protector George Augustus Robinson about Aboriginal languages. Robinson expresses surprise. Charlotte reflects, 'The tongue of our natives was soft, euphonious,' to which Robinson replies, 'Exactly, madam, music. But not language, not meaning, not what you could use. The birds made more sense.'[8] This view was, I later discovered, still held in parts of Australia in the 1960s. Set on the French side of the English Channel, 'Memory as Desire' began the transcription of my own sound history as it opens with the murmuration of starlings. After 'Memory as Desire' and 'What Is Your Name' – another colonial interrogation, although in a very different style – a number of radiophonic compositions[9] – 'Scarlatti', 'Remember Me' and 'On the Still Air' – explored migrant sound geographies: 'What is it like to live in the echo of other places? Hemi-sphered, half-voiced', the lover asks in 'Remember Me': 'Can't you write back?' and 'Whose voices do you use in your country?'[10] This sequence culminated in 'The Native Informant', where an ideal figure from ethnography becomes a metaphor for migrant nostalgia. Listening to sound recordings of his own voice, the protagonist of this piece chiefly notices the quality of the sound recording; aware of the desire implicit in his attempt to retrieve technologized memory, he experiences a kind of breakthrough: 'Beyond the imperialismo of signs .... The always significant birds .... There must be an art of sounds. ... A translation of movement....'[11]

These radio auditoria for echoes also had their colonial precedent. Echoic mimicry favoured a baroque mode of expression. The Renaissance polymath Francis Bacon pointed out, 'To make an Echo report three, four, or five Words distinctly, 'tis requisite that the reflecting Body be at a good distance.'[12] Imagine this as a geographical phenomenon; consider the multiplication of echoes achieved by the time words were received in the Antipodes. Add to this a situation where the distant colony desires to be like the other country; it would produce a cultural phenomenon comparable with that 'Echo upon Echo', whose repetition gradually dying away Bacon compares to the multiplication of images when two mirrors face each other: 'many such Super-reflexions, till the Images of the Objects fail, and die at last'.[13] In 1990 I incorporated these ideas into a sound installation called 'Mirror States',[14] but their anticipation is found in the First Fleet correspondence of George Worgan who, writing to his brother-in-law in England, imagined communication across such a distance physically, as one word pushing against another – 'I shall let fly each with such an impulsive Velocity against its leader, that by the time the last gives its Impulse, the whole will have received such an irresistable Velocity, as to make their Way against the resistance of Rocks, Seas and contrary Winds and arrive at your Street-Door with a D--l of a Suscitation.'[15] Time and again, existential and expressive parallels opened up between colonial first comer and contemporary migrant: Worgan's fantasy was likely inspired by his admiration for Scarlatti's helter-skelter sonatas.[16] The piano's sustaining pedal gives an impressed force to notes; the harpsichord's expressiveness depends entirely on the impulsive velocity of one note let fly after another.

Amplifying echoes of *The Road to Botany Bay*, the radiophonic experiments multiplied their own echoes. *The Sound In-Between*, where the script of 'Mirror States' appeared, proposed a sound geography. Going beyond the dramatization of 'the verbal gestures of first contact, the stumbling mimicry of the other person's speech', found in earlier radio, and explored in the early 1990s through multichannel museum installation, I imagined a sonic transcription of the information contained in colonial maps, an auditory landscape echoically known and mapped. Repatriated to the places that produce them, foreign echoes could contribute to a new acoustic ecology. I began the practice of reading in caves and listening to the sounds that came back. The installation practice also generated 'Echo upon Echo': in 'Columbus Echo' (1992), written and sound-designed for the new Acquario di Genova, a poetic criollismo evoked the interlingual coastlines of empire; 'The 7448' riffed on a passage from Marco Polo, creating an archipelago of journeys and voices that included a reprise of Ulysses's journey.[17] The keen interest of composer Luciano Berio in these glossolalic excursions produced a folktale multiplication of the Odyssey. Meanwhile, in Sydney, in soundscapes for Hyde Park Barracks and the new Museum of Sydney, a kind of integration of colonial amplifications and migrant poetics occurred. In scripts like 'Named in the Margin', 'Lost Subjects' and 'Cooee Song', lost voices are aired and, it is suggested, the qualitatively different histories borne on the sound of the voice. Comparing the colonial mindset to the atmosphere of the museum – both excel in the elimination of noise – 'Lost Subjects' suggested that a politics of renewal involved the migrant sensitivity to the *terra incognita* of language, a 'sneaking preference for *la fonetica poetica*'.[18]

In the following decade another change occurred, scarcely perceptible at first, eventually a landslide: the cultural, institutional and technological settings supporting these sound histories eroded and at length collapsed. 'Arguably', Virginia Madsen writes, 'without the opportunities afforded through [the radio programs] *Surface Tension* [1985–1987] and *The Listening Room* [1988–2003] the substantial and complex work of Paul Carter may not have emerged at all – his performed and recorded sound corpus predominantly involved A[ustralian] B[roadcasting] C[orporation] commissions'.[19] The appetite of cultural institutions for complex polyphonic soundscapes was also on the wane. In a *late* radio work, 'The Letter S', I reflected on the sense of an ending, but, characteristically, to do this I went back to the beginning, to another observation of Watkin Tench: the Gadigal of Sydney Cove, he wrote, could not pronounce the letters 's' or 'v': 'The latter became invariably w, and the former mocked all their efforts.'[20] On another occasion, he wrote, 'When bidden to pronounce *sun*, they always say *tun*; *salt*, *talt*; and so of all words wherein it occurs.'[21] 'The Letter S' was also a return visit to my own sound history: embedded in it is a passage from 'Lost Subjects':

Voice 5: Stop!
Voice 6: The natives never use the letter 'S'.
Voice 7: 'Top', they said, 'the man.'
Voice 5: And that was how Death came in.
Voice 6: Slurred speech has no place in the theatre.
Voice 7: But in the Museum it echoes.[22]

'S' in this work stood for *la fonetica poetica*, and I associated the new dumbing down of sound art with the reinstatement of the imperialismo of signs. Yet the absence of invention in 'The Letter S' was also striking: as a reworking of earlier scripts, it implied that history was cyclic. The vortical theory of historical change and renewal championed by Giambattista

Vico, influential in the formal experiments of Joyce and Beckett, had been ironically deployed in 'Underworlds of Jean du Chas', the last significant work for *The Listening Room*. But now the hiss was being taken out of history, the imperial schema anatomized by Dickinson shown to be a sham:

Actor 1: Kiss. Kiss Criss. Cross Criss. Kiss Cross. Undo lives' end. Slain.
Actor 3: And?
Actor 2: Again. Again, again, again.

An optimist, I seized on the evaporation of radio into digital listening platforms as another creative dispersal; my own immersion in public culture had already migrated from radio to public art; I looked forward to revisiting the radiophonic work in the context of a new dramaturgy of public space. After all, their echoic multiplication of voices was also a kind of choreography. I embarked on a new project called 'Siren Sonata', inspired by what I took to be the unfinish of the older radio piece 'Scarlatti'. The idea was to deepen the unfinish, to restore, as it were, the parts to their different origins and out of this second fragmentation to convene a new meeting place. The mise-en-scène of this experiment would be broadly speaking the city, and the measure of the work's resolution, a new kind of resonance detected in the feedback between the organization of the work and the physical arrangement of the locale. Something like atmosphere would be experienced, a milieu riding the chaos. The new post-radio function of radio would be the maintenance and amplification of the kind of communication we had championed in our acoustic art works. It would retrieve the original revolutionary purpose of putting the noise of dissidence back into history. Beyond the broad mediation between the inside and the outside of the work – the weaving of the fictive plot or score into the everyday rhythms of the street, which is the dramaturg's responsibility – I conceived of the artist's role in terms of media tactics, as the re-territorialization of listening and the fostering of a kind of rhythmic synchrony or collective entrainment. But ahead of that, I realized I needed (again) to retrace my steps; for no acting scripts of writing for radio existed; where they had been installed, the institutional soundscapes hardly survived. Equally important, the raw studio recordings, the archival sounds – in short, all the materials for composition – were scattered and needed to be recovered and reassembled.

So I undertook that acoustic archaeology in the hope it would not prove to be an autopsy. In the event it proved to be neither but, instead, the partial resurrection of an entire buried web of auditory memories, sound histories and acoustic ecological speculation. I was surprised to discover an echoic mimicry *within* that body of work – a continuous shuttling back and forth between creative writing and production and the publication of essays about the dynamics of cross-cultural communication, about the poetics and politics of different theories of language's origins, the phenomenon of absolute pitch or the importance of noise in the echolocative orientation to place. But rich as this amplification was, it also exposed a maze of connections that remain unmapped. There was the question of origins – my father's voice, mediated through sound cassettes when I moved to Australia; my mother's ear, all that she and I heard in the last few weeks of her life; and behind both the constitutive ocean of birdcalls coming up to me in the spring mornings of childhood, definitely heralding the advent of a new sonorous body. There was the enigma of echoes carried across generations and traditions, as when an impulse from beyond the grave seemed to have moved me to follow the poets to Spain, later becoming an adept of Scarlatti, almost scoring my life to his imagined pacings. In Australia there was the point–counterpoint construction of a new provisional voice, one based on melancholic as well as productive mishearings and displacements – *prolude*, the neologism, is one street yield of that period, an overheard Anglo-Italian improvisation that accurately captured an identity in evolution whose calling was not necessarily musical but embraced all kinds of play. Then there was the ontological revival of word and song in my son's entering into the house of speech. There was the unique dialogue with the poet and producer, Martin Harrison; and abiding through all these decades of echoic vicissitude there were the birds speaking to me, as if little or no translation across borders were needed.

Amplifications is the outcome of this return. Vico once wrote a *peri-autography*: a writing around the work, it cast his autobiography as a discussion of the intellectual threads joining up various projects into a larger pattern.[23] But it also presented his books, essays and speeches spatially, as corresponding to a geographical arrangement. To go back over what he had done was to go back to the places where they had been made – a kind of personal *ricorso* in whose reflective light the work was seen more truly, not through a reduction to essentials but, quite oppositely, through the discovery of echoes of echoes, meanings accumulating around the creative event itself. So, in a modest way, the radiophonic works were, or became, historical in their own right, events leading to other events. In any case they delimited the scope of *Amplifications*, establishing it as a back and forth between description and reflection, between personal experience and cultural analysis, between self-reflection and a genuine curiosity to find out what has been thought on this topic. It is therefore a meeting of many voices, sometimes the clearer for that, sometimes more confused. But both effects are intended: like the alphabet, understanding ends in rumour and sleep. A writing around the ear, *Amplifications* is best described, perhaps, as periotography.

ONE

Charms

In his book *Wild Life in a Southern County* Richard Jefferies describes the winter congregation of starlings: 'On approaching it this apparent cloud is found to consist of thousands of starlings, the noise of whose calling to each other is indescribable: the country folk call it a "charm", meaning a noise made up of innumerable lesser sounds, each interfering with the other.'[1] Growing up, an auditory charm of this kind enveloped me. The call of the birds must have come first, although my earliest memories suggest a concomitant disposition, as if my ear was attuned to their jargon. Certainly, I found it natural to assume they called to me particularly, to note their comings and goings, and orient myself to garden, orchard, path and field edge according to the signals resident in these different habitats. From the beginning, as it were, the approach was as important as the noise: if an atmosphere of birdsong seemed to ring the horizon, then the pathways I struck into its midst began to differentiate innumerable lesser sounds. Shortly after, there must have been a dawning conviction that the birds spoke to me. I detected a summons to hearken to their message. It could not be translated into a human tongue, nor was I absorbed in creating a world of private correspondences. The call was inherent in the performance, an act of spacing and timing that found me in its presence.

In the twenty-first century the social connotations of noise are generally negative. But the definitions of noise are as variable as its causes. Jefferies uses the word in a neutral way, and the amplification of his term to encompass the innumerable sounds produced by a host of different species is reasonable. I was immersed in an ocean of noises: sounds that were speaking to me impurely. Interference with their perfect reception was not due to outside intrusions: the intermittent hush of passing cars, the occasional electric telephone bell or the romping cadences of Barwick Green floating up the stairs from the kitchen radio. These signatures of the time and place belong to another story. Interference signified simply the impression of an indistinct congregation of sounds into a massing ensemble whose component parts could not be securely separated. It was as if all the frequencies of the radio were broadcasting at once, with the vital distinction that this composition was a movement form, the sea of sound I waded into when I went birdwatching a bow wave of birdsong breaking round me and trailing behind in forming and reforming wakes. It was also the sound I lay down to that seeped through the rafters and crowded mice-like along the littorals of sleeping and waking. This was consoling, and if I had been able to articulate why I felt consoled, I might have spoken of presentiment, a sense that the meaning of my existence was not to be confined by the prison house of language but 'in the change of years, in the coil of things'; as the poet sang, I could look forward to 'the clamour and rumour of life to be'.

Growing up a parish away from where Jefferies made his observations, we called the great flocks of starlings *murmurations*. The location of these varied from one winter to the next. A clue to their imminence was the sight of parties of between twenty and a hundred birds flying low and purposefully across the bare fields. Converging from all points of the compass, they must have been tuned in to a collective intelligence. At the appointed meeting place, they amalgamated spontaneously, rising like smoke from beyond the spinney, their waving columns opening, closing and reforming, black Northern Lights. Other species migrated into and out of the neighbourhood according to the seasons. But the starlings, whatever the foreign origins of their augmented numbers, behaved like a region unto themselves. Such murmurations – named as much for the great rush of air caused by thousands of wings beating as for their excited calls – staged a kind of creation. Their Moebius-strip formations could for an instant be the ruin of a pavilion. Next, rounding invisible corners, rearing up and splitting apart, they seemed deployed against an unseen enemy. Such choreographies took back the landscape, mimicking a topology before topography; they disclosed an elemental unity lost with the property division of the land and its enclosure. Then by a miracle of instantaneous chain reaction, they were gone, poured over the edge of the wood – the concentration of their noise conveyed to us as the purest interference.

They respatialized the creation of significant sounds. Their molecular involutions anticipated the shifting pitches of *Ionisation*, the hyperbolic paraboloids of the Philips Pavilion (where, again, Varèse perceived spatialized sound) – a cultural genealogy discovered later, of course. As a bird term, 'murmuration' is said to have been popularized by W. H. Auden and Mervyn Peake, but in my grandfather's vocabulary the word had already been domesticated. In any case the onomatopoeic allure of the word offered me a key to a new world where sound and sense remained fused. In its more general sense, 'murmuration' referred to a movement of the lips without speaking. It described sub-vocalic sighs or low cries emitted almost without thinking, a kind of self-absorption made audible, as if an inner speech were haemorrhaging. Scaled up to the congregation of the crowd, it was to spread a rumour. Through its onomatopoeic veil, I heard the natural origins of language: the 'ur' sound, its repetition marking the beginning of rhythm, the affixed 'm' and the 'mur' sound capturing, like our word 'mum', the double movement of sound towards speech and the withdrawal from speech – not everyday language, then, not exactly, but, maybe, the possibility of an origin of poetry in certain sounding kinetic gestures. These noisy formations spoke their name; the poet's invention – the imitative rendering of their moving masses with a word – was a translator to their country.

Like crowds, murmurations were exceptional. Ordinarily, the starling appeared on its own, a denizen of chimney pots as the bird books might say, its untidy straw nest tucked under slates. In this guise, too, it appealed to me. Unlike the sturdy song thrush, the avuncular blackbird or the melancholy robin, the peculiar avocation of the starling was mimicry. Perched atop the television aerial outside my window, or on any nearby prominence, it merrily channelled the neighbourhood: embedded in a matrix of chirrups, squeals, whistled glissandos, chicken clucks and other sub-vocalic cries were tinny quotations of birdsong, the pheasant's klaxon, the dunnock's cobwebbed dew cadence, even the convolutions of the avuncular blackbird. In my childhood I have few memories of human speech; excavating later strata of communication I find evidence of territories of silence, and I suppose that they lie over yet deeper reaches of emotional desertion, forgotten or unregistered not least because they left no audible impression. An early intuition of isolation was 'Solus'. Mercifully left unwritten, this *apologia pro vita sua* cast me as an Elizabethan actor: the reader was meant to find the paradox of an actor doomed to have no company on the stage autobiographically poignant. Had I known it, the Irish meaning of the term would have offered a richer theme: *solus* (of sound: 'clear'). In any case, the dialogues people appeared to conduct was a mystery: how did one get into that country? On what frequency was small talk conducted?

Here the starling was a leader in communication. Busily improvising a patter entirely made up of borrowed phrases, he seemed bristling with self-confidence. Not for him the labour of solitary composition; he found his inspiration on the street and in the field. His meta-communication was disarming; an ironist without cynicism, he riffed on the sound events of the everyday with all appearance of sincerity. He chirruped and chattered about talking as such; turning the dial of his radio, indifferent to what he broadcast, he imagined a community, a universe of voices assembled in a new arrangement. Parodying the sound identities of the neighbourhood, he nevertheless stood up strange and serious: his uplifted head like a rapt avian orant, his spangled throat with its little beard quivering, he opened his lemon-yellow dagger beak in prayer to the sky. But such reveries were easily abjured: at the smallest distraction, he was gone, speeding like an arrow down to the fields. The starling overcame the problem of not having a language of one's own. 'Poetic reverie revives the world of original words. All the beings of the world begin to speak by the name they bear. ... One word leads to another. The words of the world want to make sentences. ... The poet listens and repeats. The voice of the poet is a voice of the world,' the wise Gaston Bachelard had written, but the starling seemed to disprove this.[2] The theory of proper names was deconstructed: What had the blackbird's song really to do with the blackbird when it could be played on another instrument? (And besides, learnt from other blackbirds, wasn't its traditionalism unimaginative rather than archetypal?)

The starling had evidently learnt how to listen with interest. He gave back to mimicry a good name by insisting that accurate reproduction was not inconsistent with innovation. Like the ancient rhapsode stitching together a new song from old fragments, strictly speaking, inventing nothing, he improvised a composition that borrowed what was around and harmonized it. It was as if he had trained himself to be unprepared, the composition and the performance being indistinguishable. Everyone might hear something of themselves there, but hear it strangely mingled. A voice that belonged to no one might be recognized, a primordial echo of the place from which the voice came, coming back to us through the interference pattern of the starling's song. In his unframed responses I had the impression the country as a whole heard its own voice. Bachelard's phenomenology need not be set aside. As the wheeling murmurations seemed to recover ancient volumes of air, where trees, clouds, wind and light articulated a single volumetric manifold, so the individual rapper on the rooftop reversed the historical fragmentation of sound into species, into narrow melodic territories and unique timbral identities, giving voice instead to what allowed these acoustic coastlines to lie together, the pan-sonic hollow, the all-resounding oceanic in-between, the auditorium in common that alone made sense of place.

How conscious these reveries could have been when I was thirteen or fourteen I leave to the psychologists. What is indisputable, though, is that the first poem I can remember committing to typewriter was called 'The Starling'. I can visualize the typeface more easily than the words but from the circumstantial evidence I believe this poem would have represented a modest manifesto. In the absence of stories in the family, a phenomenon to come back to, I was an eager collector of other people's stories, whose manner of speaking (as much as the matter) I parroted as accurately as possible in my handwritten transcriptions. Old people's stories attracted me the most and when I elicited their personal histories of life and death, their *manner* of speaking brought distant events close, individual nuances of accent, expression or gesture communicating what words could not convey. 'Intonation: an intention which has become sound', Marina Tsvetayeva wrote to Pasternak, and in the modest kitchens and parlours where I conducted my precocious home town interviews I already knew that an accurate transcription would be inaccurate unless it could also put into words the intentions behind the words.[3] To mimic their narratives accurately, to bring over their intention, demanded an attention to context. There was no speech without another, even if one of the interlocutors was not necessarily human. The chimes of the mantelpiece clock, the lifting of the latch, the cat purring: these incidental noises counterpointed what was said, solicited and heeded as stage directions that saved the voice from coming out of the absolute void.

'The most horrible thing in the cosmos', another great Russian poet, Anna Akhmatova, used to say, 'is absolute quiet': 'She loved the noise coming from the courtyard: one person beating rugs, another calling the children home, car doors banging, dogs barking … . She laughed at those writers who tried to isolate themselves from the sounds of life going on beside them.'[4] These sounds of life infiltrated speech. To convey the living voice, the enactment of what was intended, this sounding environment had somehow to be notated. What was said was surrounded by what was heard, and mediating this relay was listening, an attention to what cues the rhythms of the sound universe might offer. The story told of old times but was telling because of the auditory consciousness that shaped it. It may be that my sample was biased. As Methodists, my relatives and neighbours favoured inward speech, and approaching the denouement of some painful recollection, they were likely to trail away into what I would now call murmuration. The current of feeling still rolled the smaller pebbles along the bed of the stream but the surface of the water had resumed its glasslike smoothness: audible in this withdrawal was an essential rhythm, something like saying the rosary. As the tide of speech withdrew into this cave of inwardness, something like a hum or distant ululation was detectable, affecting but not to be written down.

Here, definitely, was a poetic task. If the ear witness was not to be an auditory voyeur, he had to find a way to make the murmuration articulate. Not only a rhythm, pulmonary, coronary, but also contrapuntal with the sound environment that was our constant companion, a *basso ostinato* climbed up and down the hills and vales of feeling, shadowing conscious thought like a whispering *daimon* or double. Akhmatova, it appears, knew how to hum:

> When she was 'composing poetry' the process never let up for a moment: suddenly, while someone was speaking to her, or she was reading a book, or writing a letter, or eating, she would half-sing, half-mutter, 'hum' the almost unrecognisable vowels and consonants of the incipient lines, which had already found their rhythm. This humming was her outward expression of the constant vibration of poetry, which the ordinary ear cannot detect. Or, it was the transformation of chaos into poetic cosmos.

Naturally, my awareness of what was happening when I observed the descent of speech into the surd state of murmur was far less developed but I like to think that even then I would have recognized my vocation in Alexander Blok's poetic summons 'to free sounds from their native anarchic element; to bring these sounds into harmony and give them form; to introduce this harmony into the outside world',[5] even if I assumed that the sounds in question were environmental rather than human.

These reveries on the border of speech coexisted shortly with more scientific registration techniques. The advent of the early portable reel-to-reel tape recorder promised to resolve the problem of rendition by simply securing an accurate sound image. The budding poet continued to face the challenge of reproducing speech patterns in a way that conveyed the intention of the sound, but was freed of the responsibility to play oral historian or ethnolinguist. The difficulties of making decent sound recordings were legion, and instructive. Despite the promise of isolating and preserving individual bird songs in the manner of a Ludwig Koch, the microphone placed in nature made audible for the first time *noise*, not in Jefferies's sense, but in a new non-contrapuntal way. In relation to our binaural attention and capacity to discriminate between hearing and listening, the microphone hooked over a fence or crouched in the mouth of a window is low-fidelity. For, unlike the human ear, it can hear but not listen; hence, any interruption to the pure signal associated with its perfect recall sounds like noise. Its high-fidelity rendition of wind, of rain on nearby leaves, of the intermittent hush of passing vehicles or of human voices indiscreetly raised in its neighbourhood, together with its indiscriminate registration of any unpredictable sound event, identifies as noise all that comes between us and what we intended to hear. Noise in this context is not simply any sound signal that interferes with the clarity of the recording: it's bad timing as the same interfering sound might, a few minutes later, become of interest, and then, instead of fearing its interruption, we await it, counting the silence and soliciting it to speak.

One day my father communicated certain folk intelligence garnered at the local pub: a nightingale was to be heard singing at Knighton Crossing. Our expedition to hear this remains in my memory as one of the more poignant gestures towards a communication that, really, to the end of his life, remained unconsummated. Late enough for the lanes to be shrouded, although the twilight lingered, he drove me into the Vale of the White Horse to the estimated spot; we rolled down the windows and settled to listen. However, no plangent cadences issued from the coppice or lane. Other sounds were registered but that famous songster remained dumb, a failed solicitation that I came to regard superstitiously, as if the way to my father's heart was doomed to be a dead end. Recording a nightingale came about by an entirely different route, and at a different place in the nearby Thames valley. But the results were poor. Our singer exhibited the behaviour described by the naturalists (not, I thought, without a hint of disapproval), for he skulked deep in the woodland canopy, making it difficult to guess where he might sing. We had the staccato click and clatter of raindrops on leaves to contend with, as well as the wind's oddly dry crumbling against the microphone. As it turned out the recording equipment had its heart murmur, a low oceanic hiss, as if the machine pressed a stethoscope to its own revolutions, and the result was noise.

Another reason why the nightingale's territory was contracted so that a formerly common summer visitor was now classified as an infrequent visitor may have been distantly related to the cultivation of inner speech mentioned earlier. A few years later, during my university education, I visited St Alban's Abbey (as we knew it), coming across a plaque to a certain monk Sigar, a man so pious that he had feared the abundance of nightingales in that richly wooded part of Hertfordshire would disturb his devotions: 'He, therefore, made supplication that they might be removed "lest he might seem to rejoice rather in the warbling of birds than in the worship whereunto he was bound before God".' If you walk the by-ways of ecclesiology, you find that this was by no means an isolated instance of phonophobia. The absence of nightingales in Otford is, apparently, ascribed to Thomas à Becket, who is said to have been so disturbed in his prayers by the song of a nightingale that he commanded that none should sing in the town ever again. This was no idle rhetorical flourish: John Amundesham, who occupied the hermitage where Sigar had once lived, reported that in his own day nightingales not only never presumed to sing but never appeared within a mile of the place.

This is a rich theme to descant on. One wonders what means Sigar used to extirpate the nightingales (and why, taking a leaf out of Ulysses's book, beeswax ear plugs weren't a simpler alternative). But, seriously, what prevented the house sparrow, the starling, the temple-haunting martlet and a dozen other noisy species from falling under this ban? It must have been because the nightingale spoke to him. Otherwise, why would he fear that his nightly murmurings would get mixed up with the nightingale's warbling? Unless there was a jargon in common, confusion was unlikely. But the source of the distraction could have been subtler. Sigar detected in the sound an intonation, and in the intonation an intention. Consider the opposite case: a voice is heard on the other side of the wall loudly soliciting your attention. Its irregularity is as irritating as its presumption. You might feel like doing what Arthur Schopenhauer did about the coach drivers in the street cracking their whips: 'I denounce [them] as making a peaceful life impossible; [they] put an end to all quiet thought. ... No one with anything like an idea in his head can avoid a feeling of actual pain at this sudden, sharp crack, which paralyzes the brain, rends the thread of reflection, and murders thought.'[6] And then, unable to endure this sound pollution any longer, you go next door determined to put a stop to it – only to find that the source of the disturbance is a parrot in a cage. In my experience, in that moment the irritation collapses: realizing that there is nothing behind the voice, no intention embedded in the cry, it is reclassified as environmental noise and, dissolving into the general background noise of the neighbourhood, becomes humdrum.

Evidently, the summons Sigar heard in the nightingale's song survived the parrot test. But Sigar surely had more serious distractions: 'Early medieval abbeys were alive with sound. ... Like heavenly bees in their hives, monks were not silent in their industry. Their lips were always busy with the buzzing of prayer and praise.'[7] Presumably this outward humming interfered with Sigar's interior piety. Amidst this speculation, 'The Cuckoo and the Nightingale' provides the best guide to the voices of those days. One May morning, the lovelorn narrator enters a bird-loud glade and falls into a swooning fit that finds him able to speak in avian tongues: 'Methought I wist right well what these birds meant, / And had good knowing both of their intent, / And of their speech, and all that they would say.' Through a question the cuckoo addresses to the nightingale, he gives a striking example of his new-found skill. Why, asks the cuckoo, does the nightingale sing 'ocy, ocy'?[8] And the nightingale explains, in effect, that the word 'denotes the doom, which the nightingale imprecates or supplicates on all who do offence to Love'. I have some difficulty reconciling the sobbing 'piu, piu, piu' of the nightingale's song with the 'ocy, ocy' of the poem. Perhaps expressive repetition is the point: John Keats heard in the same cry 'Away! away!' and 'Adieu! adieu!'

What theory St Alban's monks might have held about the language of birdsong, angelic or diabolical, was not what interested me – whether or not, say, Sigar was familiar with the proposition that 'in its original state of pristine grace', the language of the Old Testament prophet Enoch was 'actually the Language of the Angels, and it was given forth as a lexicon of birdsong'.[9] The early evidence of environmental schizophonia shocked me. The 'romantic and impassioned' fourteenth-century mystic Richard Rolle may present a striking contrast to this melophobia. Yet, he, too, is chiefly the advocate of *inner* song. Hermits find 'clatterings distract them who are set amongst many, and but seldom suffer them to pray or think'. Alone, however, 'a heavenly noise sounds within them, and full sweet melody makes the solitary man merry.'[10] This 'sweet ghostly song' 'accords not with outward song, the which in kirks and elsewhere are used. It discords mickle from all that is formed by man's outward voice to be heard with bodily ears; but among angels' tunes it has an acceptable melody.'[11] His well-known identification with the nightingale was, spiritually speaking, apprentice work: 'In the beginning truly of my conversion ... I thought I would be like the little bird that languishes for the love of his beloved. ... It is said that the nightingale is given to song and melody all night, that she may please him to whom she is joined.' But now he knows better: his song will be immortal soul music, 'the organ-like and heavenly song of the angels'.[12]

Already, inner piety, the unscripted language of the holy heart, was conceived oppositionally, as a flight from the sound environment. An 'inward religious address' was cultivated that withdrew from the legacy of Babel, amongst whose babble was the innocent jargon of the birds. If 'by being so internalised, the revealed word of God is set to become the realized language of humans',[13] then the transcendental sound poetry of nature is demoted to nothing more than a primitive chatter. Gilbert White's identification of three distinct species of *Phylloscopus* warblers on the basis of their songs is often cited as evidence of his talent as a field naturalist. Far more remarkable, though, is the fact that the chiffchaff, the willow warbler and the wood warbler had not been distinguished before. Their songs are sharply different: in a more auditory culture, the chiffchaff's call would surely be celebrated as marking the advent of spring.[14] The silvery cadence of the willow warbler is widely heard on the outskirts of the town. Only the wood warbler is more easily overlooked. How can this collective hearing loss be explained unless as acoustic alienation fostered by the Church? Asked to attend to the birdsong around them, the people where I grew up rarely admitted to hearing more than a distant environmental tinnitus.

Modification of the sound environment occurred by way of a kind of spiritual pesticide; enclosing the inner voice and hedging it off from the polyphonic babble of coexistence contributed to the first silent spring.[15] Other enclosure acts, though, were more benign, and tended to musicalize the countryside rather than demonize it. While the nightingale recording failed because of an ancient banishment, my attempts to capture the 'dawn chorus' met with relative success. While the nightingale skulked in the bushes, the chief choristers of a misty spring break of day lined up to be heard, a phenomenon made possible, I supposed, by the eighteenth-century Enclosure Acts, which had trellised large parts of the immediate countryside with hedges, providing in this way natural staves on which the birds perched like notes. The stars of these performances, the song thrush, the blackbird, the chaffinch, the wood pigeon, the robin, a distant yellowhammer perhaps, have been canonized in literature and song, as the native informants of the place, but their rise to power is likely to have been due to their adaptability. These were the tenant farmers who managed to hang on when access to the commons was stopped, who negotiated employment and an ell of land under the new order.

The enclosure of the fields created new performance arrangements comparable with those emerging in the contemporary London concert hall, where a growing spatial separation between players and audience was bringing about a new way of listening. Instead of talking through the performance, theatre-goers were now expected to keep quiet. Silence conspired with the musicalization of sound to further eliminate noise. The artificiality of the situation was reinforced by the increasingly elaborate scenography. Such developments had their parallel in the countryside. 'Enclosures', William Marshall wrote in 1800, 'in general small and the hedges high and full of trees', create the illusion that 'the eye seems ever on the verge of a forest, which is, as it were by enchantment, continually changing into enclosures and hedgerows'.[16] This new visual aesthetic held an appeal to the ear as well. The harsher protection of property rights, reinforced in our area by the merciless execution of the Game Laws, created new divisions on a ground in a musical as well as landscape sense.[17] An indistinct surrounding of birdsong, at best sibylline, at worst little better than satanic seduction, was being converted into orderly choirs, assembled like angels. The trees, fields and rivers were being organized like stage sets, readied for pastoral revival. In mercantile faith, the country house corresponded to a tabernacle, and the avenues of trees framing it were like choir stalls. Tenanting the borders of these estates, recruits to the 'dawn chorus' practised an auditory picturesque. Glossing over the violence of expulsion, theirs was also the sound of silence.

Perhaps the way was prepared for these consolations by the prior evacuation of sound inaugurated by the likes of Sigar. I think of the remarkable passage in Thomas Traherne's *Centuries* (dating perhaps from the 1660s): 'In a lowering and sad evening, being alone in the field, when all things were dead and quiet, a certain want and horror fell upon me, beyond imagination. The unprofitableness and silence of the place dissatisfied me; its wideness terrified me; from the utmost ends of the earth fears surrounded me.' Growing up in the comfortably knitted-together north Berkshire landscape, I find this 'field' difficult to visualize. Even in the midwinter, groups of fieldfares hopped over frost-fringed furrows, rooks tenanted the bare elms and the trickle of the robin's song was audible from the field edge. Traherne's sense of abandonment may be spiritual, but it expresses itself as auditory agoraphobia, a feeling of disorientation caused by the fact that the place does not call him. Somehow, Traherne manages a turning about in the heart, finding,

> I was a weak and little child, and had forgotten there was a man alive in the earth. Yet something also of hope and expectation comforted me from every border. This taught me that I was concerned in all the world: and that in the remotest borders the causes of peace delight me, and the beauties of the earth when seen were made to entertain that I was made to hold a communion with the secrets of Divine Providence in all the world.

How much easier 'hope and expectation' would have been if, by a miracle, the 'remotest borders' had metamorphosed into hedgerows. The place would not have been silent; communion would have been sensory.

From our point of view, the field arrangements of farm and country estate were exceptional. Our habitual territory was the patchwork of allotments and market gardens that fringed the town. My grandfather trained as a gardener at a 'big house' in the district, and when he acquired a few acres of his own, created at its heart a miniature lawn and herbaceous border reminiscent of what he had seen as a boy. But this landscape devotion, useful as it was for Methodist Circuit group photographs, hardly survived his passing; and when a few years later the land was sold and the apple trees grubbed up, we looked out into a *terrain vague* of nettles and thistles and kestrel-haunted meadow grasses. This brought home the general condition of the countryside left over after enclosure. Near Cricklade, north-west of Faringdon, William Cobbett observed the impact of the Enclosure Acts:

> The labourers seem miserably poor. ... Their wretched hovels are stuck upon little bits of ground on the road side, where the space has been wider than the road demanded. In many places they have not two rods to a hovel. It seems as if they had been swept off the fields by a hurricane, and had dropped and found shelter under the banks on the roadside.[18]

If, by some Pierian metamorphosis, these folk had been changed into birds, their natural congeners and descendants would be the vagrant flocks of linnets and goldfinches, attracted to the seeding grasses. A *charm* of goldfinches, *Country Life* liked to say: the glitter of their wings and the liquid twittering of their call seemed two translations of one idea. As for the song of the linnet, it was described as 'poor'.

As an autobiographical theme, the call was always a call to come. In the dark wood of communication surrounding the only child like a distance, bird calls were paths. The golden threads they let out were not to be gathered up by walking, however; coming to me like arrows through the air, they provided flight directions. My early aerialism has left no dated memory; it belongs to the realm of Mnemosyne – whoever go to her well leave 'the temporal frame and listen to the bubbling, which has gone on without interruption since the beginning'.[19] When I climbed up onto my bedroom window sill, and spreading my arms stepped out into the air, I was not providing a 'precedent' for the present. Circumstantial evidence of this dangerous propensity is the memory I retain of curtain rods nailed vertically across the window to prevent my falling out. But it is only a vivid impression; afterwards, no one was able to confirm it. Richard Church describes comparable levitations, but his aerial adventures date from a later epoch of childhood: mine began as soon as I could stand up.[20] Later, my lazy circuits over gardens and rooftops retreated into flying dreams. Further sublimations occurred in my writing: the eido-kinetic intuition is the sense that regions are movement forms composed of innumerable curvilinear *trajets*. My ambition has not been to escape the ground: it has been to alight. When I have mentioned this, I am surprised to find many aerialists in my position.

It seems to me, looking back, that the point and cadence of birdsong provided my first movement grammar. When they broke off their orchestral tunings-up or overcame their giggling, the starling, one or in twos, or shooting out of the stubble in twos and threes, made off with astonishing certitude somewhere. This was to me always a wonder of their flight in general, the preview of perches to come, the predestination of avian destinations. Even the wood pigeon lazily cutting out crescents against the stormy cumulus knew where to come back to – so that the mastery of complicated phrases, the deft variations on a theme, were joined from the beginning with mastery of another mode of flight, and my realm was as securely skeined with shimmering filaments, the singing lines that marked out a sonorous region, as the morning bents were a dew-threaded maze of cobweb. Punctiform, curvilinear – and, together with this original fusion of sound with the arrangement of space, a correlative ground bandaged with dock, dandelion, daisy, pimpernel and burdock. Negotiating these perimeter paths I could not have imagined a future of trapeze wires suspended tightly over the void; if I traversed borrowed ground, it was with the trusting confidence of the youth in the tomb painting, who blindfolded feels his way into the dark, in thrall to the *aulos* player a few steps ahead.

The absence of stories is part of this story in more than one sense. As far as one can judge from the silence, in my father's family any discourse that could be construed as personal was frowned upon as unholy pride. In the context of human suffering, to parade personal vicissitudes was a vice. My parents may have described themselves as lapsed Methodists; in reality, it was Methodism that lapsed when its leaders supported Britain's declaration of war. They remained committed to the pulpit's theory of service and the practice of charity. Submission to the will of those in need required a willingness to talk their language: in listening to a neighbour pour out his woes, self-abnegation was exercised, as no comparable disclosure of pain was expected. So at the heart of a kind of chatter was a deepening reserve, a determination to keep mum about the things that mattered. Years after, I analysed the typical structure of those reports my father *did* care to share: referring to people in the town, they invariably described the punishment of hubris. The mechanism of their downfall was egotistical temptation, a turning away from the social contract represented by service. Ideally, a lifetime of service went unnoticed, unpronounced. So I think it is a mistake to attribute my aunt's 'O, we don't talk about that' response to almost any question of family history simply as English reticence. To speak up was, from the position of Pauline service, a misuse of language.

Ours was not a household of laughter and singing. Even writing, the public speech of the inner voice, remained mute. It is an oddity of my upbringing that I have had to learn my parents' views posthumously – from a modest stack of books shelved in the 'front room', which, since my birth, had remained unopened. The literature of their precociously early political formation – Fabian Society pamphlets, G. D. H. Cole's prescriptions for post-war recovery and Christian socialist tracts – breathes a sweet reasonableness. Promoting communitarianism, it strikes a note of meritocratic condescension. The glue of the new classless society will remain good manners, deference to merited authority coexisting with the emancipation from class-based injustice. This very English political compromise had its own history: among the interwar books stood my father's schoolboy selections from Cobbett, where the grand theft of the Enclosure Acts was described, its direct legacy our family's living conditions. But Cobbett was no revolutionary: he roundly denounced the social and economic waste visited on the village by incompetent landowners but he did not want the demolition of the countryside. His radicalism was conservative, like my father's. These unobtrusive publications had suffered a double silencing: the first opportunity to find their significance died with my parents; then, beginning to leaf through them, a new uncertainty arose. What parts of these publications especially spoke to their private, political devotions? Without their guidance, I was doomed to read there what I wanted to find.

To read the books properly, it was necessary to read their reading. But my parents were not given to underlining key passages or writing marginal comments. In all that literature I found but one trace of reading, the folded down corner of a page in their copy of Aldous Huxley's *Eyeless in Gaza*. The volume was my mother's gift to my father on the occasion of their engagement on 12 April 1941 (when my mother was only twenty): I may have read more at the same age, but she had read less to greater purpose. On the marked page begins the reformed socialite Anthony Beavis's great speech in favour of Unity, the vision that sums up all he has learnt. Unity will grow out of quiet and it will return to quiet. 'There can only be an attempt, as one goes along, to project what one has discovered on the personal level on to the level of politics and economics.'[21] Any good socialism springs from what has been discovered on the personal level, a formulation that recalls James Joyce's 'inner world of individual emotions'. Armed with this conviction, it must go out into the world of multiplicity (but for which the goal of unity would have no historical purpose). However, the peace ultimately won transcends personal expression: 'peace as a dark void beyond all personal life, and yet itself a form of life more intense, for all its diffuseness, for all the absence of aim or desire, richer and of finer quality than ordinary life'.[22]

Doctor Johnson once made the observation that respect for the soul's inner dialogue with God placed Methodists outside the law. But from another, outer point of view, that of the police, their self-conscious quietism could be seen as *compliance*. Later, when I started to investigate the disappearance of local speech and lore, I learnt that spiritual inwardness had its social counterpart in the regulation of singing. In his book *Folk Songs of the Upper Thames*, Alfred Williams (who collected songs before and after the First World War) attributed the extinction of the southern English folksong to the Victorian police who 'looked upon song-singing as a species of rowdyism'. Their threats of prosecution forced inn-keepers 'as a means of self-protection, to request their customers not to sing on the premises, or, at any rate, not to allow themselves to be heard'.[23] In winter the blackbird may sometimes be overheard singing to itself. Now, a similar kind of singing was being urged in the pub. Puritan branches of my family would undoubtedly have supported this development, asserting that the pleasure they took in hymn singing was different, a disciplined harmonization in appointed places, serving the greater glory of God.

Appointing sound to special places was another expression of a prejudice towards sounds in general. Historically, the cultivation of the inner voice was identified with political as well as spiritual emancipation from the here and now. A line can be drawn from the conception of inward speech attributable to the great Puritan George Fox and its restitution of direct discourse with God, unmediated by outward forms, to the call in certain quarters, notably among the seventeenth-century Levellers, to unlock the lands. To displace the great landowners, and to redistribute the earth fairly amongst those prepared to work it, meant breaking down barriers, dissolving the property-based hierarchies that had produced men's enslavement. Hence, in this sense, my grandfather's creation of a garden inside the garden, a little Paradise at the heart of a working orchard and market garden, did not simply ape the enclosures of the powerful: it recognized that true emancipation was not territorial; it involved a turning inwards rather than outwards. The croquet lawn, the summer house with its circuit of lime trees, the rose trellised arbour and herbaceous borders expressed an interiority that was, paradoxically, boundless: there was no end to the meditative space it extended. Neither my grandfather nor my father was a Puritan, notable in G. K. Chesterton's opinion, for their indifference to place; nevertheless, as a *locus amoenus*, the garden transcended its place.

Chesterton's general contention that a Puritan indifference to place was associated with a disdain for art certainly accounted for the silence that had fallen over our town and the surrounding villages, where, by the time I was growing up, musical instruments were rarely heard and the human voice primarily issued from the radio. Just as 'the true Puritan was ... clear that no singer or story-teller or fiddler must translate the voice of God to him in terrestrial beauty', so the Puritan felt only antagonism 'to the whole idea of spiritual blessings being mediated through saintly persons, and of any sacredness attaching to special localities or to material objects'.[24] Alfred Williams, the railway factory worker, folklorist, village chronicler and translator of Sanskrit texts, a remarkable man by any measure, was our local witness to these changes. First, he was a passionate listener. His early verse contains a number of poems directly imitative of the songs of familiar English songbirds. '"'Tis young, young, young, / The morning, morning," sang the blackbird,' which suggests that avian whistle speech, like human whistling speech, depends on repetition and phrase elaboration to communicate even the simplest concept – an observation that reinforces my impression that the bird Williams copied was the song thrush not the blackbird![25]

In *Folk Songs of the Upper Thames*, Williams reported that his native informants could not remember the words of the songs unless they sang them. By repeating them until they could fill in the gaps, they were able to give him the information he wanted. But, astonishingly, I used to think, Williams made no effort to record the tunes. My assumed superiority was, I see now, mistaken. It could be compared with the ethnolinguist's lament for a lost language, which fails to consider for whom the language was spoken and to whom it belonged. Williams recognized folksong's disappearance but refused to lament it. He rejected any folk song revival movement because, I think, he recognized that the songs he had heard belonged to a particular time and place. More, by the time he heard them they were already unheard. This is important: even he, in hearing them, was opening a tomb in search of sign, and he did not want to repeat the transgression. On this ground his decision not to record the song melodies makes sense. He paid tribute to the sonorous body without artificially extending its life. He sought to respect a falling silence without imposing on it the further silencing of sound recording. In this way its unheard melodies might continue to resonate.[26] Williams, like my father, knew that unspoken speech is not silent. It has nothing in common with the 'silent' radio voice, which, in its repression of the sound of breathing, descends from a Greek notion of listening to the Logos (in silence). The inner voice intuited here does not belong to history: without precedent or posthumous existence, it cannot be prolonged or stilled – the bubbling spring of Mnemosyne the noise of modernization tries to drown out, it flows again as soon as you listen.

What *is* heard is *embodied* sound. As Martin Harrison suggests, sounding may not have the purpose of transmitting a message (which can be done silently). To situate concept and sound together is to make a particular kind of 'world sense' predominate over all the other kinds of perception of which the mind is capable. To be conscious of being deaf and dumb is to sense that 'the sonorous body' has been straitjacketed and deprived of movement; and this is a deprivation that has nothing to do with the communication of concepts. As Harrison writes, of sonorous experience – 'Concept only occasionally occurs in sound, bent and diffracted through its impermeable medium. Larynx and ear swim through it, not as unitary instances of conceptual convergence but as myriad shoals of receptivity and emission engulfed multilaterally in its unstable environment.'[27] That sense of hurt or offence that my father exhibited when verbal performances seemed to him to try to render what should remain unspoken indicated his agreement with what Merleau-Ponty describes when he speaks of ideas that would not be better known to us if we had no body and no sensibility, for the 'little phrase' cannot be grasped or translated: 'Each time we want to get at it immediately, or lay hands on it, or circumscribe it, or see it unveiled, we do in fact feel that the attempt is misconceived, that it retreats in the measure that we approach.'[28] The sonorous body lives and leaves with the body, as I found when my father died without conveying any impression that there were still things to say.

The sonorous body identifies the voice with gesture or physical attitude. My father's rich bass tones I associate with an armchair or other nook; their quiet sonorities were ideally delivered while sitting back. The physical characteristics of speech, timbral resonance and pitch variation are for experts exercises in auto-auscultation: they express the anatomy of mouth and throat. Many voices are ventriloquial: seeming to come from somewhere else. The ventriloquist and his dummy recruit an inner voice for outer dialogue, although the stage effect curiously resembles a parrot talking to itself. All vocalizations are, in any case, mimetic, originating with nature, or mother's and father's voice. Intonation is our vocal thread to the ancestors; part of the intention expressed in a peculiarity of phrasing or a habitual breath punctuation is a desire to animate the dead; a turn of phrase can conjure up a ghost, their presence stamped in the breath pattern. But our conversations with the disembodied are not confined to childhood echoes: like starlings we channel all that is in the air – in those days, radio personalities. The whistling heard in the street was likely to be a media reminiscence; there is a spot in memory associated forever with Gluck's *Orfeo*, a pavement in Marlborough Street where one day its divine melodies appeared to me as palpably as smoke drifting up. The rhythms of walking invariably stir other rhythmic associations; throughout my life there have been passages in the city where the haunting refrain of *Che farò senza Eurydice* surges up.

The parrot's ventriloquism holds a clue to how the sonorous body makes sense of the world. Consider the theory of poetic production Shelley puts forward in *A Defence of Poetry*. Despite its Platonic overtones, his notion that 'humans are somewhat similar to Aeolian lyres (wind harps) – "Man is an instrument over which a series of external and internal impressions are driven"' – is sensationalist rather than idealist. He *says* that a special harmonizing principle, the imagination,

> reveals to those more finely attuned new thoughts ('new materials of knowledge, and power, and pleasure'), previously unapprehended relations between old ones, and allows them to perceive the good that Shelley asserts to be inherent to the relations between existence and perception, then between perception and expression, happens to be the imagination, to which he also refers as the 'creative faculty', 'faculty of approximation to the beautiful' or the 'poetical faculty',

but this is an aesthetic preference not an ontological necessity. Suppose the Turing test for artificial intelligence were applied to Shelley's claim: if 'sensation and thinking are ontologically similar', who can say a parrot doesn't approximate the beautiful best? Shelley's poet tunes the world: anyone may harmonize vibrations, translate impressions into thought, but 'poets have a special attunement to the world that allows them to produce good translations of reality which will stand the test of Time by constant reinterpretation'. But suppose that neither world nor sounding body is tuned or tunable but (like the Aeolian harp) is simply a bundle of strings that 'because they have different diameters ... vibrate in response to a common wind velocity at different frequencies'.[29]

In *Parrot*, I told the story of the wise old Amazon parrot called Ya Lur from Guatemala, who turned back a group of evangelizing Seventh Day Adventists by mimicking their unimaginably bad Oklahoma-accented Mayan. 'Ya Lur was famous far and wide, and highly coveted by the villagers for the astounding fact that, unlike most parrots, she was trilingual and had a memory like a recording machine.' Her role was 'secretarial answering machine or voice mail' in the compound. But she always played back what she had been told with interest, her message being 'mixed in with playbacks of women walking past our compound arguing about the price of tomatoes, complaining about their neighbours'. On top of 'repeating any gossip whatsoever ... redoing the dialogue using different silly parrot voices for each person', she would 'even imitate our laughter during her fine performances, repeating in turn even what we said about that. She was an auditory mirror.' No wonder the villagers regarded 'a Lur speech as meaningful, mysterious, and coming from the world of the Deities'. She was the medium by means of which 'Society' came into being, the Voice that gave them voices by giving back to them their Voices differently, with all the noise of circumstance.[30] In relation to Shelley, Ya Lur was able to tune in because she was anti-Pythagorean: no ghost of harmonic attunement inhibited her hearing.

Negative acoustic impressions also leave their trace in the sounding body. Dialogue with my mother was undialectical; impatient of dissent, any articulation, even of agreement, risked provoking a growing frustration, for she, too, found extended dialogue difficult, and, like Anthony, repeated, 'United in peace. In peace ... in peace, in peace.' Such universal consummations lay beyond rational discussion; to consider what actions they implied was moral procrastination delaying her wish fulfilment. Schoolboy attempts to historicize her vision (why, after all, *did* the world continue to go to war?) were framed as dissent and treated as personal treachery. My rehearsals of socialization were rebuffed and my communication, both spoken and written, grew guarded. A habit of considering all the points of view, a duplicity that some have found duplicitous, dates from that period of emotional subjugation. Such discursive registrations of psychic distress swam in the myriad shoals of receptivity of the sonorous body. The sound history of our domestic interior continues to haunt me. An unsourced squeal comes out of the night and I hear the chair leg scraped over the floor as my mother pushes back her chair. After half a century, it remains the note of imminent rejection. And when we had retired to our habitual reveries, and the noisy silence of traffic, drunken voices in the street or the buried life of the television resumed, I longed for nothing more than the resumption of her voice.

Noise was twinned with silence. Sigar's schizophonia, amplified over the centuries through a cultivation of interiority that withdrew thought from the sounds of life going on beside and around, fostered a double deafness. We were born into the silence of absolute noise because we valued the unheard melodies of the heart over those angelic communications audible in the atmosphere. My birth was framed by atomic blasts in Nevada and Kazakhstan. RDS-3, the first Soviet air-dropped bomb test, was detonated on 18 October 1951. 'Released at an altitude of 10 km, it detonated 400 meters above the ground.'[31] The Buster-Jangle nuclear test code-named 'Easy' exploded at the Nevada test site on 5 November 1951. A witness recalled,

> When the bomb exploded, the flash was 100 times brighter than the noon-day Sun. After the flash we could turn around to observe the mushroom cloud. No sound was heard immediately but one could see it coming, it came toward us at approximately 760 miles per hour, the speed of sound. The sound and overpressure from the blast looked like a giant heat wave. It was like a wave one can see on a hot day driving down a road but a lot bigger. When the wave reached us there was a giant boom, a deafening sound that would make a person's ears ring. Then there were smaller booms that echoed through the mountains a few moments.[32]

Geographically distant, absolute noise was culturally local. Besides, I grew up close enough to Swindon where Alfred Williams had worked as a young man. In *Life in a Railway Factory*, Williams described his own experience of schizophonia. Nothing, he said, could be compared to 'the din produced by the pneumatic machines in cutting out the many hundreds of rivets and stays'. In the Great Western Railway factory this produced absolute noise. But it also produced its conceptual opposite: wordlessness, silence.

> Language fails to give an adequate idea of the terrible detonation and the staggering effect produced upon whosoever will venture to thrust his head within the aperture of the boiler fire-box. Do you hear anything? You hear nothing. Sound is swallowed up within sound. You are a hundred times deaf. You are transfixed; your every sense is paralysed. In a moment you seem to be encompassed with an unspeakable silence – a deathlike vacuity of sound altogether. Though you shout at the top of your voice you hear nothing – nothing at all. You are deaf and dumb,

and stupefied. You look at the operator; there he sits, stands or stoops. You see his movements and the apparatus in his hands, but everything is absolutely noiseless to you. It is like a dumb show, a dream, a phantasm.[33]

The aural assault described in this passage seems to embody the threat that modernization presents to traditional acoustic ecologies. In later poems, Williams characterized the tragedy of the First World War as, in part, a deafening of men's sensibilities. His folksong research undoubtedly stemmed in part from a desire to preserve knowledge of how, as Barry Truax puts it, 'the soundscape functioned prior to industrialisation and electrification'.[34]

The sound of the Cold War and its civilian avatars was the thunderous horizon of my growing up. Strategic Air Command arrived a few miles north of Faringdon at Brize Norton in April 1951. New runways were constructed, as well as new accommodation and weapon-handling facilities. By June 1952, twenty-one Convair B-36 Peacemaker strategic bombers of the 11th Bomb Wing were in the air. From Faringdon Folly their resting fuselages were visible glinting in the sun. When the beehive stirred, an auditory tremor swept up the Thames Valley. Later, Concorde was tested there, completing our anaesthesia. We were located in what is technically called the 'carpet', 'the region on the ground ensonified by the part of the sonic boom that propagates directly downward from the aircraft to the ground'. (A 'secondary boom "carpet" is the region on the ground ensonified by the boom that initially goes upward from the aircraft but is refracted back to the ground by winds in the stratosphere above the plane'.)[35] Activity in Concorde's sonic boom trail came to a standstill, its angelic visitation producing an out-of-body experience. Visual sublimity was twinned with auditory death, a *blinding* noise producing unspeakable silence. A recent Brize Norton noise and vibration report identifies 'rollers', a kind of false alighting where the aircraft touches its wheels onto the ground and rolls down the runway 'before accelerating and climbing again into the circuit', as 'the worst case scenario for noise'.[36]

Later, I twinned the sound of military power with music revival. In 'Remember Me' (1986) I explored the echo of war in the 1951 Mermaid Theatre production of Henry Purcell's *Dido and Aeneas*. A collage of texts juxtaposes NATO war games in rural England with the rather picturesque representation of war in baroque music; Bernard Miles's desire to revive an Elizabethan England has its political counterpart in the Labour campaign against Winston Churchill ('Still fighting the Battle of Blenheim'); in Kirsten Flagstad's performance of Dido's 'Remember Me' is remembered the arrest of Kirsten's husband by the Nazis. The noises of recent European history not only amplified the emotional resonances of *Dido and Aeneas*: they intruded practically, influencing how it was remembered. There are different accounts of the recording history; it seems that four unsatisfactory attempts (15–16 October, 30 November and 1–2 December 1951) were made to record the work live at Miles's first, improvised Mermaid Theatre in St John's before it was successfully recorded, with a somewhat changed cast at the Abbey Road Studios in February 1952.[37] One speculates that the difficulties of recording live were exacerbated by interference from external noise.

In a curious sequel, the sound of 'Remember Me' is also twinned with silence in the American war drama miniseries *Band of Brothers*: 'The camp sequence in episode nine ends with Nixon walking through the camp, supervising German civilians who have been tasked with burying the dead. Nixon walks to an instrumental, all-strings version of "Dido's Lament" ... played at the slowest pace of any triple-meter music in the score.'[38] Shortly after the release of *Band of Brothers*, 9/11 occurred: 'Made before the start of the so-called Global War on Terror and televised cross a profound juncture in American history – when the nation turned abruptly towards sustained military engagement overseas – *Band of Brothers* was dated from its third episode.'[39] So, bubbling up from the well of Mnemosyne, Dido's lament was also a lament for that.

To say that Purcell's setting of Nahum Tate's words simulates the art of Mnemosyne is to say that almost miraculously it reintegrates the act of remembering with the act of perceiving sound or listening. The aria anticipates a Husserlian phenomenology of time consciousness. In this theory of temporal bridging,

> The tone 'just-past' is not a tone that was perceived before and is now remembered, but is a tone that is perceived, as it is 'retained' in its being-past, ie, 'Its being-past is something now, something present itself, something perceived.' And even the tone 'yet-to-come' is not a tone that will be perceived later and is now expected, but a tone that is perceived, as it is 'pro-tained' in its being-future, ie, its being-future is something perceived as well.

In terms of a neuroscientific theory of cross-temporal contingencies (binding of cognitive processes over time), an expanded temporality, where remembering and forgetting are *not* antithetically hinged around an instantaneous 'now', produces the view that 'the two temporal integrative functions of the prefrontal cortex, short-term memory and set, work together in *reconciling the past with the future*'.[40] It is this reconciliation that Dido desires when she asks that Aeneas's 'wrongs create no trouble'. The call to 'remember me! but ah! forget my fate', is 'temporal integration' as emotional resolution.

So the sound of silence is really the sound of *silencing*: when Agnieszka Holland uses Dido's lament at 'the emotional peak' of *In Darkness*, the reference is not to a Holocaust that happened in the past but to something perceived in the present. This may be as close as audiovisual production can come to auditory consciousness, the timeless sense of immersion in the living stream of time. Mediated sound may already lie outside 'the sonorousness of ordinary perception'. According to Martin Harrison, technologized sound is semiotic in principle – 'The timbres of such sounds, their pitch patterns, the ways in which they are sequenced as ordered sonic events, their addressiveness, even the markings of their beginnings and endings ... are all presented as semiotic attributes ... sound as "sound" does not simply exist, it has to be brought into being.' In contrast, for the sonorous body, 'A sound does not really stand for anything ... and neither does the listener somehow stand back from his or her listening in order to do a sound-specific, topographically marked-out and temporally bounded "act" of listening.'[41] In the wake of another Holocaust, the Great Irish Famine, George Petrie wrote, 'The "land of song" was no longer tuneful'; instead, during and after the famine 'an unwonted silence' prevailed.[42] Sound in the world does not come out of silence; it is its silencing that is heard.

I once asked my father to remember the sounds of childhood. He wrote me an inventory of sounds that were no longer heard: the chink of the blacksmith's hammer, the hollow tolling of milk churns, the clip-clop of hooves. These are the sounds of an England pitched between the hideous roar of jets and the retreat into an inner quiet. Together with other middle-distance sounds, including the church bells, the street vendor's cries, the shrill crowds of children in the playground, these are the valued noises of the traditional soundscape. But what does their disappearance mean? In an odd way their silencing is where their sounding begins.

In his communicational model of a well-tuned environment, electronic composer and sound theorist Barry Truax also privileges listening over hearing. Acoustic information derived from the environment through attentive and 'background' listening is the information that provides the 'environmental context of our awareness, the ongoing and usually highly redundant "ground" to our consciousness'.[43] Agreeing, Hildegaard Westerkamp also explains that 'the information we take in as listeners is balanced by our own sound-making activities which, in turn, shape the environment'.[44] At the same time, though, the soundscape model is incomplete. It presumes that listening is always listening *for something*. It forms part of a communicational chain, and even if it is a sophisticated way of monitoring the environment and tuning it, it assumes that its function is to process and transfer information between individual and environment. Historically, it presupposes loss and silence: as Truax says of the traditional soundscape, 'We are only able to conceptualize that balanced relationship as an ecological one because we have since lost it.'[45]

One April, after I moved to Australia, my father wrote to me:

> Had such a lovely experience this afternoon. I'd been digging the garden next door and, after about an hour, sat on the old wooden seat for a breather. It's a lovely warm day and I dozed off. I awoke to blackbird

song – and there he was perched on some old bean sticks, only a couple of yards from me. I sat quite still and he continued to pour out his silver stream for some minutes. I'd never before heard birdsong so close and it was a thrill I shall remember.

In understanding a non-ecological listening, one where depth of identification matters rather than the equable awareness of sound in the round, every detail of this episode speaks to me. Bill had been digging someone else's plot. This was the free labour of love, the gift of service without desire of return. A habitual practice, it existed outside the pastoral capitalist economy, in its small way enacting Marx's utopian vision for a society where, recognizing that a 'totality of needs' could only be met when each became a 'means for the other', there is a free exchange of 'social labour'. My father does not record the kind of exceptional spot in time beloved of middle-class novelists: he describes 'sensuous human activity, practice' that, at least in Marx's lofty formulation, is historical, 'a conscious self-transcending act of origin'.[46]

This is how Bill finds himself next to the blackbird. The sounds of his neighbourhood did not impinge on a listening subject: they were heard through the exertion of hoeing or planting out or digging, the after note of which is the returning consciousness of external sound associated with taking 'a breather'. These ambient sounds were *pulsed* according to the slowing heart rate audible in the ears. There is a crossover between the inner rapids of surging blood and receptivity to the inspiration of the outside world. The blackbird also rides this chiasmatic tide. Necessary to its approach is sleep. Practically, my father's inertness as he dozed bred trust, but what counted (on waking) was the sensation that the bird sang *as if he was not there*. An embodied listening occurs that transcends the practical body of labour. It is unmistakably an experience in nature of 'conscious self-transcending'. The blackbird gives its labour as freely as the gardener has given his – which is not to deny that the blackbird may benefit from worms dug up when the soil is turned over! The blackbird's communication means nothing: it speaks to the sonorous body. It is not the familiar song that is remembered, but the thrill of harmonization, a sensation that the blackbird is so close that (almost) it sings out of his hearing. Harmonized, subject and object are integrated, and, as a result, the remembered sound will never slip into the past like a sound recording or become a 'precedent' for, say, recognizing the blackbird's song in the future.

Growing hard of hearing, my mother most regretted not being able to hear the birds. Sometimes she would ask me to walk to the bottom of the garden and report back what I had heard. The retreat of human voices appeared to bother her less, as she tended to blame any communicational deficiency on her interlocutors. If only they would just speak up, she complained, echoing what her father used to say when *he* grew hard of hearing. I think the onset of deafness may have been earlier than we imagined: the control she sought to exercise over my affections may have had a physiological basis; my withdrawal from her may have been perceived as auditory. Her lack of conversational engagement may have had the same cause, her repetitiousness masking the fact that she couldn't hear us. Later, there were unsatisfactory trials with hearing aids, which, belying their name, made it possible to listen to the sound environment, but impossible to hear anything. Radio and television voices could be turned up to full volume; this made ancillary conversation impossible, even drowning out the telephone; so another cone of silence enveloped her. My mother kept a house for a guest who never came, a stranger who would receive her hospitality without question, with whom communion would be immediate and absolute. Real guests fell short of this ideal; they could not always receive what she wanted to give; they sometimes desired to transmit something back.

I try to imagine all the words in her life, and wonder whether she was the victim of a cruel hoax. The reward of resisting external speech, the Quietists said, was an internal communion. An 'unheard melody' would compensate for those 'heard melodies' that caused intellectual, emotional or sexual confusion. But instead of this she simply heard less and less. At the back of this Methodist drama was always the example of the purer Puritans, Fox's desire to retrieve the Word from words, the Image of God from signs. Or as Fox wrote, 'Soe all these names were spoaken to man & woman since they fell from ye Image of God: & as they doe come to bee renewed againe upp Into ye image of God they come out of ye nature and soe out of ye name.'[47] Fox experienced that rapture in which the Word was freed as the Sound that destroyed mere sounds, as the inward voice that was immune to deafening. Hence, at his Lancaster indictment in 1664, he prayed before meeting his judges. And his prayer was answered: 'The thundering Voyce said: I have Glorified thee: & will Glorifie thee Againe: & I was so filled full of Glorie yt My head & eares: was ffilled ffull of Glorie: & then when [the] trumpets & Judges Came up Againe, they all Apeared as dead Men under me.'[48] In this case it is the makers of noise who are reduced to phantoms and ghosts. The thunder of inward prayer is like Alfred Williams's factory: it drowns out every other sound – with the difference that it embodies a different kind of hearing. But I don't know what Margaret listened to in the long sessions with the crossword puzzle or what sensory dissociation she endured when she glanced up to see the birds feeding noiselessly at the bird table.

When the sonorous body truly hears, it is illuminated. Fox experiences these times of intense prayer as enlightenment. To receive power is to receive light: 'Stopp not ye eare against that which letts you see'.[49] This light 'which comes from Christ which never changis'[50] is the Word embodied. If we detach it from its theological frame, it is the same light of which Merleau-Ponty speaks: 'At the moment one says "light" ... there is no lacuna for me; what

I live is as "substantial", as "explicit", as a positive thought could be.'[51] Or, again, 'When the silent vision falls into speech, and when the speech in turn, opening up a field of the nameable and the sayable ... when it metamorphoses the structures of the visible world and makes itself a gaze of the mind' – due to the 'reversibility which sustains both mute perception and the speech',[52] it glosses Fox's sensation that for him also in the mute world of the body 'all the possibilities of language are already given to it'. The intermediary of this revelation in everyday practice is Holy Writ, whose repetition with interest produces a physical impression rather than a clearer conception. My grandfather, Fred, was a Methodist lay preacher. My father used to accompany him to villages on the Circuit of local churches, one Sunday piping up: 'Not the Prodigal Son again, dad'. A family anecdote notable for its rarity, it suggests a repetition without inspiration.

'When a piece is copied over, by someone else or even by the author himself, that person must re-experience himself during the act of recopying; otherwise the piece loses all the rightful magic that was conferred upon it by handwriting at the moment of its creation, in the "wild storm of inspiration"',[53] explained Velimir Khlebnikov, in 1913, and what applies to handwriting also applies, presumably, to preaching. At the moment of its creation, handwriting is a movement form. The person who re-experiences himself in copying it is like someone listening to a melody: no letter is left behind as the hand having writ moves on: the future writing is 'pro-tended' in the characters themselves and the rules governing their relationship, and, for the same reason, the evolution of the line retains what has gone before; so with preaching, which is, etymologically, to speak before, to make prophecy. Of Fred's oratorical powers I know nothing but that his popularity as a preacher suggests fluency; perhaps, one speculates, domestic taciturnity was swapped on these occasions for a declamatory power that made the old story come alive so that all of its parts were seen at once, the relation as a whole. Fred was a remarkable man; without the advantages of rank, education or wealth, he served for long periods as chairman of the local Rural District Council. A letter of condolence received after his death stated that he regarded his finest public achievement as the provision of piped water to the town, a service contributing to improved sanitation, improved health and in general relief for the poor.

To quote from the manifesto of a Russian Cubo-Futurist poet may seem out of place in the context of discussing the Methodist culture of north-west Berkshire. The late phenomenology of Maurice Merleau-Ponty is also clearly imported from elsewhere and disturbs the chronological stratification conventional in autobiography. Gaston Bachelard's 'psychoanalysis of objective knowledge' is more beguiling because it is workmanlike: weeding out facile metaphors, to replace them with more profoundly materialized concepts, it pursues the same goal as us when we replace the consumer of the auditory picturesque with the sonorous body actively making sense of its world. Such retrodictions of encounters in other countries of the mind illustrate the return of the prodigal son, the one who (as my father once said in response to a historical question) would not need to ask these questions if he had stayed at home. But they also illustrate the involuted logic of reflection, which is not stratified but interfolds all convergent data of consciousness. However dispersed these images may be in time and space, in the vortex of creative re-enactment they spiral together, concentrating inwards to give to the *punctum* in memory amplification and volume. To pretend otherwise, to cultivate a sentimental primitivism in relation to the community of Primitive Methodism in north-west Berkshire, is consciously to censor what flows into the mind when the past is re-experienced and becomes present again. I come from somewhere else; if I didn't, there would be no return.

Besides, as regards a sound history, the birdwatcher from Kazan, who translated bird calls into words, and anticipated 'transitional' species, is far closer to me than John Betjeman or Philip Larkin. His experiments in sound symbolism and language creation may throw more light on the unspoken future fantasy inscribed in our religious practices than any image of church going ('I take off / My cycle clips in awkward reverence') however brilliantly observed. After all, we were also strangers in a foreign country, intent on conversion. A recently posted note on Esau ('Charles') Martin identifies Lucy as his youngest daughter: my grandmother, Fred's wife, was born (according to this entry) in 1875 and in 1911 she was listed as a dressmaker. Charles Martin is listed amongst the great Methodist preachers of the Primitive Movement. Faringdon was the 'natural' headquarters of the Methodist Circuit. The pioneers of the movement came from the south-east, and used the language of conquest and conversion to describe their progress up the Thames Valley. They regarded the Thames as demarcating an ultramontane region, wilder than the villages lying to the south. They consolidated their position in Faringdon by taking over the Coxwell Street region. It is clear that 'we' positioned ourselves as a progressive wedge in hostile country; we were 'local', but, as good converts, outside the Anglican fold. And like any invading clan, we justified our incursion with the promise of salvation. My own name invoked the man of God who 'was caught up into Paradise and heard inexpressible words which a man is not permitted to speak'.[54]

I used to regret that neither my father nor my mother told me stories. But it is clear that no stories had been told to them. Hadn't Methodism delivered them from village superstition into the prospect of a civil, socially reforming rationalism? This process had been going on for over a century. I find that in the 1860s George Sturt was already turned out of that picturesque cradle of oral lore. 'I got from nowhere any of the local country superstitions, and the reason was almost certainly that my mother's own mother had a strong religious disapproval of such ideas.'[55] The aesthetic cult of the village (Alfred Williams, W. Hudson, Richard Jefferies) is connected with the breakdown and disappearance of this organic tradition: like the idealized soundscape of the acoustic ecology movement, it is a product of loss. Solus-like fantasies of autochthonous growth need to be placed in such contexts. The village equivalents of Paul's lifting to the Third Heaven belong in a historical order. Their expression is chorographic, related to the human geography of the neighbourhood when it is learnt through 'circuiting', a Methodist-like beating of the bounds, whose knowledge is of movement forms, not a compilation of picturesque views. Within that creative region, other discriminations are necessary. They can depend on nothing more tangible than the state of the weather, and what might follow from it – a visitation, say, by rare birds, vagrants thrown off habitual migratory paths by perplexing gales or geomagnetic anomaly. The breakthrough when it comes is a call from the beyond, but mediated by angels, always imprinted with a place and time.

Illumination or enlightenment is always a revelation of the particular. The 'enchantment' Bevis experiences in Richard Jefferies's eponymous novel occurs when light individualizes what is in view: 'Every atom of sand upon the shore was sought out by the beams and given an individual existence amid the inconceivable multitude. ... The light touched all things and gave them be.'[56] My Road to Damascus experience inverted this perspective, impressing on me my own individual existence. Over half a century later, I can still fix the place: its physical coordinates no longer exist, but it occupies a secure location in my personal geography, one that always pointed to the future. Standing on the path outside the garden wall and near the shed – the branches of the plum tree were bare – light broke through the high estate of clouds. Routines of revelation were our meteorological advantage: the shuffling of altocumulus, habitual across the border of winter and spring, produced apertures through which the sun broke in, watery, at first tentative, then growing in strength and taking hold of the field. An everyday revelation, then, except that moment coincident with a window opening in the soul: I noticed the light and stood to attention, as one surrounded and called. I climbed outside myself and saw myself, uniquely there in this time and place. I grew tall like a tree; I spread wings like a bird. Experiencing the second birth of individuation, I realized I was part of nature and it called me to speak.

The sound of the world that transcends silence is synaesthetic: it fuses seeing and hearing in an act of calling. As Harrison remarks, with regard to a sound, 'It is hard not to want to be going somewhere'; and when we do this, 'arriving at a completeness within a sonorous occurrence', there arises 'as a necessity ... a possible object'.[57] In this sense, the hearing described in this book is also *kinaesthetic*: it is perceived as a movement form, an indication or oriented chisel in space. The particular moment radiates to comprehend 'the remotest borders', as Traherne writes. Overcoming silence, migrating to 'the utmost ends of the earth', the sonorous body experiences a new communion with the world. But how might this perception, directional and circumambient, be translated into poetic expression? How is the calling to speak answered? How is the indicated communion achieved? There has to be a mechanism for translating between the individual trembling of the attentive watcher and the mass movement of the place as a whole, a rhythmic alignment or entrainment that lets the solitary watcher and listener feel embraced and carried up. Like a starling that responds to the crowd impulse when it joins thousands in a wheeling murmuration, they must possess an intuition of congregation that mobilizes the local place as a creative region.

In 1919 Edmund Selous hypothesized that the multiplex consistency of starling movement pointed to the existence of telepathic communication between birds. Riley, informs us, though, that

> recent experimentation speaks of murmuration as a phenomenon of anisotrophic co-ordination and adaptive mimicry. The theory is that each bird within the group is able to mimic the movement of the surrounding birds to an optimum number of five or six. It is suggested that the murmuration takes the form of a wave-like undulation because this mimicry is numerical rather than spatial. The starlings copy the movements of their surrounding birds over the reach of a series of unequal distances.[58]

On this argument, local enumeration practised generally produces murmuration; likewise, magnetized iron filings reorganize locally to produce a recognizable overall pattern. The equivalent in communication is a Chinese Whispers effect, a mimetic reproduction of what is audible in the immediate neighbourhood that incorporates the earlier history of the call's transmission from elsewhere in the murmuring region. Sewing its patchwork song from overheard scraps, the starling is not simply an environmental sound playback machine; like a radio telescope looking back in time when it probes deep space, its vocal *enumeration* makes *ancient noise* audible again through a new arrangement.

Sound histories are political. Moments of disclosure presuppose a prior history of enclosure. The drama of revelation depends on a frame; even temporary windows in clouds remind us that the charity of light is associated with exclusion. Sounds were associated with the transcendence of place because we inhabited a post-Enclosure environment. As regards the movement form, the freedom it implies depends on transcending boundaries. Sturt, for example, associates reintegration with a pre-enclosure freedom to ramble. Deprived of voices, places speak to him. The uneven extent of ground stretching out on every side becomes lately numinous: 'The road was solitary, but always alive and full of light.'[59] By our day even the roads had been closed to pedestrian traffic. Not only Cobbett's agricultural labourers but the ordinary inhabitants of the town were swept off the approaches to and from the marketplace, and crept against the walls to find shelter as the lorries thundered by. In the 1970s the campaign to divert heavy commercial traffic from the centre of the town recalled us to something else. The physical drowning out of speech caused by the volume of the traffic (in a double sense) was symbolically, and perhaps operationally, tied to the elimination of a space where the public interest could be represented.

Although the By-Pass Committee calling for the restoration of the public domain was probably unaware of it, a precedent had been set a century earlier: Early in 1873, 'some labourers held a meeting in the village of Littleworth, near Farringdon, in Berkshire. An ill-conditioned farmer made a fuss, and so three leaders of the meeting were summoned before the Bench for having caused obstruction on the Queen's highway.' Joseph Arch, founder of the National Agricultural Labourers Union, who recounts this story, comments, 'This was all nonsense, and only a pretext, because the Primitive Methodists had been in the habit for years of holding meetings on the very same spot.'[60] Arch on behalf of the Union decided to challenge the decision: 'We arranged to hold a test meeting towards the end of March, and it was a tremendous one. On the appointed day the big Market Place at Farringdon was more than half-filled with labourers.' As expected, the superintendent of police arrived to inform Arch that the meeting was illegal. However, the organizers had taken particular care 'to keep the crowd all round us compactly, so that people could walk or drive about easily' and the meeting refused to disperse.[61] The ensuing legal battle produced a tactical victory for the Union. But strategically the market place remained contested ground. As late as the 1945 general election, my father recalled, the local 'lord of the manor' sought to ban public speech there on the grounds that *legally* the marketplace belonged to him.

My father was named for Charles Martin, his grandfather on his mother's side. There is a family tradition that Charles Martin was Joseph Arch's 'right-hand man'. In an obituary article for the Methodist Recorder my cousin writes of his mother (Bill's sister and my aunt): 'A maternal grandfather worked closely with Joseph Arch, fighter for the rights of the agricultural worker.' I heard this too, and while I have been unable to find documentary support for the claim, it is not implausible – Joseph Arch was a Primitive Methodist preacher, like my great-grandfather – and the anecdotal parsimony of my family suggests that this association survived to enter family lore because it held exceptional significance. I seize on it because it suggests how my sound history grows out of the material conditions of production in an orthodoxly Marxist fashion. Perhaps Esau first met Arch at the Littleworth meeting, staged at 'the very same spot' where he regularly preached. Equally, or in addition, the association may have stemmed from a broader cultural alignment:

> I do not believe that the mass of peasants could have been moved at all, had it not been for the organisation of the Primitive Methodists,

a religious system which, as far as I have seen its working, has done more good with scanty means, and perhaps, in some persons' eyes, with grotesque appliances for devotion, than any other religious agency.[62]

Arch grasped the nexus between freedom of speech and a place to speak from. The long-term effect of enclosure, where, as Eric Hobsbawm has written, 'a relative handful of commercially-minded landlords already almost monopolised the land, which was cultivated by tenant-farmers employing landless or smallholders',[63] was a rural population that suffered in silence: 'The men were murmuring and muttering the countryside round, but they wanted a voice; they spoke low among themselves, but they were afraid to speak out.'[64] Arch's solution was practical: return the land to those who work it productively and in one fell swoop you solve the problems of under-productivity and poverty. 'When a man has three or four acres of land, I ask how many of such men, when they come to the downhill of life will become chargeable on the parish.'[65] Smallholding security also engenders a desire for self-betterment: 'Let the labourer raise himself by the land. I teach the farm labourer never to be satisfied while there is a chance of advancing in life. ... We should increase within his mind a just discontent for every year of his life, to make himself a better man, and go one better every year he lives.'[66] Inevitably, one expression of self-advancement will be public speaking directed towards political reform, among whose outcomes may be counted the 1882 Allotments Extension Act, the 1884 Parliamentary Reform Act and the 1894 Local Government Act (which established rural district councils).

Closing up the tiny house where my mother lived until her death in 2004, I plucked almost at random from those neglected shelves Winifred Holtby's *South Riding*. Flying across the Atlantic on my way back to the underworld, I started to read it. At once it carries me back to the early 1930s, and to the admirable Phoenix-like energy of a new generation emerging from the disillusionment of the First World War determined that it shouldn't happen again. The capacity of Holtby's people to reconcile traditional affections and a progressive politics, their English combination of deference and independence, their stoical public-spiritedness proof against every small-minded obstacle – these were the forming virtues my pre-parents incubated with such astonishing precocity in the mid-1930s. Sarah Burton is the defiant, practical idealist that my mother might, in different, more egalitarian circumstances, have become. I can see in the figure of Carne, the other side of my father, the 'natural gentleman' who, though contemptuous of the abuses of the class system, represented far better than the farmers and landowners who hemmed us in, the values of 'country life', with its respect for order, its leaning to continuity, its sense of social obligation.[67]

The revelation of *South Riding*, though, is not its general values – consistent with those explicated elsewhere on the bookshelf in their economic

(*The Economics of the Hour*), political (G. D. H. Cole) or ethical (G. B. S) applications. It is the fact that the author shows these principles in action, in the active ordering of a community. Fictional access to the public face of my parents' world comes from Holtby's simple but highly original device of tying her human stories to local government decisions. The apparently dry stuff of committee decisions is shown adding to or diminishing the sum of human misery. In Holtby's idealized shire, to enter local public life is to play a creative role in the amelioration of ordinary, individual lives. To participate in local government is more than an exercise in 'practical Socialism', it is to discover the nexus between personal lives and public realms, the direct reciprocity between decisions made in the public interest and the quotient of social suffering. To supply clean water, to assure shelter, to offer educational opportunity – these public initiatives are still directly connected to the experience of poverty, ignorance and their attendant abuses. So does a public life spring out of first-hand knowledge, and a reform of the objective conditions of existence secure a greater chance of personal fulfilment. In this utopian moment emotion and reason were still constructively entwined, and the minutes of the Rural District Council could be as filled with practical poetry as any pronouncement of the Oxford-educated laureate.

In this spirit my father and his father entered local politics, serving on a council that, as Holtby says, its Conservative members regarded as 'apolitical'. The Rural District Council was a modality of public life that uniquely mediated between the old order and the new, between inherited values and acquired ones. It was a forum where one could engage with the old mythos of an England ruled by class interests, tempering it with the 'secularism' of a socialist reason that yet remained loyal to England's culture. The England my Methodist and socialist upstarts wanted was a politically more representative and socially more just version of what already existed. True to their horticultural calling, they saw this as a process of weeding out what stifled growth and cultivating what was sturdy and likely to prove fruitful. Their political praxis and their method of cultivating the land exactly mirrored each other. The garden was one expression of cultivated order; another resides in the unread minutes of long-forgotten council meetings where, presiding over such apparently dry matters as rent rises, changes in rates, road improvements, transport needs and the provision of schools and hospitals, an attachment to place was nurtured that grafted on the improvements due to public hygiene and access to schooling. But these debates, too, I never heard.

If the silence my father kept about his views has made their recovery speculative, how much harder to penetrate is the absence of any direct testimony from my grandfather? But I am optimistic: if Bill's views could be cut together from passages in books on my parents' bookshelves, then an informed guess about Fred's views can be derived from the writings of a figure whose influence (political, social and personal) was clearly immense.

Fred, like Joseph Arch, understood the tripartite relationship between speaking up and having a place to speak from. Arch spoke from a propertied position. Living in his father's house, and inheriting it, he writes,

> I could stand up and look the whole world straight in the face, for I owed no man anything, not so much as a copper farthing. My plot of land was no waste field either; it was a fruitful garden if ever there was one; every square foot of it was tilled, planted and watered, and I raised more fruit, flowers, and vegetables off it than I had any use for.[68]

Fred's achievement in this regard was greater. After the First World War, by means never discovered, he acquired five acres of freehold land. Then, inside its working orchard, he designed a garden for congregation. A photograph exists of a Methodist Circuit Meeting, taken on the lawn around 1930. The lime grove and the elm tree stand in the background – and the summerhouse whose fascia sports antler-like decoration. The fellow sitting on the grass left of the somewhat ecclesiastical-looking group is Fred; the schoolboy on the grass is my father.

After these great meetings, the Methodist ministers and their families dispersed to their different towns and resumed their duties on the different local circuits. The organization of the Methodist movement and ministry involved an intermediate mobility. Ministers were neither sedentary nor vagrant but routinely travelled round the district, preaching at appointed local chapels. In the early days, following the manner of the Wesleys, preaching occurred on the village green, or wherever a verge, corner or other irregularity had escaped the Enclosure robbers. The parallel with Australian Aboriginal governance is striking to those who have studied it: the Scottish settler and Aboriginal advocate, James Dawson, described the 'great meetings' of the regional clans in south-west Victoria, periodically convened to conduct business, resolve differences and share information. Enclosure produced vagrancy and, consequently, tighter laws against trespass; in Australia, an administrative obsession with herding Aboriginal people off their land into 'villages' produced a new class of outcasts. 'These white slaves of England stood there with the darkness all about them,' wrote Arch of the rural dispossessed.[69] Likewise in Australia, the new labour was free to those who would purchase it because it was landless; if the slave stood still, it was because the reward of movement was arrest and imprisonment.

Corresponding to this intermediate mobility in the human sphere were the regular comings and goings of the birds, whose migratory behaviour was classified sociologically. Sedentary or resident species and regular migrants, sometimes called summer visitors, might be antithetically inclined geographically, but both respected the integrity of the territory: whether the swallow spent the winter at the bottom of a pond or hawking south of the Atlas Mountains was really of no account. What mattered was its devout *campanilismo*, its unfailing annual return to the same barn, nay, the same nest. More challenging to the stability of place-based studies (whose exemplar remained Gilbert White's *Natural History* of Selborne) were those species that were cosmopolitan, whose populations regularly absorbed immigrants from the continent, and which were themselves known to leave the district and stray as far as northern France. Non-breeding vagrants, irregular and passage migrants, as well as rare to very rare visitors, could be explained by exceptional weather patterns, and had the exotic charm of the powerless; but occasional, unpredictable crowd irruptions and invasions leaving in their wake remnant colonies were anomalous, fascinating and vaguely disturbing.

A further guide to the distribution of species (and songs) was their preference for different habitats. As all of these, lake, wood, hedgerow and meadow, lay under public ban, observations necessarily occurred from the edge; picturesquely out of reach, the middle-distant behaviour of birds was a two-dimensional drama of covering and uncovering. The danger of exposure extended to the schoolboy birdwatcher, whose curiosity drew him into the fields, and along field edges, to spinneys, coppices, the overgrown marches of farms and the borders of secret ornamental lakes or wilder woods reserved for hunting. These walks were also intermediate, solitary circuits of northern Thames Valley woods and fields regularly undertaken for the scientific benefit of the community and consciously prosecuted as unseen incursions into foreign territory. The census of souls I undertook was essentially auditory: the woods that concealed me from the gamekeeper also kept the birds largely out of sight. The lucky coordination of call, a flight path captured in peripheral vision and a momentary parting of branches in the foliage offered a glimpse of the perching birds, but those that flew low, or crept, or stirred at dusk, like the nightjar, were known solely by their give-away calls. These were the natural itinerant's longed-for moments of connection and revelation; their appeal was emotional not statistical. My family would have had no truck with Magic Methodism, but to stand concealed on the edge of a forest clearing or private lake, intent on identifying an unknown sound, induced an altered state of attention comparable to going into trance.

Territorialized and objectified in this way, song forfeited its sonorous body; indicating only itself, it was a slave to habitat. It is odd to realize how the bird books on which I feasted my eyes in the gloom after these days out colluded in this post-Enclosure prejudice in favour of fixed places. Witherby's Handbook, digested in P. A. D. Hollom's one volume *Popular Handbook*, prided itself on illustrating British birds in their natural habitats. So far as the perching birds went, which included owls, crows and kites as well as the thrushes, warblers and finches, this meant devising an elaborate scaffolding of dividing twigs, gnarled branches and curiously leafless stumps. The craft put into the depiction of different arboreal habitats – the cirl bunting amidst the oak, the serin in the alder – had a curiously primitivist aim. Its pastoral illusionism was designed to conceal the nexus between ornithological knowledge, Enclosure and the Game Laws, for most of the scientific contributors to Witherby were landed shooters. To effect this historical legerdemain, it was desirable to convey an atmosphere of settled antiquity. M. A. Koekkoek's finches evoked misty May mornings in country parks and on the fringes of farms with a sentimental ardour that far surpassed what was necessary for identification purposes. Such elaborateness of setting suggested longevity of residence and loyalty to place. The birds depicted in the *Popular Handbook* were dependants who, year in, year out, could be relied upon to sing fixed tunes from appointed perches.

In this context, a throwaway remark of Australian writer Hal Porter – 'The skylarks have no politics' – turns out to be badly misjudged. Evidently colonial intruders in Australia, they were associated in England with land that had not been enclosed. To find them in abundance, the easiest way was to climb White Horse Hill. Offering a bird's eye view of the Vale below, Faringdon's corallian ridge stretching north-east towards Oxford beyond it, and in its lee the misty gloom of the Upper Thames valley, the White Horse reproduced my earliest flights in the kestrel's bow-bend or the skylark's ascensions. I remember how, in September, I cycled to the base of the Hill and climbed to the Iron Age camp behind the Horse to study the diurnal migration of birds across my district. The naivety of attributing to flocks of meadow pipits flying south-west across the brow of the Hill migratory intentions now seems palpable. Most likely they swooped back to earth out of sight round the Hill. Swallows were on their way somewhere but flew at a height where the Hill's elevation made no difference to their direction. At the time I was in thrall to a new breed of 'young ornithologists' books, promoting ingenious observational stratagems and inductive boldness. In contrast with the older literature, these guides presumed a mobile, extra-territorial observer; they cultivated a hyper-visualization commensurate with surveying larger and wider places, with a corresponding diminution of interest in the diagnostic value of birdcalls.

As part of the National Trust estate, White Horse Hill had been stripped of any legends associated with it. Assessments of its significance were severely archaeological. The identification of the figure as a horse went unquestioned, even though to the untutored eye it more plausibly resembled a bird-headed cat. It took me a long time to find out what it meant in my life. My chief association from those days is kinetic: I remember one day running from the horse's tail down the steepest slope towards Dragon's Hill, running, leaping, in the full flight and power of my heedless ten years of athletic power; and not noticing the farmer's single stranded electric fence, bursting through it, as if breasting a tape; and nothing could enclose me or rein in my flight; and somehow absorbing the shock of the metalled road that rides the saddle between figure and Hill, scaling the verge beyond and fast striding, as if sailing, across the waving grass until, up, my momentum absorbed, I landed breathless and poised where St George was said to have slain the beast, a correspondent breeze rushing outside while the blood beat in my ears. In this flight was recapitulated my early aerialism. Standing on a rounded summit in the South Downs, W. H. Hudson wrote, 'I can almost realise the sensation of being other than I am – a creature with the instinct of flight and the correlated faculty; that in a little while, when I have gazed my full and am ready to change my place, I shall lift great heron-like wings and fly with little effort to other points of view.'[70] I had the same sensation.

In 1857, a great meeting was held here. The occasion was an attempt to revive the custom of regularly cleaning, or scouring, the White Horse. Traditionally, the 'ceremony of scowering and cleansing' had been the occasion of 'great joyous festivity'.[71] But the custom had lapsed: 'What with the Catholic Emancipation, and reform, and the new Poor Law, even the quiet folk in the Vale had no time or heart to think about pastimes.'[72] One victim of this 'transition time' was singing. 'We have ceased to be a singing nation,' someone says, in conversation with Thomas Hughes as they take supper in the great booth: 'Songs written for the people, about their heroes, and, I believe, by the people. There's nothing of the sort now.'[73] The sequel proves him partly wrong, as Hughes records a number of songs performed in response to this challenge. But they are only sung because an exceptional circumstance has brought people together. Hughes recognizes this nexus, and a function of his book *The Scouring of the White Horse* is to promote a folk revival. As the chief threat to customary lore (and law) comes from the Church, he prints a lengthy defence of feasting written by a sympathetic minister. But its advice that dancing and singing are only 'lawful and right in the sight of God, and fit things to do when we are rejoicing before Him',[74] is, in effect, the same condemnation of pub singing recorded by Williams. The year 1857 was the last time a large crowd congregated here to sing and dance.

TWO

Returns

In June 1991 I visited the provincial city of Lecce in Puglia in the heel of Italy. It was not my first visit but this time I travelled with sophisticated sound recording equipment. One evening I set up my Sennheiser stereo microphone on the balcony and started to record. At this time I had been living for nearly ten years in Melbourne. I had become fascinated by the problem of living in a new country, where, it seemed to me, the stable points of reference that make communication possible were missing. I retrodicted this feeling of being all at sea onto the colonization of Australia, and started to collect evidence that early explorers recognized this aporia in their cultural resources, and in response developed and applied poetic techniques to bring the place into being and to make it speak their language. Articles on this topic attracted attention, and I was invited to co-script a theatre work called *The Horizon Papers*. I clearly remember the day I pulled out of that collaboration, and retreating to the historical collection of the State Library began to write my own work, later broadcast as 'Memory as Desire'. In the same year I wrote a second radiophonic work, 'What Is Your Name', which derived much of its material from early white attempts to 'get' the Aboriginal languages. The following year *The Road to Botany Bay* was published with the subtitle 'a spatial history', a neologism that aimed to foreground the kinesthetic foundation of *relating*, in the double sense of description and orientation.

Lecce is a city of returns and echoes. The choreographic genius of its curving streets and elaborately arabesque baroque facades is distilled in the gesture of the Solomonic or barley-sugar column, common in the churches. In *Baroque Memories*, an anti-novel largely written in Lecce that summer, a narrative structure is devised that imitates the column's double spiral. Instead of evoking characters with stable and evolving identities, I conjure up a world of doubles, where coming into being in a new place (always as ephemeral as a whirlwind of dust arising out of the plain) is a matter of chance encounter and echoic mimicry. Intended is a description of the migrant condition where, in the absence of a language in common, a discourse of sounds in-between is improvised. Sounds in-between can develop into languages, as they do in pidgins; but they can also be the basis of singing together. Because they are mimetic, they allow the sound environment to be a secret sharer in the conversation. A feature of the speech performances generated in this way is that they may have no other purpose than to defer the ending. The interest of the exchange, in the sense of *inter esse* or what exists in-between, may be indexical rather than symbolic: what will count most is the sculpting in real time of a hollow, or place of shared coexistence, rather than the translation of linguistically pre-scripted information.

The *vico* I record is an urban auditorium; its narrow walls amplify the voices below and lift them to the rooftops. A *passeggiata* is in progress, flowing in and around doorways, pausing where the *invecchiate* huddle in chairs. Children play at ball or run in and out; swifts scream overhead against the lilac afterglow of the *tramonto*; television sets are just audible from kitchen interiors; a dog barks on a nearby rooftop pergola. Listening through the headphones, I become aware that the order and distribution of sounds is not random. It appears to be produced by subjects who listen to one another and consciously thread themselves into the collective sonic fabric. Children cry out, women pause mid-sentence, then resume passing the time of day; someone arrives or leaves, their passage surrounded by a cloud of greetings and partings; a scooter cuts in and revs to a halt; the silence in its wake fills with the murmur of conversation resumed; a dog barks, a swift, unusually low, screams past, decisive as a conductor's 'release' after which the tempo relaxes. Then, like crickets when the shadow moves away, the street comes alive again: a radio burst from an opening window, a clatter of dinner plates; a child's name called into the dark may trigger it. Underneath this spatio-temporal distribution of sounds an oceanic urban intonation is audible: unidentifiable rustlings, squeaks, tremors, as if the city is clearing its throat.

When I made a radio documentary based on the Lecce sound recordings, the title I gave the piece ('Tuned Noises') reflected the difficulty I had in defining the phenomenon I had observed. Neither musical nor linguistic analogies could respect the fact that the compositional principle at work was, almost paradoxically, deliberately unconscious. In *The Soundscape: Our Sonic Environment and the Tuning of the World*, Murray Schafer advocated studying 'the effects of the acoustic environment or soundscape on the physical responses or behavioural characteristics of creatures living within it'.[1] Implied is an ideally steady-state relationship between human speech and the 'soundscape' or 'sonic environment'. The effects referred to here are certainly unconscious. If they climb into consciousness it is because of a physiological necessity to notice them – Sigar's nightingale or Williams's railway factory are extreme instances. However, their climb into consciousness, while it may be negative – Schafer's 'flat-line' industrial sound falls into this category – is unlikely to be positive unless it can be musicalized. In musical terms attunement may be contrapuntal (a kind of echoic mimicry) or involve some form of harmonization (a certain pitch equivalence). But either response is consciously conceived and executed. Similarly, to talk of a *dialogue* between speech and environmental sound implies an intention to communicate with one another. In Mozambique human honey hunters and avian honey guides may talk to each other about realizing a common goal, but this must be exceptional. In general, the collective auditory awareness that has to be hypothesized to explain the Lecce phenomenon seems entirely without aesthetic or discursive ambition. Beyond its continuous self-production it has no purpose.

Tuning is a misleading metaphor, but to classify the sounds in the street as *noises* is also problematic. The assimilation of what is undoubtedly a creative auditory region to musical or rhetorical terms endows it with a false intentionality. But to regard it simply as elevated noise (even if noise is evaluated positively) may be to ignore the sense in which it *is* a self-sustaining communicative structure. Sound psychologist Albert Bregman argues that auditory perception demonstrates a principle of 'psychophysical complementarity'.[2] Contesting the Gestalt view that perception is determined by forces of attraction and segregation inherent in our mental representations (whether or not they correspond to reality), he hypothesizes that the way we analyse sound evolves directly out of the character of a sound environment where 'noise' in the neutral sense of 'unwanted sound' is ubiquitous and silence unknown. But the problem is complicated. Most sounds we encounter are neither pure sinusoidal tones nor sudden noise bursts (the two classes of sound favoured by laboratory research). Such familiar noises as a car passing, a jar falling and breaking, the clatter of knives and forks, the sound of paper rustling, not to mention all the environmental sounds collected in Lecce, are neither periodic nor random. Bregman writes, 'We have no knowledge of how to characterise these sounds or of how they are classified and organised by the auditory system.'

In a structuralist view of meaning formation, the problem of auditory integration does not really exist. The Russian–American theoretical linguist Roman Jakobson cited neurological studies that showed, or apparently showed, that the right ear (or left hemisphere of the brain) was attuned to speech sounds while the left ear (or right hemisphere) specialized in recognizing non-speech sounds – music and environmental noises – to make the case for a fundamental distinction between speech and non-speech sounds. According to these studies, while the right ear is able to 'perceive speech sounds within real, meaningful words, synthetic nonsense syllables, and even in speech played backwards', the left ear specializes in 'all other auditory stimuli, such as musical tones and melodies (both unknown and familiar), sonar signals, and environmental noises such as a car starting, the sharpening of a pencil, water running and oral emissions apart from speech – coughing, crying, laughing, humming, yawning, snoring, sniffling, sighing, panting, or sobbing'.[3] Applied to vocalization, this distinction produces some counter-intuitive results. The linguist, Edward Sapir, for instance, thought that the 'sh' sound, made by channelling air between the tongue blade and the front of the hard palate, could be classified *either* as an unplaced sound (imitative of waves withdrawing from the beach) *or* as the fricative diphthong at the beginning of the word 'shout' (where it represented a speech sound whose meaning resided in its grouping with other speech sounds). The articulation might be identical, but only in the latter case did it convey 'significant concepts'. Theorizing *two* 'sh' sounds, Sapir doubted whether, even though they sounded alike, they were 'the same type of physio-logical fact'.[4]

While this analysis – which provided a dramatic framework for my 1987 radio work 'Remember Me' (subtitled 'Music for the right ear') – may pave the way for computerized speech, it throws little light on what might be called the 'Lecce phenomenon'. More recent studies of auditory perception reported by Stephen Handel suggest that localization studies quoted by Jakobson embody a mechanistic or vertical, rather than holistic or horizontal, view of brain function: while vertical organisation may be characterised as genetically determined, as it is localised in distinct neural structures, and for this reason is computationally autonomous in the sense that the output is automatic, and for this reason is not affected by other cognitive processes,[5] a stress on the horizontal processes of remembering, judging, comparing and associating might reveal the same cognitive processes acting on all sorts of input – visual images, speech sounds, odours, musical sounds.[6] Given that our auditory apparatus is attuned evolutionarily to dynamic events in a noisy environment, rather than to the pure, isolated sound events of the laboratory experiment, it seems likely that horizontal processes are essential to its successful operation. Given that auditory information is context dependent, that what may be interpreted as noise in one context, as music in another, may, in yet another, be the bearer of 'significant concepts', it seems unlikely that a neat division of functions across hemispheres will be ecologically efficient.

However, the effects audible in the Lecce recordings elude this kind of analysis. As a human murmuration, they represent the formation of a crowd consciousness that depends on a ventriloquial capacity to speak in relation to somewhere else. They may be said to notate a movement form, a crowd formation whose figures arise directly from interaction with a constellation of ambient sound sources. Rather than convey a concept or an emotion, they stage the ever-present desire of communication as such. A model of auditory cognition is needed that respects the outwardness of echoic mimicry. Conventionally, hearing voices is treated as a disorder. Handel concludes that both hemispheres have the potential to take on the typical functions of the other and differences between them may be less due to the speech non-speech distinction than to the complexity.'[7] This aligns with psychologist Julian Jaynes's argument that both hemispheres of the brain understand language and the hallucinatory voices schizophrenics experience may represent 'a partial periodic right hemisphere dominance that can be considered as the neurological residue of nine millennia of selection for the bicameral mind'.[8] In pre-Homeric, bicameral times, outwardness might, after all, have existed: sounds might have 'spoken'; birds might have been mistaken for the voices of angels.

This is a more liberal model of auditory consciousness. It allows for the cognitive assimilation of noise. However, it throws no light on how these noises are used to produce or compose communicative structures or resonant group atmospheres. To understand this, Canetti's concept of the 'acoustic mask' deployed in his plays, is a good starting point. As a stylistic invention, it consists not only of 'the words and sentences a character is in the habit of speaking' but of 'the sound and modulation of the voice as well as the gestures accompanying the utterances'.[9] An intention is, as Tsvetayeva said, communicated through the intonation, but an unconscious, even contradictory, desire may also be expressed. Groups, too, possess acoustic masks. People speaking a foreign language, for example, appear to perform a speaking pantomime, even though the words cannot be followed. In *A Torch in My Ear*, Canetti described a conversation between three or four people overheard in a Viennese café:

> The interplay among them produced the most surprising effects. The voices paid no attention to one another; each started off in its own way and proceeded undeviatingly like clockwork, but when you took them all together, the strangest thing happened; it was as though you had a special key, which opened up an overall effect unknown to the voices themselves.[10]

But what is that 'overall effect'? Canetti imagines it in terms of betrayal or self-revelation: the conscious content of the words masks an unconscious desire. This may be murderous (as when a crowd forms, speaking with one voice). Alternatively, it betrays a desire of communion, even when there is nothing to say.

Inhabiting a multilingual society, Canetti was always sensitive to the way communities hid themselves from one another; even within one local population, access to, or exclusion from, language defined relative insider or outsider status. His observation of the emergence from 'voices that paid no attention to one another' of an 'overall effect unknown to the voices themselves' may seem paradoxical. Extended to include the 'interplay' between human and environmental sounds, though, it is no different from the Lecce soundscape where each sound has its fixed identity but together they produce an effect, which I tried to explain on the analogy of tuned noises. In the non-human sphere it can be compared to the double life of the starling: not only a determined individualist but also haunting flocks whose formations it can neither detect nor design. The 'interplay' essential to the translation between individual and group does not occur in language but through the sounds in-between, which are available for reassignment – they hover between languages and lend themselves to migration from voice to voice. The means of this interplay is an echoic mimicry that defines communication performatively.

In Australia, it emerged that a baroque gift for elaboration might be a reaction to an existential silencing. Where the master language wore an impenetrable acoustic mask, all sounds potentially signified. The identity that emerges echoically through their imitation faithfully mirrors the subject's migrant condition – a situation Antigone Kefala describes in her 1984 publication, *Alexia: A Tale of Two Cultures*. To explain the lack of affect that typifies ordinary Australian English speech, Alexia relates a story:

> Basia, her only friend at school, also a Refugee from another Old Country, said that, many many years before, everyone on the Island had been forced to swear an Oath of Silence, and to speak only when absolutely forced, and even then to use the minimum of sounds, and if possible only a few, such as 'mg', or 'ag', which were forever repeated in sorrow, regret, surprise, admiration, entreaty, contempt, mockery.[11]

Listening to these instances of phonic iconicity, Alexia 'imagined that there must be a great and subtle complexity in these sounds and that her ear was not attuned to them, so she kept listening, listening, as one would to the cry of birds, hoping to discover the key to their language'.[12] Kefala's irony is directed towards the laconic nature of Anglo-Australian speech. But there is another irony of which, I imagine, the author was unaware: historically, Aboriginal languages had been listened to in the same way. At the end of the nineteenth century, the self-styled philologist Hermann B. Ritz, not content to 'connect the original speech sounds with definite psychic states and processes', maintained that in Tasmanian languages, 'all things were distinguished according to two ideas, namely, rest and motion. The liquid consonants expressed motion, and all the others, rest.' Had Basia consulted Ritz, she would have learnt that the combination of 'liquid + guttural', represented by 'mg', signified 'motion + rejection'.[13]

The Victorian philologist Max Müller, whose theory of 'phonetic types produced by a power inherent in human nature' Ritz followed, rejected the idea that the original constituent elements of language might have been interjections or imitations.[14] But not everyone was so phonophobic; and, again, advocates of an alternative 'bow-wow' theory of language had their migrant avatars. James Dawson, who studied Aboriginal languages in Western Victoria in the mid-nineteenth century, was convinced that many of the words collected in his Vocabulary were 'onomatopoetic' in origin. Dawson could produce over thirty birds and animals where 'the native names have been applied ... in imitation of the peculiar sounds they utter'.[15] He knew this appeared to contradict Müller's theory but concluded sturdily, 'The facts are against him.'[16] Take bird names like *Kueetch kueetch* (the rosella), named for 'its cry'[17] or the onomatopoeic invention *Kuurn kuurn kullut*, imitating the colonist's cockerel, and translated as 'call for daylight'.[18] Is this sensation that the environment *speaks* in its own name greatly different from the experience of the post-war migrant, who, arriving at the Bonegilla Migrant Reception and Training Centre, spent the day given him 'to settle down' going for a walk, reporting, 'My ears were getting used to the sounds about, and smiled [*sic*]. Cattle mooing, dogs barking, sheep, bees, birds, well THEY were "talking" as in Europa. It lifted my morale'?[19]

But what did they talk about? This seemingly guileless example speaks directly of the legacy of the Enclosure Acts, one of whose overlooked effects, the Hammonds point out, was the historical isolation of the poor. Social, as well as economic, divisions also widened; the 'companionship of the classes', celebrated by William Cobbett, broke down. Finally, territorial divisions (the laws against trespass) saw the 'commons of England [and Scotland]' renamed the poor or, worse, 'the mob'.[20] The *sound* of this systematic rural expulsion (and its corollary, usually involuntary emigration to the colonies) was bleating sheep, a sound, as Australian poet John Shaw Neilson realized, twinned with the silencing of the land: 'The land is all for thee, O woolly sheep. / For thee we grapple with the undergrowth, / For thee we spoil the glory of the trees. / All bloom must perish on thy heritage – / Begone, O singing birds and humming bees'.[21] Our ear witness's soundscape is an image of loss. Exiled to Siberia, the agrarian reformer Alexander Herzen expressed his nostalgia for his 'childhood home' in a sound image:

> The shepherd cracks his long whip and plays on his birch-bark pipe. I hear the lowing and bleating of the returning animals, and the stamping of their feet on the bridge. ... Then the voices of the girls, singing on their way from the fields, come nearer and nearer; but the path takes a turn to the right, and the sound dies away again. House-doors open with creaking of the hinges.[22]

In this remarkable image, sheep are the sound of coming home; but their return is premised on someone else's leaving.

The cows, dogs and bees may have 'talked' as they did in Europe but, apart from an immigrant skylark, or vagrant starling, the birds of the bush certainly did not. Legal exclusion from the squatter's paddocks, together with the clearing of native vegetation, caused a kinaesthetic immersion in the sound environment to be sublimated into a middle-distance auditory picturesque. Instead of considering sounds as productive of new senses, they were musicalized; instead of responding to the call, the listeners inserted a division between transmitter and receiver. In line with developments in the European theatre, they fell silent. And nature, extending the orchestration of the 'dawn chorus' into the evening, performed: 'We shall now be saluted by a very unique concert, which is so far professional that the performers are well acquainted with their parts. We will approach the banks of the nearest rivulet; and at once perceive that these musicians are frogs, who each in a different key sing merrily all night long.'[23] 'Emigration', as another Adelaide resident rhapsodized, 'provided it be to a good climate, brings its delights as well as inconveniences, and summer suns and cloudless skies, are better than gaslights, and Paganini himself need not be regretted, when you can listen to the music of rural sounds'.[24]

The apotheosis of this acoustic mask ('mask' because in the guise of aesthetic approbation it betrays a fundamental sonic alienation in which sound, instead of generating an echoic composition of self at that place, is musicalized) is an experience that disturbingly anticipates the sonic utopias imagined a century and a half later by the acoustic ecology movement: 'The incessant chirping of myriads of locusts lost its monotony, and with the rippling of waves upon the distant shore, blended every sound together so sweetly, that for a time the travellers listened in admiration to the most delightful of all melody, – the harmony of nature.'[25] Invoked is an ideal listening position that anticipates the focus of the microphone. Listening to everything nothing is heard; removed from the means of its production, the receiver feels no obligation to respond to what is heard. The rhetoric of harmonization disguises the refusal of an environmental contract of the kind that Neilson invokes when he invokes 'the glory of the trees'. More disturbingly, when the same auditory picturesque is applied to speech, a human contract is refused: 'The sound of their chattering in their camps was pleasing. Many of their words were exceedingly tuneful because of the prominence given to vowels. ... Pretty were the vowel cadences and sweet was the effect of rude dirges and chants carried by the wind over hill and creek and scrub to a distant listener.'[26]

Colonial history is a sound history. A recent study of Pitjantjatjara sound classification explains that

> categorising and naming according to sound ... is consistent with the concept of 'naming' a place by using the correct stylised sung form of the name; with 'singing' a person by manipulating his behaviour through the medium of song; with taboos on the use of the name of the dead, since their spiritual power can be tapped through the correct sounding of the name.[27]

Further, 'the sounds of the environment in general provide a different form of classification. There seem to be two main aspects. One illustrates value judgements about sounds. The other is an extensive vocabulary of onomatopoeic words which suggest that much of the classification of surrounding environment is done on the basis of the sounds produced by the thing named.'[28] On this basis, the writers hypothesize that 'one of the most critical elements in classification by Aboriginal people, probably throughout Australia, is sound, both musical and environmental'. The term meaning 'to sing' in Pitjantjatjara can mean 'the act of causing an event through the power of song'.[29] But, evidently, to the distant colonial listener it had the opposite meaning; musicalization was Malthusian theory for the ear; the cadences dying on the air foretold the dying of a race. When Ritz reduced the Tasmanian languages to two ideas (motion, rest), their speakers had either gone away or been put to rest. Somehow the speakers had connived in their own silencing; for whatever 'prominence' vowels might have had, they were dismissed – according to Ritz, the vowels in Tasmanian speech 'are liable to be changed at will' and are therefore 'so unstable as to be of no importance for our demonstration'.[30] A source of spiritual power was translated into lack of forethought, or cultural fecklessness.

Looking back at the first decade in Australia, I find the chronological sequencing weak. Experiences are organized helically, as ascents and descents that are homologous with the circuits of earlier events, although adding to their interest or modifying their significance. The temporality of that time is acoustic, rotatory, immersive and echoic in its movement form, rather than perspectival, possessing a purposeful forward direction. Migration magnified the amplitude of the oscillation between places: the chiasmatic interplay of remembering and forgetting, the mirroring of going away in coming back. Thomas De Quincey wrote of experiences that connected themselves with his life

> at so many different areas that, upon any chronological principle of position, it would have been difficult to assign them a proper place; backwards or forwards they must have leaped, in whatever place they had been introduced ... belonging to every place alike, they would belong, to the proverb, to no place at all; or (reversing the proverb), belonging to no place by preferable right, they would, in fact, belong to every place.[31]

David Malouf characterized *The Road to Botany Bay* as the imagination's 'true discovery of Australia', but the secret sharer in that journey was a way of relating found earlier in Venice. That book was a *geographical mask*; structured as a return journey, it betrayed a desire to make whatever place I was in belong to every place.

In the name of silencing, colonization silenced its own violence. 'The spot which had so lately been the abode of silence and tranquillity was now changed to that of noise, clamour and confusion,' David Collins wrote of settlement at Sydney Cove, but he added, 'After a time order gradually prevailed everywhere. As the woods were opened and the ground cleared, the various encampments were extended, and all wore the appearance of regularity.'[32] Contrast such acoustic clearances with the sumptuous sound picture that Gabriele D'Annunzio paints of the Festa della Senza in *Il Fuoco*, '*Lo strepito di un'acclamazione sorse del traghetto di San Gregorio, echeggio pel Canal Grande ripercotendosi nei dischi preziosi di porfido e di serpentine che ingemmano la casa dei Dario*'[33] or (of the crowd's clamour in the Piazzetta), '*La folla nera e densa nella pausa ondeggiando, i vani delle logge ducali si riempevano d'un confuse romorio simile al rombo illusorio che anima le volute delle conche marine*'.[34] An enlightening comparison, or just a bizarre juxtaposition born of my personal history (I first visited Australia after living in Venice)? In any case, the fantasy of a Venetian Australia persisted: in *The Lie of the Land*, the *intoned* 'dot-and-circle' Papunya Tula painting practice is compared with the *tonal* values achieved through the Venetian art of *macchiare*; both reimagine geography as rhythm, noting a humming ground whose intonation resists imperial history's linearization of time and space, its confusion of order with silence.[35]

Writing to T. S. Eliot about Guido Cavalcanti, Ezra Pound insisted that the Florentine poet (and contemporary of Dante) 'thought in accurate terms … the phrases correspond to definite sensations undergone'. It seems to me that, despite their seductive fluency, D'Annunzio's phrases have a comparable forensic precision. An urban setting is evoked where the sounds of nature wear a human mask: a *strepito* (noise, din, clamour) metamorphoses into an *acclamazione* (a crowd shouting its approval). In the resonant chamber of the city, the tumult is amplified, circulated and absorbed. The Grand Canal's cool surface causes the sound waves to slow down, to refract and seem amplified. Angularly reflected off the Ca' Dario's hard, bright walls, the echoes falls into phase, an impression conveyed by the repeated guttural consonant *c* (riper*c*otersi, dis*c*hi). The pulse softens, dying into the rhythmic slap of waves against *fondamenta*. The architectural apotheosis of a city composing itself sonically is the Ducal Palace, the supreme auditorium for echoes. In a reversal that recapitulates the myth of Venice borne ashore Venus-like, the transitory rumour of the crowd's ebb and flow produces the illusion of the sea, its mighty swell and eternal whisperings. In returning the voice of the crowd, creating an 'overall effect' unknown to the individual, the 'organising principles' of the Palace produce 'involutes', as De Quincey called them – 'far more of our deepest thoughts and feelings pass to us through perplexed combinations of *concrete* objects, pass to us as *involutes* (if I may coin that word) in compound experiences incapable of being disentangled, than ever reach us *directly*, and in their own abstract states'.[36]

In Botticelli's *The Birth of Venus* the goddess drifts ashore in a scallop shell. D'Annunzio, though, compares the cavernous reaches of the Ducal Palace to a conch shell. In developing his concept of memory involutes, De Quincey borrowed the language of shell taxonomy. Introducing *The Conchologist's First Book*, Edgar Allan Poe (no less) defined the 'involute' as a rolled hollow without a spire. A volution wound inwards, in contrast with the revolute, it is but one type of univalve interior architecture. Most univalves, he explains, possess a 'Pillar, or columella ... that process which runs through the centre of the shell in the inside from the base to the apex of most univalve shells, and appears to be the support of the spire'.[37] The 'spire consists of all the whorls of the shell, except the lower one ... termed the body of the shell'.[38] The 'whorl is one of the wreaths or volutions of the shell'.[39] Hence, an 'involuted spire' characterizes 'those shells which have their whorls, or wreaths, concealed in the inside of the first whorl or body, as in some of the Nautili and Cypraea'.[40] In late 1896, working on *The Interpretation of Dreams*, Freud tried to explain to Fliess the *polychronicity* of memories. 'Our psychic mechanism has come into being by a process of stratification, the material present in the form of memory traces being subjected from time to time to a *rearrangement* in accordance with fresh circumstances – to a *retranscription*,' he claimed: 'What is essentially new about my theory is the thesis that memory is present not only once but several times over, that it is laid down in various kinds of indications.'[41] It appears that De Quincey anticipated this palimpsest conception of memory. In addition, he had ascribed to it a distinctively vortical structure.

A sound history may have a distinctive form, homologous with what it remembers. A particle theory of sound's generation might visualize D'Annunzio's scenes as a kind of Brownian motion, where a multiplicity of sound trajectories bounce off obstacles, collide with one another, and produce an incessant, random variance. Referring to the evolutions of dust glittering in a sunbeam, the Roman poet Lucretius wrote, 'You will see a multitude of tiny particles mingling in a multitude of ways ... their dancing is an actual indication of underlying movements of matter that are hidden from our sight.' Transposed to the realm of sound, this might produce what is called random walk noise.[42] Alternatively, a wave theory of sound propagation might take account of the motion of the air to visualize the movement of sound as a turbulent complex composed of eddies, whirlpools and spirals. A combination of these two propagation models might produce a composite figure, a revolving involute, a self-generating and self-sustaining spiral of matter. Eagles riding thermals describe this figure. The swirling snowflakes may also exhibit it – 'Nor it flewe not streight, but sometyme it crooked thys waye sometyme that waye, and somtyme it ran round aboute in a compass,' wrote Roger Ascham, to illustrate to the apprentice archer the turbulent medium into which the arrow was fired.[43]

As a figure of attention, analogous to the act of listening that is open, erotically inclined to receive the call, the echoic involute discovered in the whorls of Venice in the mid-1970s is re-transcribed, rearranged *in accordance with fresh circumstances*, a number of times in the next twenty years. An early re-transcription is a polemical engagement with Florentine poetics where, in an immodest critique of Dante, I set out to define the 'definite sensations' represented by Guido Cavalcanti's *spiritelli*. Discussing passages in the sonnets where snow falls out of a clear sky (*aria serena quand apar l'albore / e bianca neve scender senza venti*) and the approach of the *donna* causes the air to tremble (*Chi è questa che ven, ch'ogn'om la mira, / Che far tremar di chiaritate l'are?*), I imagined the *spiritelli* as lovers' glances.[44] Each spirit could be visualized as a single snowflake; their thickening swirl was the lover's sensation of being overwhelmed (blinded). Fifteen years later, in 'On the Still Air' (1990), the first involute of Cavalcanti and Dante was involuted into a second poetic debate (between T. S. Eliot and Ezra Pound) and both further involuted into private and public stories of betrayal personally experienced in Rome.[45] Then, in *The Lie of the Land*, the Brownian motion of the *spiritelli* surfaces a third time, *rearranged* again – Ascham's snowstorm is quoted to illustrate a distinct sensation – 'as if numberless photons dashed through the air, their friction producing a tremulous light. The air, galvanised by the gaze of desire, is a force field. The same psycho-kinetic sensation is evoked in the line "*e bianca neve scender senza venti*" where, of their own volition, flakes of light fall, palpable, errant, quickly melting.'[46]

In my notebook for July 1990, I find a passage that explores the analogy between birds in flight and winged words, and between both and musical notes. The frame of the composition disclosing the pattern in the movement form might be the sunbather's contemplation of glittering waves:

> 'On The Still Air' is composed as a complex of movements within a frame or across a screen. Here the silver gulls and crested terns drift along the foreshore, strung out in a glittering single file, and then alight, bunching into a dense, reflective flock. The slow waves trickle up between them, sweeping an edge of shadow, concentrating it into a ribbon of darkness. Beyond the beach the breakers continue to zip up white waves breaking. In the air starlings gather sound and let it out again in slow reels. So the composition is built up from densities and rarefactions, intervals of notes that are like gnats in the forest sunlight, minutely picked off one-by-one by a bird that plucks them like strings. Focus and unfocus, the singular event and the kinetic mass, the object and its halo of glitter: it is the same dialectic in 'On The Still Air', where punctilious detail alternates with enigmatic rumour. What do the messages from the Brigate Rosse mean? The tension between meaning and movement is like the relationship between poetry and music, and to reproduce it is not to represent it but to embody it technically.

J. Godwin in *Harmonies of Heaven and Earth* claims that Trecento motets were of this experimental kind: 'The motet typically sets a phrase of plain chant, without its words, in a regularly repeating rhythm, and adds to it to higher parts, moving somewhat faster, each of which has its own text. The texts may be Latin and religious or French and amorous. Often one of each kind, sung simultaneously. … No one listening to a motet for the first time can disentangle the two simultaneous poems, or perceive the underlying rhythmic structure.'[47] It is as if Cavalcanti and Dante aimed to dignify the simple monophonic songs of the Troubadours in a comparable way. In fact, a theme of their *canzoni* will be the ambiguity of the sense, the endless rising and falling between different registers of meaning. About the punctiform *mise-en-scène* of the *donna* there is an almost architectural glitter produced by the structuring momentum of the metre – remember Pound's claim that music for their songs is completely missing, as if the poems contained their own musical instructions. The genius is to hide these instructions in phrases that also signify so that the pattern in the rhythm and the meaning of the poem emerge simultaneously.

All night a stiff wind off the land played the truncated cypress outside the window: an imitation ocean. In the morning the waves were double waves: as they broke a film of spray unzipped upward and swept back, a reverse wave, mirroring the forward motion. The backward wave, echoic, ethereal, simultaneously contrapuntal or perhaps simply winding back the recording, seems to fold time back on itself and produce a 'standing motion', a mesmeric cycling on the spot, where, possibly, in the haze of the spray a figure could form, a stable ghost, a persona or voice walking on the waves.

Such involutes are like murmurations. Perplexed combinations of sense data, choreographed into whorls, they are creative forms, figures of becoming at that place. They are heard as noise, seen as the trembling of light or, in an integration of the senses, as a kinaesthetic undulation of the crowd. The 'overall effect' corresponds to what Jean-Luc Marion calls a 'saturated phenomenon' which 'refuses to let itself be regarded as an [abject] object … precisely because it appears with a multiple and indescribable excess that annuls all effort at constitution [assimilation to an abstract concept]'.[48] You cannot see all the snowflakes, although you apprehend a forming vortical order in their apparently random individual trajectories. You cannot hear all the starlings, or all the voices of the crowd, but a distinct *romorio* or acoustic mask is discernible. You are conscious of congregation, a principle of amalgamation that bridges the gap between Solus, the individual, and Rumour, the crowd without reason. In this sense the impurely named 'tuned noises' of Lecce were both a discovery and a return. Self-awareness is needed to make sense of them. After discussing auditory perception, it is necessary to listen to oneself listening, to hear the oceanic ambience organizing itself into significant forms. Something new is found and formed, but wrapped around the older *columella* of an involuted memory structure whose constitution is essentially auditory. Lecce was new but heard there was a compositional *ritornello*, an echo, an amplification of an older creative principle.

In Melbourne I read how the country came into being for a second time. The Woiwurrung-speaking Wurundejeri said that the creator deity Bunjil grew angry with the people he had created because of their wicked ways. To punish them, he entered their encampments 'cutting men, women and children ... into very small pieces'. However, the pieces did not die but 'moved as the worm (*Tur-ror*) moves' and great whirlwinds (*Wee-ong-koork*) came, changing them into flakes of snow, which were carried over the earth: 'Thus were men and women scattered over the earth.'[49] In 'Mirror States', a sound installation proposed on the Yarra River between what are now World Trade Centre Wharf and South Wharf Promenade, a sound composition was proposed that imitated the 'great whirlwinds'. Four eight-storey banks of loudspeakers – two on the north side, two on the south side – were to demarcate a 'discrete acoustic territory'. The script stipulated that the four sets of loudspeakers should be in dialogue with one another, the voices appearing to move back and forth across the river and circulate clockwise or counter-clockwise; over the duration of the work, the sound was to spiral upwards from one storey of speakers to the next. A notable feature of the design was a secondary counter-dialogue between the local urban sound environment and recordings of that environment, achieved by interspersing the noisy revolutions with windows of silence when the outside noise (normally masked or filtered out) became suddenly audible. The elemental phrases and noises composing the whirlwind were conceived as acoustic 'snowflakes'.[50]

And then, a couple of years later, 'Tuned Noises' was reporting a similar structuration of urban sound in Lecce. Another whorl in the involute, it demonstrated not only the persistence of a theme laid down in earliest childhood but its surprising generosity of invention. The metamorphosis that occurs in a yet later work for radio is whimsical proof of this. 'Underworlds of Jean du Chas' (1998) is a Beckettian drama set on the banks of Lethe. Hollow and his son Didi recline in deckchairs scanning the waves; perhaps something will happen. (Icarus will plunge out of the sky or Charon's barge arrive for them.) They have in their luggage a portable cassette player and a collection of tape cassettes. As the cassettes have not been indexed, the only way to find what they are looking for is to pick one cassette after another at random: 'Hollow stops the tape, noisily ejects it. Scrabbles in his heap of cassettes for another, as he does so muttering "A lifetime of voices and not a name or number among 'em".' In Scene Three Hollow inserts a new tape, commenting, 'These stolen sounds he left his son.' But the new tape is unsatisfactory, as the stage directions explain: 'The recorded portion of the tape has come to an end. The blank tape continues with its characteristic hiss and regular minor clicks as it unwinds from one spool to the other.' Listening to the noise, Didi reflects, 'How they eat each other up. He wastes to nothing: he grows fat. Then everything is reversed, Hollow.' 'Revolutions of life!' Hollow replies sardonically, to which Didi adds, 'Father and son'.[51]

'Underworlds of Jean du Chas' riffs on a hoax lecture that Beckett (fresh back in Ireland after two years as *lecteur* at the École Normale Supérieure) delivered to the Modern Language Society of Dublin around 1930.[52] In parodically rococo French prose, Beckett announces to the world the proto-Malone character of Jean du Chas, founder of the mythical art-movement Concentrism. The lecture is an exuberant satire on academic pedantry: 'Underworlds of Jean du Chas' naturally follows its lead. Pedants, writers and scholars have a constitutional need to dethrone their fathers, to contradict the authorities. Beckett had described Jean du Chas as leading 'one of those horizontal lives' (see Scene Four). I decided that Jean had in reality led a *vertical* life. His art movement, Concentrism, was conceived by one used to seeing radiating circles of water. The *formes circulaires* that Jean championed were not those painted by Robert Delaunay: they were the ordinary experience of a *diver*. Yes, the real Jean du Chas was a diver – he said himself that Concentrism was a prism on the stair – obviously, the Orphic address that Jean delivered to his followers (in Beckett's lecture) contained cryptic instructions for a high-dive into the abyss of the unknown (Scene Fourteen).

Jean du Chas shared Beckett's own birth-date. He was an alter ego: a fictional persona. Another author who spoke personally through masks was the inventor of heteronymies, Portuguese poet Fernando Pessoa, who, in notes published after his death, characterized himself as 'the centre that doesn't exist': 'In the water spinning float all the images that I have seen and heard in the world. ... I am the nothingness around which this movement spins.'[53] In Jean du Chas, centrifugal and centripetal forces struggle for possession of the soul. When the concentric circles begin to spin, a void is opened up (a centre that does not exist) at the heart of the vortex. Plunging into it, Jean finds the 'images' of his life (the selves he might have been, the lives he might have led) whirling round him. But the rotatory motion that gives him access to the depths proves to be centrifugal; what he hopes to grasp is flung apart. The Second Plunge of 'Underworlds of Jean du Chas' reports a number of these rotatory forays that end in disappointment – Jean's childhood trip to the Ruhr and the escaping miners' pigeons (Scene Eight), his adolescent initiation into the wonders of the spider's web (Scenes Nine and Ten), his first 'rapture' (Scene Eleven) and the centripetal love affair (Scene Twelve) – 'Why did you run away? Why did you not look? The little void is turning, growing; it is spreading out its arms to engulf all of Europe.

It is us. The train is reversing out of the station. The train has not arrived. It will not take the plunge.'

When it became clear I was settling in Australia, my father began to talk. Among the sound materials in *my* collection that I used in the original production of 'Underworlds of Jean du Chas' were cassette tapes, audio letters my father had recorded and mailed to me. Beyond the obvious pathos of distance, the messages forwarded over a period of years had a poignancy all of their own. Finding it difficult to satisfy the hunger of the cassette tape's sixty-minute recording blankness, my father resorted to various stratagems. One was to assemble topics he thought might interest me over a period of weeks. A sound diary emerged, a sequence of stops and starts and a collection of anecdotal episodes, which might or might not develop a larger theme. Another was simply to send the tape when he had nothing more to say, leaving part of the cassette blank. While there was nothing unusual about such self-editing, I was struck by Bill's self-consciousness; almost invariably the beginnings and endings of each recorded section attracted their own commentary. A significant part of the audio letter was the producer explaining his own editorial decisions. Most striking, though, was his actor's awareness of script, for invariably, and perhaps more than once within an individual communication, my father would announce that he had nothing more to say. This absence of story became in his telling a topic in its own right. In fact, when I listened back to the collection of tapes, I had the impression that their major theme was saying goodbye.

A more perplexing aspect of these audio transactions was my father's regularly repeated instruction to *record over* his recording. Perhaps there was a cultural context for this: historians are scandalized by the BBC's reuse of recording tape, leading to destruction of programmes now regarded as 'classics'. Evidently, the capacity of recording technology to record anything is a two-edged sword; while it captures more, it creates more to discard. In its extreme form, Bill left my mother to record the news on the other side of the tape; in this case, his side of the tape was left empty with the sole instruction to use it as we wished. In the event, a complete withdrawal from speech was staged, and the conversation consisted in responding to a silence. But the intermediate situation was more common; allegedly in the interests of economy, the same cassette tape was to go back and forth between Australia and England, each new communication erasing an older one. Had this instruction been followed, the posthumous legacy of our exchange would have been curious: a drawer-full of my voice recordings would have been found but no oral trace of my father. Nowadays, this injunction to talk over his voice would be interpreted as passive aggression, a classic example of doubletalk where the seeming willingness to communicate disguises an emotional withdrawal. But his self-erasure expressed a different wish, a desire to keep alive the illusion of conversation occurring in the present. He knew that the recording signified the echo of absence. By resisting the perpetuation of the voice he refused to live posthumously. I heard in the call to silence him a quiet summons to come back.

The Lecce recordings coincided with a telephonic summons from the composer Luciano Berio to attend him in Radicondoli. He was planning a new music theatre work, a version of Homer's *Odyssey* inspired by Vladimir Propp's *Morphology of the Folktale*. The idea was to use Propp's theory of narrative functions to elicit the 'fundamental structure' of Homer's epic poem, and to repeat this six times. Each repetition would be a return to the same material in a new way; each new story would preserve the arrangement of the story but transpose it into a different set of circumstances. I remember that in responding to this brief, I pointed out that Berio's plan had to some extent been anticipated in the largely lost post-Homeric 'Epic Cycle',[54] where the essential relationship of father and son spawns new genealogies and geographies. Odysseus marries again. Telemachus acquires half-brothers; they fight among themselves or gang up against father. New adventures are undertaken. Wars and amours occur resembling those in the *Odyssey* but in different countries. The classical rhapsode responsible for these plot variations was a patchworker: cutting local materials to shape to fulfil the narrative functions, he preserved the story's fundamental structure. If you laid these later variations over the original Homeric plot, I suggested, you got a surprising result: every key moment (or 'function') in the story generated a number of might-have-been alternative scenarios.

We drew up a 6 × 6 grid. Vertical columns corresponded to archetypal 'functions' of the Proppian folktale: initial disorder, crisis or loss; appearance of hero to right the wrong; hero's adventure with trials; hero's successful recovery of what has been lost; triumphant return and re-establishment of order. Five functions are recognized; the sixth, Berio's innovation, was the episode of return itself when, reconfigured as a hinge moment in which, as in *Finnegan's Wake*, the hero is compelled to begin again, the whole cycle starts over once more. This sixth empty function, which in an almost Spinozan fashion draws attention to the illusion of free will or psychological motivation, reclassifies the work as metanarrative: displaying its own structural scaffolding, it engages in a debate about the ethical basis of aesthetic representations. What does it mean to be moved by an evident fiction? To personify this paradox, I suggested we call the work 'Outis', after the answer Odysseus gave Polyphemus. 'Who goes there?' the one-eyed giant cried; and the hero replied, 'Nobody'.

Populating the functions horizontally was potentially to review the entire history of storytelling, or that part of it known to us. Even if our character template was strictly based on the characters in the *Odyssey* that propel the functions forward (Pallas Athene, Poseidon, Odysseus, Telemachus and Homer himself – on the analogy of the 'empty' sixth function, we also factored in an 'empty' trickster figure), we were, culturally speaking, spoilt for choice. Further permutations and complications arose when the 'local' force field of associations was taken into account: if, as Berio wanted, in one cycle Poseidon assumed the persona of the Kurogo in Kabuki theatre, we had a trickster on our hands – 'If the scene is set in snow, the Kurogo is then dressed in white. If the set represents the sea, he may be needed to push boats around, in which case he will be tastefully dressed in blue and white stripes.'[55] If, in another cycle, our no-man/everyman/trickster character was represented by Grock, 'the last of the great clowns', what scripting was needed? When Berio was a child in Oneglia, Grock lived next door. Berio and his school friends used to steal the old man's oranges –

> Then later – I was 11 years old – I had the chance to see him in performance at 'Teatro Cavour' in Porto Mauricio, I realized [his genius]. During that performance, just once, he suddenly stopped and, staring at the audience, he asked: *Warum* (why). I didn't know whether to laugh or cry, I wished I could do both of them. After that experience I haven't stolen oranges from his garden anymore.[56]

Was Grock challenging the future composer to invent meta-theatre? The 'adventures of a simpleton among musical instruments' offered Berio a theme perfectly adapted to his genius. It did not need 'winged words'.

Among the transformations of 'Outis' there was one I particularly liked. In the Fifth Cycle our no man appears as Big Ears. A story that had at its heart a mishearing (Polyphemus misinterprets Ulysses's reply to his question – a reply that is itself echoic mimicry deployed strategically) and a refusal of song (Ulysses blocks his sailors' ears with beeswax to avoid hearing the Sirens) invites reflection on the perils of first contact. In those days we both found inspiration in Giambattista Vico's poetic etymologies – Vico, whose theory of historical *ricorsi* or returns had so appealed to James Joyce, managed in the *New Science* (1725) to derive the entire history of Western civilization from an etymological archaeology; suggesting that phonically similar Latin words concealed semantically suggestive clusters, Vico anticipated Freud. However, where Freud used word associations to weave a guiding thread to the labyrinth of the individual unconscious, Vico examined the history buried in the older strata of the collective psyche. Anyway, in the spirit of Vico, I paid much attention to the etymology of Odysseus/Ulysses and his many Homeric epithets. When these were teased out, we seemed to be dealing with an entire population of characters rather than a single hero. Thus, among many characterizations, Odysseus is not only *outis* but also *polymetis*. *Ou* is the negative of fact and statement, and *me* the negative of will and thought. *Outis* means 'no one at all', while *metis*, which means 'the same', also carries the sense of craft, resourceful wit.

Using cunning intelligence, *metis*, he denies himself to prove true to himself: in short, the personification of the creativity that lies in the ambiguous relationship between sound and sense.

Among the semantically suggestive sound-alikes hovering around Odysseus/Outis was Otis, the Bear, named for his big ears. Ulysses possessed exceptional hearing. Perhaps this was the particular danger that the Sirens posed; as geniuses of the shore they whispered to him of a different destiny, one that his keenness of hearing alone discerned. Because he hibernated, the bear was associated with Shaman philosophers like Pythagoras, who famously visited the Underworld – Odysseus's ten-year wanderings far from Ithaca could be similarly construed. Bears are associated with bees, and when the buzz of bees is turned into golden speech with rhetoric and poetry. When 'honeyed words flow from their lips', princes 'can easily reconcile a dispute with soft persuasive speech'.[57] As an exceptionally gifted listener, Ulysses was like the Kara Kirghiz oral singers described by the nineteenth-century Tsarist official, V. V. Radlov:

> He has a whole series of 'elements of production', if I may so express it, in readiness, which he puts together in suitable manner according to the course of the narrative ... [these] consist of pictures of certain occurrences and situations, such as the birth of a hero, the growing up of the hero, the glories of weapons, preparations for battle ... characteristics of persons and horses ... the art of the singer consists only in arranging all these static component parts of pictures with one another as circumstances require, and in connecting them with lines invented for the occasion.[58]

In short, as an improvising minstrel modelled on the classical rhapsode, our Ulysses was putting the morphology of the folktale into practice long before Propp had thought of it. And the happy parallel with our situation was even closer:

> The procedure of the improvising minstrel is exactly like that of the pianist. As the latter puts together into a harmonious form different runs which are known to him, transitions and motifs according to the inspiration of the moment, and thus makes up the new from the old which is familiar to him, so also does the minstrel of epic poems.[59]

This, as I understand it, is close to the technique of the *Sequenzas* where Berio, through a kind of return to what is familiar makes a new transcription according to the inspiration of the moment. Insofar as 'Outis' was a new *sequence*, the Big Ears Cycle invited a new transcription of Berio's 1984 work with Italo Calvino, *Il re in ascolto* (The king who listens).

The collaboration was drawn out and ultimately unsatisfactory. If I may risk a structural reason for this (as no competition of talent is intended), our attitudes towards the found word and its poetic interpretation were too similar. Another way to say this is that Berio's demands on language were contradictory. He asked for 'winged words', borrowing Homer's formulation, but reserved their poetic development to himself. Ascribing to me the role of cultural native informant, he expected me to provide him, Ossian-like, with folkloric phrases out of which he could weave his music-magic. I was going to dig up from the many-layered Ilium of cultural memory the equivalent of the famous, but anonymous, cries of the old London street vendors used in *Cries of London* (1976). Eco would translate my words into Italian, Berio explained. He himself would write the variations on the hero's name. Such stipulations differentiated my task from Edoardo Sanguineti's – Berio's usual librettist – who habitually wrote across languages in a style I much admired. Without the licence to explore the potential of word sounds to generate a multiplicity of voices and archetypal situations of conflict and resolution, I had no way of testing their compositional fertility both within individual cycles and across successive iterations of the fundamental structure. Unable to calculate the intra- and interlingual power of sound-alike and near sound-alike word sounds to generate theatre, I was back in the situation that had led me a few years earlier to write 'Memory as Desire'.

After this attempt at collaboration failed, I wrote 'Introducing Ulysses', a work for two human voices (Echo and Ulysses), accompanied (I asked) by Athene's 'Tyrrhenian trumpet' and saxophone. Sixteen short scenes derived the drama of *Outis* from a set of variations on Ulysses's name. 'Introducing Ulysses' itself echoed 'Cooee Song', an earlier work for actors and (electro-acoustic) echo. The essential digraphs ('ou' and 'ee') are the same in both works. Anyway, when the opportunity to collaborate with another composer arose a few years later, I proposed a work called 'Winged Words', which would frame the performance of 'Introducing Ulysses' inside the story of our collaboration, the highly condensed libretto that came out of my collaboration with Berio. Andrew Schultz, my potential collaborator, had suggested that my Echo was less a mask of Pallas Athene (as I had imagined), and instead behaved much more like 'an older man'. I endorsed this suggestion, remarking, 'in this way the figures of Berio and Me are prefigured or refigured in Echo and Ulysses respectively'. And I added, 'I like foregrounding this parallel structuring for another reason. It seems to me that Berio always refused to be the "author" of *Outis*.' That is, he always refused to be the work's 'father'. Instead he tried to foist paternity of the work on me. He asked me for winged words – inseminating seeds, if you like, that would fertilize his latent creativity. Besides casting himself as the womb,[60] in a way Berio also cast himself as the 'son' to my 'father'. Of course, this placed me in a double-bind. As power lay with him, this reversal of roles was a game played to his rules. It did not give me freedom to create a new kind of music theatre. It meant that whatever I offered him, he could pretend that it was addressed to another (the author, father). In this way, I was always addressing my work to an empty space: 'Berio refused to be at home for me, he was always wandering, like Ulysses.'[61] Is a psychologist needed to see through this? Even posthumously I sought father's approval.

The inner biography of that period is rotatory: geographical relocation created occasions to carry over preoccupations from one place to another and recognize them in new arrangements. Ideas were twirled in the hand like precious facetted mirror balls: turned around, the light of the world playing on them, their spherical mosaic reflected a polyhedral reality. The transcription of a curvilinear world as a collage of differently oriented facets communicated, perhaps, an important compositional principle. At the time of the Radicondoli residencies, Berio was investigating an obscure nineteenth-century Italian composer of dance music; daily, scores of once-popular waltz scores spewed from the fax machine. In *Baroque Memories*, this prolific but forgotten musician reappears as an urban choreographer. In an ambition to coordinate social life, Forgetson has devised a *ballet méchanique* far superior to the on–off of traffic lights. The logic is simple: everyone passes from rest to motion, everyone has to make a journey and everyone, more or less, has to come back, passing from motion to rest. Design the streets, then, as a network of conveyor belts – horizontal dumbwaiters circulate endlessly, you alight wherever you want. This theory transposes Propp's analysis of the fairy tale to the organization of crowds. Instead of worrying about the psychology of the individual, a master of urban movement forms orchestrates essential functions. It is a scaled-up, comic version of Balinese shadow-puppet theatre: 'Everything ... is in effect calculated with an adorable and mathematical precision. Nothing in it is left to chance or personal initiative.'[62] Forgetson's task is to write urban scores that produce this kaleidoscopic effect.[63]

Around 1990 Les Gilbert, sound artist and director of Sound Design Studio in Melbourne, negotiated a contract with the architectural firms Cambridge Seven Associates (C7A) and the Renzo Piano Workshop to design soundscapes for the new Acquario di Genova. The opening of the new building was to coincide with the 500th anniversary of Christopher Columbus's First Voyage. C7A were proposing an opening exhibition that told the story of colonization from the American point of view. In imagining the design and content of the exhibition, Eduardo Galeano's recently translated *Memory of Fire* trilogy was a primary point of reference. In this trilogy, Roger Davis summarizes,

> Galeano presents over eleven hundred vignettes of prose and poetry in a style that has been described as 'magical journalism'. Each piece, whether two lines or two hundred, stands alone as an individual testament to a moment of myth, conversation, observation, announcement,

correspondence and decree. Galeano does not tie them together but allows his vision of the past of the Americas to emerge as each piece is presented in chronological sequence to 1984.[64]

A non-linear, generically diverse labyrinth of found stories that allows readers to find their own narrative pathways may be unusual in a literary work; transposed, though, to the spatialization of information structures in an exhibition, it is conventional, even inevitable. Exhibitions are consumerist spaces, scarcely distinguishable from designer shop windows; the cumulative effect of the displays is often phantasmagoric. How, in this context, is the soundscape to preserve, say, the clamour of encounter, the cannonades of the collision or the long after-effect of silencing?

Certainly, it was to avoid the middle-distance aesthetic of the auditory picturesque implied by this question. When responsibility for soundscape content fell to me, I suggested we focus on the communicational tactics of first contact. First contact is never first contact: meeting strangers, all parties become newly self-aware. Where nothing is understood, listening separates from hearing. Not understanding what is said, hearing instead an ocean of sound, whose sense lies somewhere in the spectrum between environmental sound, word sound and verbal sign, the hearer becomes a critical and creative listener. In sifting this glittering suspension of sound for sense, echoic mimicry is a primary tool; imitating those utterances that sound almost familiar, or which the other party pronounces most emphatically, a kind of echolocation occurs: independent of any content, a situation of mutual non-understanding is communicated. The echoic vocalizations enable temporary coexistence to be negotiated. Of course, this is an idealization; probably no episode in the history of European imperial expansion ever produced quite this cross-cultural emotional equilibrium. On the other hand, the alternative that ignores its possibility is precisely the unilateral extirpation of peoples and cultures that C7A and Galeano lamented and sought to reverse. When Antonio De Nebrija's grammar of Castilian (the first of any modern language) was presented to Isabella in the same year that Columbus set out for Cathay, the Queen asked, 'What is it for?', to which the Bishop of Ávila replied, 'Your majesty, language is the perfect instrument of empire.'[65] In a revisionist account of imperialism, the first step is a revision of language.

The script of 'Columbus Echo' (1992) incorporated material from the old literature of European mercantile expansion collected by Hakluyt and Yule. It lifted from contemporary travel accounts (Varthema, Ramusio, van Linshoten, Dati, Diaz, Ibn Batuta and Pigafetta). It reproduced with a Glissant-like exuberance contact languages (pidgins, lingua francas, trade jargons and creoles), from the Caribbean to the China Sea, learning much about their construction and function from such notable scholars past and present as Hugo Schuchardt and Michael Silverstein.[66] I paid attention to the Genoese dialect (in a field expedition to the town of Carloforte on the

Isola di San Pietro off the south-western coast of Sardinia we recorded a dialect that retains the phonology of medieval Genoese). I considered how poets, from Homer to Camoens, from Jordi de Sant Jordi to Saint-Perse and Wallace Stevens, had represented the noise of geographical expansion and political exile. Todorov and Baudot's *Racconti Aztechi della Conquista* contained sombre accounts of auditory terrorization. I dipped into the ballad chapbooks of Spanish-Jewish exile. I wondered how birdsong was heard and notated – Clément Janequin's *Le Chant des Oiseaux* was important. I pondered Columbus's own ear: What clues to the way he heard and spoke were to be disinterred from his *giornale di bordo*? Anyway, something about this trans-lingual gallimaufry must have worked as it impressed Berio sufficiently for him to suggest we work together.

One of the great achievements of the Genoa project was its recording in Melbourne: assembling performers who commanded fifteen to twenty languages between them (depending on how their utopian language improvisations were classified), we practically demonstrated how a multilingual migrant society might open a radically new dialogue with the colonial past. The courtesies actors showed one another when lacking languages in common, the emotions they were able to communicate in invented languages they co-improvised, the translations between different pidgins (themselves already multilingual hybrids), the extraordinary choral imitations of Jamaican birds, Arabic spice marts and speaking charts recuperated a 'wild storm of inspiration', all the more remarkable because they are generated by the echoic economy of the performance – rather than any authenticating personal history of suffering. These experiments inspired the acoustic mask of 'The 7448' (1992), a radiophonic sequel to 'Columbus Echo', where four word sounds define the work's fundamental structure and functions – 'la, la', (exploration), 'tush' and 'shhh' (discovery), 'ee' and 'oo' (first contact) and 'agh' and 'om' (violence, death). The essential powers attributed to these phonemes are, *contra* Ritz, onomatopoeic. They emerge when the imitation of sound is itself imitated.[67]

The 500th anniversary of Columbus's First Voyage provoked much reflection on the tactics of first contact and their meaning. Stephen Greenblatt, for example, argued that 'improvisation on the part of either Europeans or natives should not be construed as the equivalent of sympathetic understanding; it is rather what we can call appropriative mimesis, imitation in the interest of acquisition'.[68] He added, 'A process of mimetic doubling and projection ... does not lead to identification with the other but to a ruthless will to dominate.'[69] But this is, unwittingly, to perpetuate the normative character of what Octave Mannoni, writing about French colonial Madagascar, has called *homo imperiosus* who, 'much less unhappy when he is absolutely alone than when he is afraid he may not be', fears nothing more than attachment to another.[70] In other cultures, though, including those that the Spaniards and Portuguese encountered in the Caribbean and South America, a different model of sovereignty exists, where one acquires social existence mimetically – illustrated, for example, by the bilingualism which characterizes societies like the Bororo, who are divided into two exogamic moieties. There, 'it is literally through the *other* (someone of the opposite moiety) that an individual exists socially'.[71] There, to be 'heard' socially depends upon mastering a mimetically based feedback loop between listening and speaking.

In the explosion of revisionist writing less attention was given to the sound history of conquest. Here, reading for 'Columbus Echo' produced some interesting reflections. Firstly, Columbus's reputation as a poor observer (and a worse listener) may be partly an effect of transcription – copying Columbus's journal Las Casas 'corrected' many Portuguese spellings and idioms. Did Columbus, in fact, speak a standard language? Guillén Tato notes that hardly any of Columbus's compass directions derive from Mediterranean sources; they come from the north, adapting terms used in English, French and German.[72] A healthy admixture of Greek, Latin, Arabic and Catalan terms suggests that Columbus had also absorbed the Mediterranean's cultural and linguistic diversity. It is highly likely that he was familiar with the Mediterranean Lingua Franca known as Sabir. Linguistically diverse crews, sailing between different coastal language communities, borrowed words wherever they could. Sailors' pidgin was simple in syntax and grammar, baroque in lexicon and style of delivery. Far from being deaf to his linguistic surroundings, 'the genius of Columbus reveals itself in the Diary,' writes Tato, 'in a remarkable capacity for assimilation'.[73] When Columbus informed his royal patrons that he would capture six Taino people and bring them back 'so that they can learn to speak', he was not implying they were dumb (or he was deaf),[74] simply differentiating between the practical patois of everyday exchange in multilingual situations and Castilian, the language of unilateral conversion and enslavement.

The day after making landfall (in the sentence following his proposal to kidnap six natives) Columbus remarked, 'No animal of any kind did I see on this island except parrots.'[75] The juxtaposition of the avian king of mimics with the language acquisition program is suggestive. Recalling the *Popol Vuh*'s portrayal of the parrot as a colourful trickster, Miguel Asturias notes the parrot's role as America's ambassador – Columbus returned to the Old World with a parrot, not a hawk, on his arm, 'a diplomat who to his jacket of live colours adds a tangled speech typical of the dialogue that would follow between Europeans and Americans'. The parrot's echoic mimicry, its deceptive dialogue, serves, he suggests, as a self-protective tactic. It 'allows Latin America to save itself from exoticism seeking foreigners by "counterfeiting paradises"'.[76] The now extinct Cuban red macaw played a complex role in the symbolic economy of conquest. From the European point of view, the sighting of parrots was confirmation of a geographical hypothesis. 'Parrots were the symbol of the Indies. All the medieval commentators, including

Marco Polo, had included them among the marvels of the Orient. Columbus was well aware that parrots were not to be found either in Europe or around the Mediterranean.'[77] From the Taino point of view the European interest in parrots had a different meaning. It is suggested that the Taino came to believe that the parrot was the white man's totem, the guiding spirit without which they could never have 'descended from the sky' – the Taino seem to have insisted that the Spaniards flew rather than sailed to their islands.[78] In a further twist, Columbus treated the Taino as parrots without minds of their own: 'I believe that they would easily be made Christians, for it appeared to me that they had no creed.'[79] As Sider comments, reasoning of this kind made imperial domination a self-fulfilling prophecy. Explorers like Columbus or Giovanni de Verrazano 'parroted their own fantasies of native intentionality' – or, in this case, non-intentionality.[80]

As regards counterfeit paradises, their most significant ornithological representative in Columbus's diary is the nightingale – although here, again, he or his copyist was clearly not a birdwatcher. On 7 December 1492 Columbus 'heard the nightingale sing, and other little birds like those of Castile'.[81] A few days later, on the thirteenth, conditions were again propitious: 'The breezes were like those of Castile in April. The nightingale and other small birds sang as in the said month in Spain, for he says that it was the greatest sweetness in the world.'[82] The nightingale is unknown in the New World and (as Columbus seems to acknowledge) sings in April, not December. A modern editor suggests that Columbus heard one of the American mockingbirds. In Jamaica the species in question is known as 'The Singing Bird, Mock Bird, or Nightingale'; because it sings at night, it is also called the 'Jamaica Nightingale'.[83] But the correct identification is with geographical desire, for on 29 September 1492, long before sighting San Salvador in the Bahamas, we learn, 'the breezes were very sweet and good smelling and he says that nothing was missing but to hear the nightingale'.[84] When, later (on 7 October), Columbus commented, 'most of the islands that the Portuguese hold they discovered through birds', he referred primarily to sightings – but out of sight of land, he also strained his ears to *hear* what lay ahead.[85] 'The wakeful nightingale', who 'all night long her amorous descant sung',[86] was endemic to the Earthly Paradise – whose location obsessed the Admiral: Sigar tried to block it out, Christopher, to amplify it.

In Australia, parrots also pointed towards Paradise. In 'Memory as Desire' (1986), a conversation between two couples (ex-explorer Charles Sturt and his wife, Charlotte, and ex-Aboriginal Protector George Augustus Robinson and his wife, Rose) takes place in the French harbour town of St Malo. Birds feature largely in the conversation. The murmurations of starlings streaming off the sea ('It is natural to ask where they come from.' 'Where they go to.') are discussed, as well as the onomatopoeic names of Australian birds ('There is a history in it'). However, the major *bird scene* is derived from Sturt's account of his last expedition. Like Columbus, Sturt's expeditions (1829, 1835 and 1844) had a paradisal purpose: in this case, to locate an inland sea. To fix the likely position of this hypothetical water body, he appealed to parrot flight paths. In a horizontal projection that far surpassed my migratory speculations at White Horse Hill, Sturt collated two sets of observations: budgerigars flying directly *from* the north along the shore of St Vincent Gulf (adjacent to Adelaide) and budgerigars flying *to* the west-north-west from a position on the Darling River near present-day Bourke in north-central New South Wales. Now, Charlotte explains, 'if we were to draw a line from Fort Bourke to the west-north-west', and, the Captain adds, 'from Mount Arden to the north', 'We should find', Charlotte continues, 'that they would meet a little to the north of the Tropic'. Robinson provides the logical conclusion: 'Now birds that delighted in rich valleys or kept on lofty hills surely would not go into deserts and into a flat country.'[87]

This is a *homo imperiosus* fantasy par excellence. If Sturt had spoken to the Pitjantjatjara, in whose territory the inland sea of his imagination was located, he would have discovered that the onomatopoeic name of the budgerigar subtended a community of objects and relationships, including a story about theft and reprisal.[88] In this sense, at stake was a theory of communication. 'How are we to explain the invariable flights of the bird towards a particular *unknown* region?' Sturt's contemporary, the Scottish philosopher Dugald Stewart, had said, 'For it must not be forgotten that its migrating instinct has at once a reference to a period of the season in the country which it leaves, and to that in the country for which it is bound.' Unlike Hume, Stewart stops short of attributing reason to animals but flags again the necessity of Aristotelian notions of final causes or design ('the wisdom of nature'). 'Sagacity' is instinctive but there is 'a certain power of accommodation to external accidents'. To illustrate the combination of instinctive impulse and external chance in social behaviour, Stewart cites the innate language capacity observed in the human infant as it acquires its mother tongue. Against the modern prejudice (John Stuart Blackie spoke of 'a certain horror of a sort of merely animal element in the creation of language'), the migration of birds and the acquisition of language might be determined by the same principles.[89]

'Memory as Desire' is a journey into 'the inland sea' of the colonial unconscious. Sturt typifies the imperial type who is 'confident in the face of abandonment' because he has 'converted [it] into independence'. In a development of this psychological thesis, Mannoni explains that such emotional disengagement depends on

> refusing to see one's own gaze reflected in the gaze of the other. It is this refusal that obliges the white man to substitute his own *ersatz* emotional economy. But the basis of his substitution is the sense of abandonment that occurs when, instead of a face in the landscape, he sees only a blind spot, the void of his own gaze.[90]

While this thesis imprisons the colonized subject in a mirror state decided by the colonizer, and ignores his revolutionary agency, it retains value as a description of environmental alienation. Sturt's withdrawal into reverie, as he goes over the details of his last inland expedition, substitutes a false interiority for engagement with his surroundings. In historical terms, his type withdraws from the sense of things and, in particular, refuses to listen to the calling to come invested in sound. In 'Memory as Desire', I wanted to create an audible transition from an English 'dawn chorus' to an Australian one. The choruses were to be created by the actors listening through headphones to environmental sound recordings, and imitating them. In another post-production experiment, we reproduced the flight call of the budgerigar by playing back our human voice recordings at four times their normal speed.

An 'overall effect unknown to the voices themselves' is created when each of the characters, Sturt, Charlotte, Robinson and Rose, proceeds undeviatingly like clockwork. However, I also wanted to suggest the noisy environment against which what they say is heard. Unassigned lines in 'Memory as Desire', quotations, asides, exclamations and other rhetorical in-fill fulfil the same function as the non-human sounds in the backstreet soundscape recorded in Lecce. They position the conscious conversation as the echo of a shared auditory unconscious. Unlike Sturt's metaphysical disorientation, the echoic mimicry that secretly binds the speakers together creates an orientation to each other and to the environment that is *internal* to the dynamics of the performance. In another context, anthropologist Roy Wagner meditates on the significance of *bats* in so-called totemic thought. He posits a 'genuine semiotics' where humans listen for themselves in conversation, by this echolocation, learning about the limits of communication: 'It is because sound is not meaning but the meaningfulness of direction that ... the bat [can] listen to itself as a navigational vector.' He continues,

> It is in sound's inability to merge with or directly encode the meanings attributed to language that it similarly becomes meaningful for human beings, allows them to listen to themselves as vectors of meaning through a medium that is not meaning. Those who wish to ground meaning in language are disposed to imagine the 'sign' through a magical precision bridging sound and sense, but such a coding, to the degree it were precise and exhaustive, would render impossible the 'play' or ambiguity, the irony of sound and meaning – would nullify sound's echolocative possibilities.[91]

Works such as 'Memory as Desire', 'The 7448' or 'Underworlds of Jean du Chas' were products of a distinctive radiophonic culture promoted by the Australian Broadcasting Corporation. They were commissioned by producers like Martin Harrison and Andrew McLennan familiar with the 'Neue Hörspiel' of West German radio, where the play was not an acoustically illustrated narrative but instead 'an auditory event in which all sound phenomena, whether vowels, words, noises or sounds, are in principle of the same value'.[92] Klaus Schöning, one of the leading *Ars acustica* producers in Germany, who later worked with me on a number of productions, explained the innovation of sound artists he worked with in this way:

> Their sensitivity to both the spoken word and a semantic articulation and their recording and montage techniques led to the artistic exploration and exploitation of documentary sound (*Originalton*). The subjective language of anonymous people was given a voice; things otherwise never published were heard. Recorded on tape, spoken language and quotations from the media could now be cut and spliced, analysed, composed, and arranged as a montage. In the words of Ernst Bloch, they were 'made recognizable through change'. ... Language *itself* began to be seen as a source of quotation and documentation – a repository of history.[93]

In Schöning's Studio for Acoustic Art, language was conceived as embedded in sound and arising out of it. Language revealed its 'history' – its autonomous potential for expression – when freed from the semiotic cuirass that repressed its vibration, damping down (in the musical sense) all those qualities associated with sonorousness. As Hagelüken notes, when such features of non-written language as emphasis, sound, rhythm of speech, speed and melody of speech and even vocal character are treated as semantically fertile, 'the structure of language in time can also be altered. Earlier dependence on the purely linear treatment of literature (reading as a sequence) is abandoned in favour of more musically inclined solutions, such as simultaneous speech.'[94]

Through the association with Sound Design Studio, it was possible to realize this simultaneity spatially. Using interactive art technology (computers and sensors), Les Gilbert specialized in designing sound environments for architectural spaces. Prepared sound – one of my scripts might generate eighty to one hundred and fifty separate sound events – was stored in digital files. Gilbert's role at this point was to write the random access program that established the parameters of the sound experience; sounds could be regionalized in time and space; they could be differentiated in terms of frequency and rarity; and it was also possible for a local sound event, once triggered, to generate its own cascade of surrounding events. To deliver this experience, hundreds of speakers might be hidden in the architectural fabric, while motion sensors were strategically placed along the main walkways. The overall effect was to align the sound environment to the choreography of the crowd. But the alignment was subtle, elastic, as any trigger from a sensor might be queued or otherwise delayed in its execution. It might be argued that the impression of randomness was a technological illusion. But that is fundamentally to miss the point; it is to perpetuate the fantasy that the visitor can or should be an unimplicated observer or hearer. As Artaud wrote, manipulation by another does not necessarily represent a loss of agency or a diminution of identity. The dramaturgist's adorable calculations can produce 'a double body, doubly membered'.[95] In Wagner's terms, an echolocation occurs in which, in principle, it is impossible to separate the origin of the sound from its kinaesthetic shadow.

Installed in heritage-listed buildings or new custom-built museums, permanent sound installations, such as 'Named in the Margin' (Hyde Park Barracks, 1992), 'The Calling to Come' and 'Lost Subjects' (both Museum of Sydney, 1995), challenge us, the public, to confront a collective responsibility in perpetuating historical amnesia. In *Material Thinking* I tell the story of how the last remnants of the first Government House had to be destroyed in order to build the first Government House Museum – a perfect illustration of the destructive character of historical memory when it confuses what happened with the theatrical representation of events. What T. S. Eliot's character Harry says in *The Family Reunion* seemed to us true of most museums: 'All that I could hope to make you understand / Is only events: not what has happened. / And people to whom nothing has ever happened / Cannot understand the unimportance of events.' In our critique of what Tadeusz Kantor had called 'the "Artistic" Place', where 'everything is readied for the reception of fiction and illusion', we wanted to foreground the invention, or concomitant act of production, at the heart of remembering. Kantor imagined actors not acting – 'Everything that happens on stage would become an event – perhaps a different one from those that occur in the audience's spatial and temporal reality of life, but still an event with its own life and consequences rather than an artificial one.'[96] Among the stage events imagined here – so much easier to realize in a museum where audience/visitors and spectacle/exhibition space are not separated – would be the recovery of the sonorous body. Or as lines from 'Lost Subjects' say,

1: This is the ledger of borrowings. Notice its single column.
2: Once part of a larger building.
3: Once trees propped up clouds.
1: This is the room of clouds.
2: Here are collected the breaths of the departed.
3: There are instruments that can condense their final words.[97]

The subjects lost here have been twice colonised: by land theft and spirit theft. Kantor is clear. These 'sterilised and immunized places (it is difficult to call a museum or auditorium a real place)', dedicated to the presentation of 'ambiguous acts of representation', prevent ambiguity from being embraced as a manifestation of reality.[98] In this sense, the communicational dilemma of the migrant directly parallels the actor's difficulty – Kefala's Basia or the Bonegilla detainee embrace ambiguity as an existential reality. To solve 'the dilemma of autonomy and representation, with ease', the new actors (but they might be new historical subjects) do not retreat into a Cartesian ego-consciousness. Without a nostalgia for self-representation, they embrace their professional personae, with characteristically ambiguous results: 'They do not imitate anything, they do not represent anybody, they do not express anything but themselves, human shells, exhibitionists, con artists.'[99] They are no longer actors reproducing the ambiguities of reality. They are players whose own performance is ambiguous. By this device, like echoic mimics, they bring into being a new, in-between place: 'Playing is identified in the theatre with the concept of performance. One says, "To play a part." "Playing," however, means neither reproduction nor reality itself. It means something "in-between" illusion and reality.'[100] And, as Kantor stresses, redefining the poetics of performance also alters the 'situation' of the audience-member, who is no longer one passively in 'a state of hearing', but 'a potential player'.[101]

In migrant Australia, where everyday performances are themselves ambiguous, stage histories merge into life. Literary support for this observation occurs in a self-published book *The Shoe in My Cheese*, by one R. A. Baggio, whose title immediately places authenticity under the sign of ambiguity. The challenge of acting a part (typically specified by the anglo-centric nation-state) is not, Baggio suggests, to rebel against the interpellated acoustic mask but to embellish it – by an act of echoic mimicry that parodies the literal, to create a new figure which fuses self and actor in play. Baggio reports non-Italo-Australian mispronunciations of his name:

> They named me, Rino, pronounced ideally Rino, and locally as Reen-oh; Rhine-oh; Renault; Reo; Reen, by the over sixties; Ringo by the Beatlemaniacs; and sometimes Ringe or Roscig, out of the junkmailers' computers. Our surname, Baggio, is similarly well-suited for invention, improvisation, innovation, and fantastication: Badgee-oh; Baggy-oh; Badge-EE-oh; Bug-Eye-oh; Barge-ee-oh; Buggy-oh; Bar-Joe; and the Danaean, Baggos, favoured by our dairyman on his invoices for thirty years, or so.[102]

Baggio appears to lament the lack of 'a magical precision bridging sound and sense'. But his nostalgia is ironic. In fact, it is precisely the slurring of his name that makes room for the ambiguity of play. Imitating his mimics, Baggio baroquely surpasses them. Through 'invention, improvisation, innovation, and fantastication' that parodies theirs, he does not express anything but himself – an exhibitionist.

And, as I say, the vicissitudes of identity formation in a migrant society seemed to me to recapitulate what must have often happened along the frontiers of colonization but which, never generating representable 'events', remained historically silent. In a discussion of the Lower Mississippi

Indian pidgin, Emanuel Drechsel has wondered 'whether – with different first languages – speakers of a pidgin really understood each other in conversing in it, especially in its initial and highly variable stage'. Could such a linguistic compromise, he asks, 'reveal non-utilitarian purposes such as those of a linguistic game'?[103] If it did, the success of the 'game' could not have been measured by its semantic yield. It must have been assessed phonically, *onomato-poetically*, through the echoic free variation of 'players' mimicking one another in the absence of agreed terms of reference. Even communication that is clear, straightforward and mutually intelligible may retain traces of this 'background radiation' of ambiguity. This is the implication of D. H. Whalen's recognition of 'indeterminacy as a linguistic phenomenon'.[104] Indeterminacy arises as no two speakers speak or hear what is spoken identically. This obvious fact, although grammarians ignore it, means that every speaker stands a chance of being misunderstood.[105] A 'semantic noise' backgrounds every attempt at communication. But what is its origin? Whalen speculates about divergent syntactical and lexical judgements, before acknowledging that 'there is a great deal of room for personal variation and divergent determinations on the phonetic scale [but] since the phonetic data does not lend itself to structural analysis, I will leave the phonetic level as an untried case'.[106] Nevertheless, in stabilizing potentially murderous situations phonetic free variation may have played an important historical role. Even in a less fraught migrant situation, it can, as Baggio shows, provide a useful safety valve.

Occasionally these diverse sound histories converged in a single work. 'The Native Informant' (1993) was partly inspired by the recordings I made with my Philips L3586 portable tape recorder mentioned in the previous chapter. When I came across these by chance on a home visit in the mid-1980s, I found (besides the crackly dawn chorus and the scarcely audible nightingale) recordings of my own voice. Twenty years on, machinery to play was no longer generally accessible. Recorded on a 1964 Philips EL 3586 portable reel-to-reel tape recorder at a speed of 2.4 centimetres per second, the tapes ceased to be generally playable when the recording industry adopted faster speeds – 4.75 or 9 centimetres per second. To render the old tapes audible, it was necessary (i) to play them at double or quadruple speed (the slowest speeds available on most machines at the time of the 1993 production) (ii) to dub the speeded-up original recordings on to a tape itself playing at double or quadruple its minimum speed and (iii) to play back the dubbed tape at half or quarter speed. This acoustic archaeology is described because each transcription generated its own artefactual sound; in particular, the speeded-up voice uncannily resembled birdsong.

In Australia, when I could finally listen to the transcriptions, I was disappointed. Instead of gaining access to the foreign country of my adolescence, I heard a schoolboy reciting first declension ancient Greek verbs and conducting a dialogue with himself in rather shaky French. Instead of a sound picture of a lost culture, I heard my former self speaking in foreign languages. This situation reminded me of what we had learnt during the Museum of Sydney project about the techniques the colonial authorities used to 'get' the local languages. Invariably, 'intercourse with the natives' is opened by a kidnapping. As 'What Is Your Name' reflects, violence is rhetorical as well as physical. But figures like Arabanoo, Coleby and Bennelong are not language teachers; their object is to learn the language of their captors so that they can regain freedom. Arabanoo does not want to give the English the names of things; he wants to talk back: he intends to initiate an echoic exchange that is seriously playful. The attention that First Fleet surgeon William Dawes paid to the sonorous body of language was exceptional.[107] More generally, in Australia in the nineteenth-century ethnolinguists analysed the structure of Aboriginal languages using grammatical models derived from Latin and Greek. Evidently, in this situation the 'native informant' was a construction of the white investigator: 'The ideal this mythical creature conformed to was that of the linguistic theoretician' and one recent writer goes so far as to say, 'Most of the time the lonely subject was one mystical native speaker who could even be the author himself.'[108]

In 'The Native Informant' sound histories from the Upper Thames and colonial Australia converge; but so does a sound history of technology. In some of the Philips recordings, the noise of the recording process is louder than the sound source being recorded; listening back, the 'native informant' (portable tape recorder) communicates the noise of its own feedback rather than accurate information about a lost acoustic ecology: 'It is the hiss of listening I hear now'[109] or, as Kantor would say, neither reproduction nor reality. The lost soundscape is located in the noise: hearing in it 'hiss of tyres, spool-warp, manic thrush ringing the steeple of the ear, the nightingale, the angel voice of an England England no longer hears', something is audible 'in-between illusion and reality'. But deriving atmospheres from atmospherics is not news to ham radio operators, as the classification of 'whistlers' illustrates: 'A whistler is a very low frequency or VLF electromagnetic (radio) wave generated by lightning ... perceived as a descending tone which can last for a few seconds. Whistlers can be categorized into Pure Note, Diffuse, 2-Hop, and Echo Train types.' The electromagnetic 'dawn chorus' that occurs most often at or shortly after dawn local time is a related phenomenon. The Wikipedia entry comments cryptically, 'With the proper radio equipment, dawn chorus can be converted to sounds that resemble birds' dawn chorus (by coincidence).'[110]

Throughout this period I conducted an unanswered (or unheard) dialogue with the acoustic ecology movement coming out of Murray Schafer's original 1977 publication *The Tuning of the World*. At issue was precisely the status of sound *as history*, and therefore our collective and individual agency in changing history by listening differently. Listening that hears the sound environment unconsciously composing itself is not passive: the listener is actively attuned, inclined or responsive. The corollary, at least for me, is kinaesthetic: a region that calls itself into being is criss-crossed with tracks or trajectories. It can be compared to the time-lapse image of the flights of a ball in a game of basketball or soccer, or to the Brownian motion of Cavalcanti's snowflake-like *spiritelli*. A listener who discerns this kinetic architecture is a player in its formation and reformation. In this environment, ambiguity, the bifurcation of sound and sense, can be compared to endlessly splitting and criss-crossing forest tracks. The focus of these will not be a symphony of birdsong; it is likely to be the human cry, the sound of the limits to human communication. This is the truth value of ambiguity, as Artaud grasped: its 'perpetual game of mirrors, in which a colour passes into a gesture and a cry into a movement, leads us without rest along rough paths we find hard to follow, plunging us into a state of uncertainty and unspeakable distress which is truly poetic'.[111]

To excise the history of listening from the presentation of the sound environment is to lapse into the realm of representation. To silence the spatial history of encounter, the invitation to respond contained in the calling to come, is to be deaf to one's own hearing, its location and its coming there. Virginia Madsen captures this paradox in a discussion of Schafer's proposal to locate microphones in remote wilderness zones, and to transmit their sounds 'without editing into the hearts of the cities': 'Schafer, through this "radical radio" where no editing occurs (no cuts, no wounds), is present in nature as never before. The microphone does not open a window of transparency onto nature. Rather, the microphone, and the whole machine attached to it, amplifies and heightens the sounds of nature (as well as those cultural intrusions), creating a hyperspace.'[112] Tomas makes a related point about Steven Feld's much-celebrated CD *Voices of the Rainforest*. Leaving aside the exploitation of digital sound recording and mixing techniques to simulate the Kaluli's auditory world, Tomas feels that the very presence of Feld's technology tends to make disappear what it would preserve. Its semiotic inscription 'is the product of the movement of Western technology through a foreign space'. The Kaluli 'can sing to us (to me) from track number 6 ... [but] I can never reply'.[113] In the end, Tomas thinks, writing in his Montreal apartment, 'the sounds of the Kaluli resurrect a history of colonial relations rooted in *this* as opposed to *that* space, because the compact disk has promoted a strange intermingling between a Kaluli world and a Canadian world in which the latter world serves as defining context for the former world'.[114]

Twenty years closer to the endgame of auditory biodiversity, we may assess *Voices* more charitably. In any case, the object of these criticisms is neither Schafer nor Feld (both musicians who work directly with the audio-archaeological environments they study): it is the microphone (here representing the entire drift of sound recording innovation towards the elimination of noise). Madsen's or Tomas's critique applies to the Lecce street recording: the electronic ear that overhears the scene is a semiotic reduction of the real, a disembodied transcription, its compositional suggestiveness due to being de-corporealized. Its representation of communicational plenitude is technologically generated nostalgia. Unless it helps sustain the world it records, it is colonialist in orientation. In Brandon LaBelle's terms, the microphone cannot act as the auditory conscience of the scene. Arguing that the 'geography' of speaking and listening is essential to understanding the demand of the other, LaBelle maintains that an ethical gaze involves an 'ear ethic' – the 'one-to-one pressure, the encounter of your gaze and my voice, and the responsibility of building place, between you and me' presupposes, LaBelle suggests, 'the third body or third ear, the one sitting over there, in the wings, off-screen, out of frame; the one, that is, who *overhears*'.[115] In *Meeting Place*, LaBelle's eavesdropper is compared to the 'erotic ear' in Socrates's *Phaedrus* or the one who, in Canetti's scenario, hears the 'overall effect' unknown to the speakers themselves. In the Lecce scene the third ear is diffused throughout the architectural ensemble of echoic surfaces and hollows.

But, surely, it makes little sense to demonize the microphone and its associated recording technology in this way. It is better to think of the assemblage as a transcription machine. Berio notes that for centuries 'the practice of transcription had a function analogous to that of records'.[116] One would not blame the piano for deterritorializing sound or placing the listener in a false position. If Theodor Adorno thought that, in the early recording period, the continuing practice of keyboard arrangement 'kept alive the listener's critical ability in the face of the violent anaesthesia administered by the fetishized and unblemished concert recording',[117] it was because of new listening habits. The development of the recording industry and radio are directly linked – 'Radio is a promotional vehicle for recordings.' Adorno's acoustic anaesthetic was mainly delivered via radio, 'a "lean back" technology for passive consumption', but this was not, in principle, irreparable.[118] Bertolt Brecht had urged as long ago as 1929: 'Radio is one-sided when it should be two-.' Its object cannot 'consist merely in prettifying public life' or 'bringing back coziness to the home and making family life bearable again'.[119] Instead, he suggested, 'change this apparatus over from distribution to communication. The radio would be the finest possible communication apparatus in public life, a vast network of pipes.'[120]

Usually, this is taken as a remarkably prescient vision of the internet, but if one focuses on the pipes, it could also be read as a formula for a new architecture of communication: reticulated radio or telephony for the urban collective. Whatever the case, the content of such an assemblage would not necessarily be information (official or dissident) but the amplification of noise. It is relevant that Brecht's frustration with radio arose in the context of negotiating a radio broadcast of his music theatre work *Der Lindberghflug*. As Steve Schwartz comments, 'The radio fascinated several major composers of this [twentieth] century and seemed to hold out the promise of a new kind of art, as well as a new kind of audience, outside the traditional concert venue. The "new art" never really panned out. Just about every work written expressly for radio ... owes its life to live performances.'[121] Hagelüken is more specific: 'The "open" politics of radio, championed by the director-generals of the individual radio institutions [under the Weimar Republic], contributed to this spirit of enthusiastic experimentation. But the political pressure on the individual directors with regard to their programming had already begun at the beginning of the 1930s. The first purge of radio actors took place in 1932. In the years 1932-33, the majority of leaders of the development of radio (and *Hörspiel*) in all its diversity were dismissed.'[122]

Composition is a history of transcriptions. In Berio's sense, which oddly echoes De Quincey's notion of the involute, an earlier composition is not collaged (selectively quoted in a new musical context) but, rather, opened towards the future: going deeper into its melodic or rhythmic structures a new arrangement is released for the future. A new sense is not produced through a process of juxtaposition: a return or *ricorso* is conducted that signifies a poetic, as well as political, conviction about human potential. Essential functions and arrangements are intensified. The transcription makes audible a tradition of openness, which as Eco argued, 'refers to the latent potential of a contingent and contiguous event or events to suggest multiple and textually consistent interpretive possibilities'.[123] In this connection I recall the occasion on which I informed Berio that the nightingale singing in the valley below was an *open work* – Berio had insisted that the well-known six-line transcription of the nightingale's song in Athanasius Kircher's *Theatre of the World* would supply our bird notation – but no two nightingales, I pointed out, sing alike. To quote a transcription in this way is to revert to an art of collage where 'the signifying features of the quoted material appear to remain intact, and at most a slight reterritorialisation occurs'.[124] To realize the original through a concomitant mode of production, as Edmund Husserl says, to achieve 'the revolutionary quality of "becoming intense", requires a basic elimination of Sense'.[125]

Anyway, the point is that the 'possibilities' are not imagined as endless: 'A work will establish a "field of oriented possibilities" that results from chains of associations "previously suggested by the co-text".'[126] In realizing this creative potential, the technologies of transcription can be active partners. To hear the hiss of history in the transcribed recording is genuinely to access an environmental unconscious, to recover noise's 'revolutionary quality' (in a double sense). Friedrich Kittler argues for a direct connection between Sigmund Freud's theory of the unconscious and experiments Hermann Gutzmann, lecturer in speech disorders in Berlin, was making around 1900 with telephones and phonographs. Gutzmann discovered that prompting his patients to respond to nonsense words produced *nothing but parapraxes*. Because both machines – due to transmission economy or technical imperfections – limit the frequency band of language on either end, what subjects 'understand' can differ from what they have 'heard'. Gutzmann spoke nonsense syllables like 'bage' or 'zoses' into the mouthpiece; the ear at the other end heard 'lady' or 'process'. 'A simple question brings to light an unconscious.'[127] In this sense, Freud 'takes the place of phonographic tests'.[128] Dream transmissions become in his reception meaningful psychic facts. As regards radio, a simple interruption to transmission plays a similar role: the unconscious of the medium is exposed. Transposed to radio theatre, it produces Beckett's formulation: 'The experience of my reader shall be between the phrases, in the silence, communicated by the intervals, not the terms, of the statement.'[129] What Adorno calls Beckett's 'dissonance' uses silence to expose censorship; opting out of 'an historic moment whose promise of freedom produces a viciously circular domination',[130] and whose cultural counterpart is the realist literature (and radio) of resistance, he transforms passive waiting into an active listening spiralling towards the end.

Quoting Brecht in support of his view that radio has failed to fulfil its cultural and political potential, Dan Lander has written,

> The history of radio art represents a struggle to overcome the enforcement of the arbitrary boundaries drawn by the paranoid hands of the state. These boundaries stifle creativity in many ways including the political, the aesthetic, the conceptual, the sensual and the multitude of creative imaginings that shape the various modes of expression and perception in a diverse cultural terrain.[131]

But works like 'Memory as Desire', 'Remember Me', 'The Native Informant' and 'Scarlatti' at least insist on *listening* to the radio. Subtitled 'a romance for radio', 'Scarlatti' engages radio as radio theatrical sound effect, as plot function (in Propp's sense), as erotic symbol and as medium. As medium, it questioned the 'coziness' of the listening environment, represented by being tuned in to a particular frequency. It was a poetic transcription, in this sense, of the spirit of Domenico Scarlatti's music, which his biographer and editor Ralph Kirkpatrick characterizes as noisy: he 'imitated the melody of tunes sung by carriers, muleteers, and common people', but also the 'dizzying whirl of twinkling feet, stamping heels [and] the inevitable castanets'.[132] The original 'Scarlatti' tried to imitate this free association dramatically, partly by the juxtaposition of 'found' phrases, partly by the simulation of a radio being successively tuned to different stations.[133]

I always had the feeling that 'Scarlatti' was unfinished. As more or less open works resigned to the fact that 'nothing but the main form of a work will resist the vicissitudes and calamities of presentation', as Ezra Pound put it,[134] any script of mine can be further augmented, translated, reworked and its lines reassigned. None of them can or should be mathematically precise movement scores. But, in the case of 'Scarlatti', I thought for a long time that the original transcription was structurally incomplete: Scarlatti was a great *improvisateur*; an auditory shaman of everyday sounds. However, his compositional mastery primarily depended on an astute and original deployment in his sonatas of a type of binary form. The first part of the binary structure introduces the first theme 'A', then modulates to a related key and introduces theme 'B'. Instead of a development (as in the later classical sonatas) there follows a 'mirror' (a short passage usually modulating back to the tonic key, and this passage represents a point of reflection). The second part of the binary structure is then theme 'B' followed, without modulation, by theme 'A' concluding the sonata.[135] In a crude sense, and allowing for the critical fact that dramatic form and musical form are not isomorphic, I felt that the resolution of the story remained to be found. In the original 'Scarlatti' the two characters (Paco and the Englishman) meet because they are drawn to a radio in the street mysteriously playing a Scarlatti sonata. Set in the last days of the Franco regime, when, 'waiting for the boss to die, heroic gestures had no place', they pass the time, listening to the radio. At the end Franco has indeed died but nothing is resolved: the question 'What are you going to do now?' is unanswered.

Insofar as it imitated the form of a Scarlatti sonata, the first part of 'Scarlatti' had been written but not, I felt, the second part. But I didn't know how to complete it. Kirkpatrick stated that Scarlatti's compositions generally lack a 'recapitulation', offering in the second half instead a 'restatement'.[136] Perhaps this was another clue to the phenomenon of repetition in difference. It reminded me of what Pessoa said about writing poems: 'I've written quite a few poems, / I'll no doubt write many more, / And this is what every poem of mine says, / And all my poems are different, / Because each thing that exists is a different way of saying this.'[137] Scarlatti, like Pessoa, repeats himself because he is not afraid of repeating himself. And then two events changed this. In 2011 what the Spanish newspapers described as a 'lively' debate 'erupted' over the proposal of a national commission to exhume Franco's bones; early in 2012, after a thirty-five-year silence, the individual on whom 'Paco' was loosely based contacted me by email. In the strange coincidence of two themes, one public and one private, surfacing after so long, an echo was fast approaching that invited transcription. Raised again was the question of resolution. What is a return without recapitulation? In music it might be a *ritournelle*, or refrain, something like birdsong ... but in language? A recurrence perhaps?

Among Domenico Scarlatti's older Neapolitan contemporaries was the philosopher and writer of the *New Science* (1725), Giambattista Vico. It is unlikely that they ever met: Domenico was only one when the eighteen-year-old Giambattista accepted a tutoring post in the mountain village of Vatolla, in the Cilento, south-west of Salerno, where he stayed for nine years until 1695. By the time Domenico was eighteen or so, his father, Alessandro, the renowned opera composer, had packed him off to Venice. Much later, Alessandro and Giambattista may have been acquainted through their membership of the Roman Academy of Arcadia.[138] A late baroque sensibility is common to both, exemplified by their shared interest in the creative possibilities of ambiguity but it took dramatically divergent forms: while Domenico diffused sound masses into musical molecules, and is in a sense the least florid, most anti-baroque composer of his distinguished generation, Giambattista was a cloud-builder, conjuring from the poetry of etymology a figure of history as a spiralling, revolving staircase in which 'the pertinent institutions of the past – forms of government, laws, modes of ownership' – recur. History does not exactly repeat itself, anymore than Scarlatti's sonatas recapitulate the main theme: instead a kind of transcription occurs – 'The recurrences between discontinuous but symmetrical stages of history' can be described as 'correspondences.'[139]

The Cilento, the ruggedly mountainous region that bulges out into the Tyrrhenian Sea, dividing the plain of Paestum to the north from the rocky coastline of Calabria to the south, and inside one of whose in-turnings ancient Elia nestles, offers a bewildering complex of folds; long spurs extend towards the sea, lofty, here and there rising to small peaks before suddenly dropping down to the coast; these spurs attach to a spinal cord that is itself zigzag, broken, many fingered. Ridges descending from these heights run not in parallel but converge or diverge; and the pattern of the steep valleys carved out by millennia of flash floods reproduce these fractal patterns on a smaller scale, valleys sporting their own lateral ridges, their own alluvial steps, chasms, opening plains and closing gorges. Dotted through this region of proliferating horizons are tiny villages that cling to the steep upper slopes; they consist of little more than a main street snaking up a ridge and hanging off the side of it a fishbone arrangement of side streets. Vatolla, reached by a slow winding road, is like this. The miniature piazza, little more than a belvedere, occupies a small saddle in the spine. A few steps above is the parish church – the day I was there schoolchildren in file carried Easter lilies into its cold interior. A few paces below, a city within the village, its imposingly high northern wall at an angle to the winding main street creating a wedge-shaped *largo*, is the country palace of the Rocca family, complete with medieval turrets, where Vico taught, studied and dreamed for nine years.

I visited the room where he slept and worked when his daily duties were done. Vico commanded a view in which the descent and evolution of humanity might be imagined; beneath his window sill the mosaic of tiled roofs, stone walls, tiny terraces and flowering trees comprising Vatolla;

beyond, stretching vertiginously downwards and rapidly stretching away into the blue haze, the huge basin of the ridged and intersected valley-system of the Cilento, an arrangement of horizons, a complex genealogy of topographical lines of descent, confused and confusing as to their origins, even if their underlying reason is simple; and beyond them the resolved symmetries of the Gulf of Salerno, and its twinned institutions of destruction and renewal, the peak of Vesuvius and the matchstick temples of Paestum. At night from the adjacent turret the slow *ricorsi* or apparent retrogressions of the planets might be watched. In 'the silence of the solitude' of Vatolla, as the plaque erected in the 1870s puts it, Vico spread the wings of lofty speculation, ascending to the sublime heights of his new, reflective science.

However, the streets were *not* silent, nor was Vico deaf or blind to his surroundings. The Cilento was the topographical equivalent of the etymological method that Vico would use to map the true origins of words and to show how in the course of time they had been metamorphosed into a hundred divisions of meaning, whose shrouding heights (caused by the ever-deepening erosion of usage) cut them off from their original significances. Reflection was cognate with one of the four devices of poetic logic, irony. Irony, considered positively, as an awareness of the forgetfulness of the history of words, the ever-present ambiguities inherent in their commonplace application, was a fact of communication in Vatolla. The sharp division between patrician and plebeian, between lettered aristocrat and unlettered peasant, was obvious – perhaps keenly felt by the subordinate Vico. In the *New Science* one of the chief sources of evidence for his thesis is the figurative language and superstitions of the country peasantry; in earshot of his study, Vico heard picturesque phrases, which, he thought, contained the vulgarized remains of a once-noble mythopoetic faculty. His new science of poetic reflection would return these expressions to their origin and cause them to open out a new landscape, the history of what humankind has made out of the mind.

We know about the evolution of Vico's ideas from the *periautography* he wrote in 1725, responding to Count Gian Artico di Porcia's '"Proposal to the Scholars of Italy," calling for the scholars to write their intellectual memoirs for the educational benefit of the young'.[140] A *periautography*, or writing around the self, is life writing as spatial history; it derives the narrative of a life from the works through which the reflective self evolved. The Cilento is not a picturesque backdrop: it is the model of a reflective science where *anabasis* and *katabasis*, climbing to the origin and descending to the future, are twinned. For every White Horse Hill there is a Vale. It posits that the setting of a life shapes the formation of ideas; it substitutes outwardness for inwardness or, in other terms, a poetic logic or *ingegno* 'that connects disparate and diverse things' for the axiomatic derivation of the cogito found in Descartes.[141] To walk around oneself, to survey the field of ideas, is not a singular, concentric activity: it takes the same form as the spiralling gyres of the homing pigeon looking for landmarks to find its way. The curvilinear path of the connections made in this way offers a genealogy mapped in terms of returns. But the returns see the earlier work in relation to wider origins and open futures. These *ricorsi* deepen the involute producing new works that are transcriptions, or new correspondences. *Amplifications* is *periautography* in this sense. More accurately, it is *periotography*, a writing around the big ears of listening.

THREE

Rattles

When I responded to the message my mother had left, she told me that lung cancer had been diagnosed. From that moment we began to hear ourselves talking. A silence opened up like a sinkhole in the pavement, and instead of contemplating it, we stepped round it, beginning to speak from the other side. On a doctor's say-so, we took Death's part. Crossing to the time of last things, the discourse of consolation began. As if dying could find at last our common project, we began making arrangements. The talk was hollow; it drowned out the dread. Our mutual reassurances were a *folie à deux*, reaffirming the habit of a lifetime. Perhaps always, denying change, we had made language an undertaker. The telephone is said to be an inadequate channel of communication in such situations. In a way, though, it gives calumny nowhere to hide. A silence had opened up, alluding to an experience beyond representation, from whose bourne no echo comes back. It would have been honest to honour it, even in the poignant interval retuning our ears to the normally ignored sounds of the everyday (all that does not concern us but which envelopes us from cradle to grave). But the culture of the telephone prohibits this, as it casts any pause in the conversation as a sign of withdrawal, abandonment or disavowal. So to avoid this signifier of betrayal, we blunder on betraying the truth. After a lifetime of conversations that were invariably left off angrily, unresolved, we have between us a history of silence that cannot be repeated. The telephone sees to that, solicitous in coaching us across the little contraction whose physical manifestation is a lump in the throat.

Curiously, around that time of the phone call, a distinguished novelist told me about a different experience. Someone he had based a character on was dying. So, as she was dying, the novel was read to her. She gave her blessing to his interpretation of their story, and died. It was as if she had been waiting for her story to be told. A life ends unsatisfactorily if it is unshadowed. His story was her pall. It was the voice of another she needed in order to find the strength to give up the ghost. It was the living voice that could accompany her on the downward path, interceding with the furies. Or was this operatic scenario exactly what she heard, and seeing through it, was *this* the motive of her parting? Life had been one unfulfilled story. Here was another. The author wanted to be her Orfeo. He wanted to catch her dying breath, so as to hold her hand forever. She had to settle for the dispensation of his genius. It was a consummation devoutly to be wished. My relationship to story has always been different, more querulous, causing me to ask, 'Might other voices and visitations have filled those hours?'. These last words that she heard, this echo of the stories she had told – aren't they botched from the beginning, returned to her disguised, enriched and false? In the end, the darker truth hits home for me: there was not, nor can there ever be, any harmonization. Between speaking and listening, writing introduces an absolute divide. As for the beneficiaries of this fiction, the readers are not curious about the factual circumstances. They are as indifferent to the fate of the lovers as the birds the dying hear half-muffled in the walls of the house.

Certainly, such literary transubstantiations of the echo as this eluded me. My father never heard his recorded messages to me played back; by the time 'Underworlds of Jean du Chas' was broadcast, he was dying. It would have been cruel to suggest that he survived, at least in my underworld, merely as the one who always said goodbye. It needed the good humour of good health to prevent this migrant licence from seeming simply a betrayal of trust. Similarly, my mother never heard 'Light', another radiophonic work produced in those years.[1] William Light played a major role in Adelaide's urban design; he also died there in 1839 of tuberculosis. 'Light', the work we made for the 1996 Adelaide Festival, was mental theatre realized as a site-specific sculptural installation, soundscape and performance. It took the form of a séance in the course of which the dying surveyor is visited by figures from the different parts of his life. The pageant of spirits is heard through the folds of breathing or, more accurately, the growing death rattle of Light's infected lungs. As his Double explains, 'The voice comes directly out of the chest. ... As if a puppet were lodged there, speaking in the character of Punch ... . Sometimes as if speaking through a cleft reed ... . Doubled, trebled, as through a speaking trumpet ... . As it might sound applying the tube directly to the larynx ... . Or smothered, like the ventriloquist's.' Much of this information came from Laennec's *Treatise on the Diseases of the Chest*, published the year before William Light expired.[2] Laennec describes four classes of *râle* (sounds produced in the air spaces of the lungs): sonorous, sibilant, crepitant and mucous. The sonorous *râle* 'is sometimes extremely loud and resembles the snoring of a person asleep. At other times it resembles the sound produced by the friction on the string of a bass violin; and occasionally it resembles the cooing of a wood pigeon.'[3]

This was not suitable information to convey to one dying of lung cancer, especially when Laennec adds the macabre observation that the resemblance to cooing 'is sometimes so striking that we might be tempted to believe the bird concealed under the patient's bed'.[4] To have listened to 'Light, a séance drama' in 2003 when she was struggling to breathe risked infecting *all* happy associations with birdsong: the dry, 'sibilous' or sibilant or hissing rattle was said to resemble 'the chirping of birds', as if, in a typical underworld reversal, Death's Furies had appropriated the dawn chorus.[5] But perhaps this was not the story after all: when I arrive in Faringdon, I find not the dark forest but the seashore. Margaret greets me with a description of her recent examination. Wax was placed in one nostril, like a semi-embalming. Asphyxiating, she was flooded with oxygen, surviving despite herself. I recall what she said seven years ago when Bill was dying. She dreamed she was standing on the banks of an angry swirling river; it was plucking him from her, tugging him ever more forcefully out of her reach; he was sinking into the billows, and she could not pull him back; he was waving, going under, the gale groaning, the waves whipped white threatening to suck her under. She could do nothing but cry out. Now she could not even cry out. When she struggles to speak or draw breath, it is not the lungs 'filling up' with fluid. She descends handcuffed and blindfolded into the black river. This is not Light's world but Jean du Chas's underworld: 'My ears buzz and my mouth tastes bitter. I lower myself further. I am going to sleep. I can't fall asleep in such dizziness. There is little light. I reach for the next knot. Miss it.'[6]

One day I drive my mother slowly through chrome, amber and robin-breast fields and woods to Southrop: herring-bone nave wall and a late-Norman font, where the Vices are inscribed backwards and the Virtues forwards in intricate triumph over their avenging saints. In sorrowing light to Westwell and its finger of stone commemorating two officers killed in the Great War; and to Christmas-like Burford – where something happens for which I am unprepared. My mother waits in the car and sends me forth as her sensory probe, to report back what I have heard and seen. But, instead, walking down Witney Street, I find myself at the Mill where, in the midst of the drawn-out withdrawal from my marriage, I stayed with my then lover, an event that coincided with my father's death. My emotional inertia prevented us riding the rapids of change; and it ended bitterly. I have let the time sink to the bottom like waterlogged timber. Four months before he died, she emailed me: 'My experience of being with you has heightened my awareness of the importance of sounds … . You know about the importance I attribute to your voice. Do you remember how you made me laugh, when you read aloud to me passages from Du Chas?' And about this place: 'Walking with you and making love to you creates the desire to listen in me … . You wanted us to stay by the watermill because of the sound, wanted to make love on the balcony to be able to weave a story of sounds ... . The physical and emotional story composed in a landscape that fostered your first desires.'

I don't report this to my mother, nor communicate my other memory from that time – that the telephone in the room where we stayed was padlocked. It was like a stuffed bird in a cage: a gagged telephone. Can a locked phone ring, I wondered? And if it does, the caller would never know the telephone had *made the call* and decided not to pick up. When I think of the long-distance relationships through my life, and the scope they created for emotional inertia, a locked telephone looks like a kind of solution. In its chastity belt it cannot recall us to our duties. Its *Notschrei*, which is 'not only a sound, but an irresistible noise, the kind you have to do something about', is nullified; we are licenced to forget. Its calling to come, the cry of an existential lack, which we recognize as the echo of our own distress, is quietened: 'It's like a siren, or like an automobile burglar alarm that can't be turned off, but renders all thought and concentration impossible. It's like a person who keeps calling on the telephone, begging and soliciting, someone who won't shut up and won't stop. It's both noise and a call (*Ruf*).'[7] This is why well-hearing Odysseus plugs his sailors' ears with wax. If a modern *Odyssey* were written, its hero would be called Telephonus – the answer to Polyphemus's question 'Who goes there?' would be an unanswered telephone ringing. The Windrush is channelled into the mill stream; the undershot waterwheel rotates: the rush of the millstream and cascades of uplifted water tipping backwards off the wheel are deafening. We enjoy a wordless *ricorso* or, as Jean du Chas says, 'There you have it: remembering and forgetting: the rotation method. Consider the joy'. But it only happened there once.

Nursing the sick, the carer practises environmental auscultation. It is like having a baby in the house, constantly listening for anything out of the ordinary. An unusually prolonged silence is as concerning as the cry of distress waking from a nightmare. The meaning of other signs changes: I am surprised how quietly Margaret sleeps. She used to snore. Now she sleeps like a babe. But it is not a good sign. It means her breathing is growing shallower, sluggish. I creep back down to the kitchen, the crypt-sized scene of the drama that we played out across a lifetime; always at right angles, as I sat at the table facing her while she gazed out through the kitchen window. I disturb the poker next to the Rayburn, a hangover from the anthracite period, and its little clangour immediately recalls me to Friday night washing rituals half a century ago. Margaret held out for a decade against the introduction of an automatic washing machine, using instead a gas-heated tub with a paddle to agitate the laundry. The paddle was operated manually via a drive shaft in the lid of the tub attached to a metal arm. You gripped the handle at the end of the arm and pushed it, 90 degrees to the left, 90 degrees to the right, at each half beat banging against a raised stop in the zinc alloy lid. An almost musical clang was emitted, but what I chiefly remember is the industrial drumbeat of the handle savagely wrenched left then right, hammering against the stop, my coming downstairs after being woken and helping out, the steam-filled kitchen, the momentary alliance of mother and son as together we counted out the prescribed 140 turns.

The noise we make blocks out the sound environment. Absorbed in routine activity, we build a wall of deafness. Retreating from the world, we also let down our auditory defences. Too weak to act, unable to repair the breaches, we lie down and listen: the swish of the traffic, the iambic 'de-dum' of the never repaired manhole cover when it is stepped on, the robin's sibilance scarcely distinguishable from the kettle's whistle – these sounds break in. They no longer frequent the background; they are no longer lost in the wake of our forward motion. Instead, now as we glide without volition towards the sea, they stand either side, like immortals. They inhabit a time indifferent to our narrative; they will persist as before after we have gone. These are the sounds that intrude into the funeral service or mingle with the mourners in the car park afterwards. A robin's cadence like a silver rail leading down into the dark, a song thrush's defiant resurrection chant, the philosophical complacencies of the blackbird: such pathetic fallacies as these pervade the atmosphere. Other habitual sounds that belong to this category include children screaming in a nearby playground, a door slamming, its clap amplified down a passage, the description of the Australian interior provided by caged budgerigars next door, the dismal melodic chimes of 'Greensleeves' and the impression seasonal gales give that the house is at sea.

William Light's *Last Diary* is largely confined to two topics: the state of his health and the weather.[8] Light did not have the advantage of Laennec's *Treatise* and, I argued, in *The Lie of the Land*, correlated his difficulty in breathing with the condition of the atmosphere. In 'Light', I proposed a far more fateful correlation, as appears in this passage from Act Two, Scene Three, evoking his last delirium:

> E (Light's first wife): Look at the clouds.
> Light: Suffered much from debility occasioned by the atmosphere.
> E: They are like lungs.
> Double 2 (Light has five doubles): Once I was single.
> Ten (One of the choral commentaries): No smoke without fire.
> Eleven: Again, again and again.
> Light: If there were a change.
> E: Myriads of small specks.
> Double 2: I married a wife.
> E: Long tapering columns.

Light: If, say, Tuesday, 16 April, yes, very fine weather all day. Felt much better.
Double 2: My wife took a fever.
E: Mackerel back, striae, like the grains of wood.
Eleven: Again, again and again.
Double 2: Married another, much worse than the other.
E: Cumuli come, at first like a small spot, the nucleus on which it forms tuberculated.

Apropos of maculations, the starling is among the auditory symptoms of atmosphere in Act Two, where, with reference to Light's Suffolk childhood, 'Breathing Paths become more eventful, and more enveloping. They introduce tubercular breathing patterns per the indications in the script. These patterns modulate into atmospheric phenomena, in particular, *English* birdsongs, the "chirrup" of small birds, and where the script indicates, the calls of the wood pigeon, corncrake and starling.'

The Breathing Paths mentioned here referred to linear arrangements of loudspeakers. 'Light' was installed in the park immediately to the north of the Torrens River (and diagonally opposite the Festival Centre) known as Pinky Flat. Two Breathing Paths crossed through each other, each stretching 300 to 400 metres east–west across the park. Aerial sounds were heard from speakers suspended in the trees; the *rhonchus* of Light's tubercular lungs emanated from powerful speakers planted in the earth's cavities. The roar and rumble of troubled pulmonary inspiration and expiration that made the ground tremble was derived from recordings of water blasting out of blowholes in the Canary Islands. I suppose these recordings where the rumble of imploding seawater alternated with the swarming of bubbles exploding as compressed water escaped upwards bore a resemblance to Laennec's moist râles or mucous rhonchoi – we did not conduct an audioscopy. 'When the moist râle is very large and infrequent, we can almost perceive the bubbles to form and burst. When it is numerous, large and constant, it is sometimes so noisy as to resemble the rolling of a drum.'[9] Another moist *râle*, the 'tracheal' or death rattle, may be so loud that it suggests 'the noise of a carriage on the pavement'.[10] More to the point is another observation. In certain cases, 'a slight vibration [is] communicated to the stethoscope when the site of the *râle* is immediately beneath it.' Laennec pointed out that this sensation, 'like that occasioned by the voice, may sometimes be felt by the hand very distinctly.'[11] At Pinky Flat, walking a border between the lungs and the external environment, the pathology of the place was *heard* through the feet.

One day during my first visit I return to find the kitchen in darkness. The house seems to have withdrawn into itself. I have a premonition of what it will be like when she is gone. It listens to itself: the minute shuddering of the refrigerator, the tiny groan of the window latch, a cupboard door springing open, a floorboard taking the weight. The house conducts an internal examination. Sometimes now Margaret murmurs to herself; hard of hearing, like the house she sounds her own depths. Communicated is a desire to be *heard* listening to herself. The dull thud in the back of the throat articulates the wheeze and whistle of compromised air passages. The house breathes, too: the windows wear a minute bubble-wrap of condensation; wiped away, the view remains blurry; the little birds at the bird table spectral flutters. I creep through to the front room, and find Margaret asleep. I am relieved that all is well and utterly quiet. I savour the irony of not disturbing her, of wanting her relief from consciousness. The house makes no demands on her and holds its breath. Escaping the artificial heat of the kitchen, I pause on the stair, taking in the signature damp clinging under the skin of the wall. It is always there, gathering outside the furnace-room kitchen, its moulds and vapours biding their time.

The doctor says that the tumour eating up Margaret's lungs is of the 'large cell' variety. The cancer is also in her lymph nodes or glands. Eventually, if she lasts, it makes its way to the brain. I am not sure, though, that Margaret hears any of this. Treating his visit as a social call, she conceals her suffering as a distress lest she inconvenience him. Telling me the story, I guess he imagines I will translate; but I am not his puppet, and besides do not know what to say. Instead, after he leaves, I read Shelley's 'Ode to the West Wind': 'O uncontrollable! If even / I were as in my boyhood, and could be / The comrade of thy wanderings over Heaven.' Symbolic meteorology holds out more hope than any medical scenario. Apropos, Margaret tells me this day about her *abhorrence* of puppets. This information is twinned with the sudden memory of a childhood swimming accident in Blackpool, where she nearly drowned. I surmise that the link between these items is a fear of asphyxiation. There is spiritual asphyxia as well as physical choking. Anyway, it is another reason for not passing the time listening to 'Light', in whose 'séance drama' the characters conjured up are puppets of the Medium. And yet, summoned by the spirit master's curt 'rat-a-tat-tat', these lieutenants of Death speak with a refreshing candour:

Ten: Doctor! Doctor! I am dying.
Double 2: As if a puppet were lodged there, speaking in the character
 of Punch.

But spirit possession comes in many guises. In the sequel to 'Light', 'Jadi Jadian', a site-specific performance about Light's mother made in Georgetown, Penang, we sought assistance from a spirit healer: in the 'rat-a-tat-tat' she used to accompany her incantation I heard the weekly tattoo of the washing machine downstairs in the kitchen.

The body of the earth has its cavities like ours. As cycles of time alternately calcify and melt its interior structures, it breathes. It also has its death rattle.

A textbook written in 1845 describes how 'a mass of snow' filled with an 'immense number of bubbles of air' is changed when penetrated with water into a glacier by 'successive congelations': 'The air is disengaged in the form of little bubbles, that rise and burst on the surface of the little pools of water ... with a very marked crepitation.'[12] Here 'crepitation' refers to the appearance of the surface, but it is the frozen echo of a sound event. The process of its formation is no different from the impact of pneumonia on inspiration, identified by its crepitant *râle*: the crackling is supposed to be due to 'fine bubbles' in the lung cells: 'We assume that air and fluid are in the alveoli, and that the *râles* are produced by the bursting of microscopic bubbles of moisture.' The sound 'can be simulated by rubbing a dry cigar between the fingers or sprinkling salt on a hot stove'.[13] After congelation comes melting – and more majestic symphonies:

> On a sudden, as if from some prodigious distance, there fell upon my ear the sound of musical instruments, pure and clear, but barely distinguishable. I halted and listened: there could be no doubt, there was the beating of a drum, and from time to time the sound of brass instruments. ... We moved on, and the sounds continued, becoming rapidly more intense, and soon, as we approached a deep, narrow crevasse, the mystery was explained. At a considerable depth below us, a trickling streamlet in the interior of the glacier fell from one ledge of ice to another; the crevasse under our feet played the part of an organ pipe, and the elastic mass of ice struck by the descending rill produced sonorous vibrations.[14]

Over the Alps and back in Melbourne, I am tied to the telephone. Sensitized to its summons, I overhear other calls to come. One day from an indeterminate distance there penetrates the squeaking hinge distress call of a blackbird. I would like an inventory of these unclassified sounds of alarm. They track us through the days and nights, and from continent to continent. I doubt if anyone besides myself in this suburb heeds that call. It falls below the rim of auditory consciousness, unmarked, but scratching its nail into the soft tissue of consciousness. It suggests the last of the race, the solipsism of the last language speaker. It is the cry heard when the silence falls through the traffic around the bed of the dying. It populates deserted gardens, junkyards, hospital car parks, every species of place from which the facile melodies of everyday activity have been evacuated. It comes back in, a kind of weed-sound seeding the toxic soil, inhabiting its traumatic wounds. I imagine it is the first and the last sound of life, what lasts when bird song contracts into the drivel of tubes, only its bayonet thrust gets through the curtains, the low hum of life-support systems and the accompaniment of aeroplane drone, traffic and music with the bass turned up that are the signs of wanting to live elsewhere. Later that day, cycling, I observe a dead blackbird at the side of the road and put two and two together; motivated by more than existential angst, the bird's desolate squeal responded to the news.

Responding to the *squillare* of the telephone, I fly back to England. We were making love within earshot of the Pacific when the message came through. I remember a congelation that infected the scene. The waves grew cinematic, fluttering in their restless movement on the spot, rustling like grass in the swimmer's ear. They merged with sheets zigzagged and crumpled where we wrestled with each other. It is doubly symbolic, this leaving of the light. The comparison of the passage back with Margaret's last days is unavoidable: the luggage check-in, the passport control, the passengers like patients, row after row. The giant doctor organizes palliative care against disaster; the lifejackets are, although useless, carefully stowed. There is a curious interregnum of double starlight, the twinkling geometry of nameless Russian cities below, the mistier, spread out star patterns above; then the great red eyebrow of the dawn chasing us from the east. The descent is muffled up in English cloud, light caught in its twists and folds. And then the resumed 'hum' of ground transportation, driving out of the airport's gloom. In the fields either side of the road, some time between leaving and returning, the colour withdrew and the bare hedgerow elms swim in the mist like giant fan corals. The little town has scarcely drawn breath since I left. I find my mother, I rearrange her blankets. I call my lover in Australia. The lover's discourse (even by telephone) is rhythmic. 'Everything we do with a certain skill – we do rhythmically without noticing it,' says Novalis, claiming a human or metrical origin even for machines: 'In all arts and crafts, all machines – in organic bodies, our daily tasks – everywhere – rhythm – metre – beat – melody.'[15] The kettle is coming to the boil and I make tea.

Margaret said last night that, having often wondered what Bill felt in his last months, she understood now: it is a filmic sense of people, traffic, clouds, voices and noises occurring behind a glass of irrelevance. But these screens characterize a life as well. Acts of withdrawal occur wherever something is left unspoken. I remember childhood stays with my grandfather in Letchworth; he inhabited a cottage that was the architect's answer to post-traumatic stress disorder, a bungalow called Rosemary Cottage. Caring for her father, my mother took care to preserve the illusion that nothing had changed since before the Second World War when they used to attend the Wesleyan Chapel in Harrow; in forty years her loss of faith was, I am certain, never mentioned. Where no change was admitted, there remained only the daily routines to discuss. Emotional inertia enveloped us; we swam through it like fish, pretending to talk. The mantelpiece clock maintained its solemn tick-tock, periodically whirring ahead of striking the hours. Perhaps in its temporal circularity grandpa heard his own triumph over chance. Some irregularity in his life had been overcome. In his retirement – which, symptomatically, stretched out thirty years – the psychic fibrillations of his First World War service and the later eighteen-month convalescence from a car accident (in the 1930s) had at last been ironed out; the perturbations of progress resolved into a Vico-like *ricorso*. As a railway clerk whose promotion followed the direction of the LMS line southwards, from Stockport to London, he had synchronized his working days to the rail timetable. Retiring to a place outside the national clockwork, where the regular beat of the clock secured him against accident, he enjoyed a kind of immortality.

The fishbowl face of the mantelpiece clock reflected the leadlight windows looking onto a picturesque garden. Bored and restless, I used to stare out at the soundless scene. I used to watch the blackbirds stalking the invisible worms under the green lawn, cocking an ear, hopping purposefully forward and pausing again, jabbing, jabbing the turf, hauling up the stretched out rubber band. To pass the time, I drew a plan of the lawn and marked on it the courses of the hunter birds, by these, I suppose, plotting the underground labyrinth of wormland. This dismal cloak-and-dagger routine, a kind of secondary auscultation where the blackbird was my stethoscope, suggested a struggle between fate and free will, as if mastery of either depended on a cartographic grasp of the possibilities. But what appeared a matter of chance at one scale looked inevitable when considered regionally; statistically, the worms as a whole were doomed to be extracted from their dense humus and gulped down. After my grandfather died, we kept his pocket diaries. The entries, as minimalist as those in Light's *Last Diary*, were a record of his social life; but empty pages were as frequent as those where something happened. One had the sense that keeping the diary up to date mattered as much as any train of events. After my mother died, I collected *her* diaries. Lists of daily tasks furnished their main content – until, near the end, routine now irrelevant, she started to record her feelings, as she was dying, beginning to speak up.

Outside the window jackdaws flutter heavily upwards, but the sound of their wings is withdrawn. A woodpecker grips the rotten branch of the apple tree, and performs a rat-a-tat-tat, which, being inaudible, seems ineffectual. These portents occupy the vacancy that Christmas day opens up. Mother retches, doubled up over the toilet bowl, and then, murmuring against Morpheus, falls asleep. A sound diary today includes the roar of the vacuum cleaner. As it is pushed forward, the machine lets out a sotto voce sigh and groan, which I cannot distinguish from her voice, from decades of querulous summoning sucked out of the domestic space, swallowed into the machine's maw and absorbed into its breathing pattern. The squeak of the window latch when the puffs of west wind beat against the roof seems to cry out in stifled pain, a tremulous whimper. It might be the dying agony of a rabbit trapped by the foot. The wind comes in waves, breaking slowly over our near level estuary of roof, bare trees and aerials. The thin screech of the blackbird is caught up in it, like a streak of foam. In Margaret's panting, punctuated with hardly audible squeaks, sighs and groans, I picture a limed bird beating its wing uselessly against the branch. Over the undulating line, like the shark's streak sewing a wave, the thin pipe of wheezing, sometimes 'like a prolonged hiss, flat or sharp, dull or loud', sometimes like 'the chirping of birds'.[16] The hiss of oxygen periodically released under pressure imitates the swish of passing cars: seductive, both, promising release from the here and now. Now it is my soiled hands rubbing together; the dirt dried on them is language turning to dust.

Inside the ear, the channels are silted up. Trapped membranes quiver, generating ghostly creatures and spirit speech. They are caught up in their interior forests, and lack the strength to break through the undergrowth. They are struggling into swampy ground, sinking up to their necks in mud. The sickle moon is no handle. They are following the sound of their own shallow breathing, as if it were the footfall of a messenger; but it is the outgoing of their own lives that is marked, and in whose path they follow. Now nothing exists without its characteristic note. Cups, door handles, switches, clattering unevennesses in the ground and the slide of slippered feet: they are known by their use, recognized by their signature sound. In these tiny overheard collisions the routines of life are translated. The toilet flushes, and the cars keep promising to arrive before they veer away into other conversations. These events she registers as a deaf composer might know her music: imagining them, she hears the sting of her deafness. Each shrill, whistling punctuation in the night of the ear, is recognized as the double of a sound last certainly heard a decade ago. Rowing down the tunnel of this inside world, the sky is still visible between the deepening banks; only, the birdsong has withdrawn. What is audible is the tinnitus of ancient noise, snatches of Chopin or clogs on cobbles. Dead voices come back and hold converse with her. While I listen to what can be heard, she listens to the unheard sounds. For instance, the doctor cannot get through to her, but Bill speaks to her from the chair, while her twin sister echoes her thoughts almost perfectly.

In the morning Margaret struggles to dress herself. Every exertion produces a little cry. To breathe in demands concentration, as the smallest miscalculation releases mucus back into the lungs: shallowness of inspiration struggles with shortage of oxygen. Each out-breath is also stage managed, either to clear her crepitous lungs or to mutter words under her breath. To talk when she should save her breath is her defiance of Death: the ventriloquist of her distress, she will not be his puppet. What wells up is not only bile but the spirit. There is contraction as well as expansion; the birds that chirrup in the chest's black pool remember the fields. There is not only withdrawal as the spheres of irrelevance lay siege but also a new participation in the suffering of others. As the spirit gets ready to migrate, physical confinement is twinned with a journey to 'the uttermost ends of the earth'. Now that nothing, not even the habitually repeated, will be repeated, everything is mysterious. Casual conjunctions acquire a unique significance. In the evening we watch an adaptation of Thomas Hardy's *The Mayor of Casterbridge* – with the sound turned up. News coverage of the Bam earthquake in south-east Iran follows. I want to make sense of it for her, to find some connection between Henchard's self-immolation and the collapse of a 2,300-year-old citadel. Margaret writes a cheque for the British Red Cross Bam Earthquake appeal.

We slid into the new year in a coffin of sleep. A national gust of wind blew out celebrations from Edinburgh southwards: woke before that pallor that betokens dawn, and switched on the radio. Tightrope my way into last night's clothes, and creep downstairs. Margaret sleeps like a baby these days: no snoring, and none of the wheezing that used to follow her footsteps to bed – so that after she lay down you had the impression she was still creeping along. It is raining, the penumbral gloom that is permanent at this season thicker. I clatter about in the kitchen, and bringing a book back from the front room, settle down to listen and wait. It is Bill's copy of *The Actor Prepares*. After the Second World War Bill had helped to establish the Faringdon Dramatic Society. About the time of my birth, it was not only the Mermaid Theatre that arose Phoenix-like from the ashes of conflict: under the banner 'Fairy Story of Faringdon's Little Theatre', *The Wiltshire Herald and Advertiser* told how 'a rusty, derelict nissen hut, appropriately surrounded by mounting weeds and rubbish dumps, has been transformed into a model theatre'.[17] In the 1950s, Bill produced over a dozen plays for the Society. Like Stanislavsky's actor, he prepared, underlining key passages in pencil. One of these reads, 'The external immobility of a person sitting on the stage does not imply passiveness. You may sit without a motion and at the same time be in full action. … Frequently, physical immobility is the direct result of inner intensity, and it is these inner activities that are far more important artistically.'[18] It describes Margaret, slumped and motionless in the armchair. 'All men are capable of love for all other men. But we've artificially restricted our love. By means of conventions of hatred and violence,' Anthony protests in *Eyeless in Gaza*.[19] True to the 'inner intensity' of her creed, Margaret was always moved to give alms.

Allied to fixity of opinion, such intensity can be frustrating. In the time of last things it is a last stand against Death. It positions the palliative care brigade as Death's helpers. At 3.00 am the next morning I drift up from sleep to hear Margaret calling me to help her. I find her sitting on the edge of her bed somewhere between vomiting and suffocating. What, I write sententiously to my wife in Australia, would she have done had I not been near with my elf-keen hearing? Earlier this week, despite the role it might play in allowing her to continue at home, she flatly refused to order and use a personal 'alarm' – designed for these crises. *If I am not here when this next happens*, I write. Each morning Margaret wakes in painful distress and looks greyer, weaker. She sits at the kitchen table with her head resting on her hands. I stand beside her helplessly, distracted by the little birds outside squabbling at the bird table. Her washed and rollered hair has lost its spring and looks dishevelled. Her cheeks begin to shrink; her purple-skinned ankles are swollen. Her breathlessness agitates her, and the agitation makes breathing more difficult. In the afternoon she says, 'I don't want to live any longer.' I lean in close, take her hand and begin the whispering; reassured against asphyxiation, she perks up and begins ordering hymns for her funeral. I find this posthumous planning vaguely exciting, as if we are taking the fight to death. And I understand her refusal. 'A role which is built of truth will grow, whereas one built on stereotype will shrivel,'[20] reads another underlined passage. The alarm, itself a dramatic stereotype, makes stereotypes of us. Simplifying what is to be done, it causes human sympathy to shrivel. Her refusal saves us from this false *quietus*.

It also reorients us to other auguries of change, aligned, like the garden Margaret tended, to the seasons. The safe language of travail was ornithological. It depended on cultivating a narrow cut in space – the view of the espaliered apple tree dividing the backyard from the flowerbeds and lawn and the bird table in its lee. At a scale of years and the period of returns from the underworld, it furnished mother and son with a chronology of visitations in which ultimate reunion might be patterned. Yesterday, a male blackcap alighting on the bird table recalled us to March 1965 when, in a particularly bleak spell of weather, a male and female of the species had frequented the very same spot. That was the time the snow bunting appeared. The confinement of perspective magnified the chance of the unexpected happening: an occasional troop of long-tailed tits leapfrogging along the twigs, the goldcrests that used to seep through the sapling pines, an infrequent woodpecker, the almost once-in-a-lifetime visit of a redpoll. Coming from the out of sight, demonstrating the existence of a world beyond, they were angels. The starlings ganged up to drive them out or squabbled noisily among themselves, but a thin piping dialect was audible as these visitants, driven by pressures we could not imagine, strove to string a thread across the abyss of unfamiliar terrain. Their unpredictable loyalty refracted through our identifications created a bond. Other species are newer and have yet to be added to our calendar – this winter's pied wagtail, say, that cheekily eyes her through the half steamed-up kitchen window. But its message may not be reassuring: stepping outside while Margaret is asleep, I find wagtails everywhere in our cheerless streets, like victims of seismic shock, infiltrating foreign territories because their habitat is lost.

Death robs us of time even before death: knocks on the door when the doctor comes running alternate with hours where there is nothing to do. Cancer's tempo is flexible, but maintains death's steady accompaniment throughout irregularities of pulse. Bunching time into breathless episodes, it also creates intervals of slackness where, post-routine, one might go for a walk. Retracing the solitary paths and field edges I haunted growing up will not upset the pace of change, but it clears the air, provides a kind of amplitude. The sodden earth clinging to my boots is strangely associated with 'sticky surfaces' and 'submerged orifices', as if walking could be a breathing exercise, an old incantation helping relieve the congestion. In the intervals I have been making notes for a book about creativity. It is amongst other things a defence of mud, drawing a parallel between the colloid constitution of the Cotswold and Thames region and a practice of combination – like, I now think, the way my mother selected and grouped the flowers in her garden. It is a challenge to Adrian Stokes, my guide and 'other eye' in Venice, and his identification of the good object with marble, with a limestone characterized by solidity, integrity, depth and anthropomorphic expressiveness. Another composite object is proposed, soft, dispersed, multiple and associative. Odd support for my idea came this morning, when looking at our copy of *The Natural History of Selborne*, I came across this physiographic reverie:

> I think there is somewhat peculiarly sweet and amusing in the shapely figured aspect of chalk-hills in preference to those of stone, which are rugged, broken, abrupt, and shapeless. ... I never contemplate these mountains without thinking I perceive somewhat analogous to growth in their gentle swellings and smooth fungus-like protuberances, their fluted sides, and regular hollows and slopes, that carry at once the air of vegetative dilatation and expansion.

Was there a time, White daydreams, 'when these immense masses of calcareous matter ... were raised and leavened into such shapes by some plastic power; and so made to swell and heave their broad backs into the sky so much above the less animated clay of the wild below'?[21]

This is a pulmonary theory of the earth's formation, as if chalk hills were sponges, fungal formations like the human lungs; the morphology of the clouds above resembles the 'vegetative dilatation and expansion' of the chalk slopes below. Perhaps these systems model the entire respiratory tract, including the organs of vocalization. White Horse Hill possesses in the rippling folds of 'The Manger' a structure analogous to the lateral ridges in the roof of the mouth, the rugae, said to assist speech. Organic depositions of oceanic time, the Downs are also the inorganic inspirations of the earth's uplift. Made of shells, they have an ancestry of echoes. Ultimately formed of the same matter as limestone, they took a different evolutionary path, one that suggests a different theme. Robert Graves thought that Homer's Trojan horse was a peace offering to Athene, and 'originally the same White Goddess ... a horse, because sacred to the moon'.[22] The spring on Helicon named Hippocrene was originally struck not by Pegasus but by 'the moon-shaped hoof of Leucippe ("White mare"), the Mare-headed Mother herself [Demeter]'.[23] Demeter and Athene are simply different aspects of the same matriarchal principle, as is Blodeuwedd the Owl,[24] the Sumerian goddess Anna-Nin,[25] Ariadne[26] and the Libyan Moon Goddess Neith.[27] The Methodists abhorred this kind of mythography: astrological speculation, psycho-cosmic identification, indeed, superstitious animism of any kind, were anathema to my mother. Perhaps in reaction, I seized upon Graves, writing my first tree poems under his influence. Is the White Horse really a mother figure, a primary generative womb, rather than an image of vaulting creativity? But, in Graves's poetic *vade mecum*, everything leads back to the same principle: as Cardea, ruler of the 'Celestial Hinge', or Eurynome, 'Queen of the Circling Universe', or Rhea of Crete, (who 'lived at the axle of the mill, whirling around without motion, as well as on the Galaxy'[28]), the White Goddess is also the nymph of the mill wheel to whose continuous liquid folding and unfolding we made love.

Back from the walk, time speeds up again. Margaret gets worse, poor thing, calling me again in the night. She is bringing up what she describes as red hot fluid from her stomach, and feels that it invades her lungs if she tries to rest. I hear her downstairs for hours afterwards, groaning and mumbling and occasionally coughing or retching. Sometime towards 5 am she 'drops off'. When she gets up, and insists on staggering downstairs, the performance is repeated, only she is *successfully* sick three times. An 'ague' overcomes her and she shakes uncontrollably. She has not eaten significantly since Tuesday, and now finds it difficult to drink more than a few drops. She looks 'at death's door'. I chivy round, call the doctor, nurses and collect prescriptions. Tidied up, we wait; a spell falls upon us, expectant quiet. Comes the knock at the door and, like the prince in the fairy tale, the doctor appears, bringing in his wake a tail wind of alarm, children's cries, telephone bells, the screech of tyres and ambulance sirens. He does not wake the princess from her slumber. Instead, with his needle, he puts her into a deeper sleep. The respiration comes hard, raucous, as if every ruined organ shared its suffering, as if (to someone skilled in sonic divination) you were inside Colonna's ruined colossus ('Everything was there that is found inside the natural body: nerves, bones, veins, muscles and flesh, together with every disease, its cause, cure and remedy') and could discern the resonant structure of it simply by sighing ('I uttered a loud sigh ... and instantly heard the whole machine resonating').[29]

Hedging its bets against an uncertain diagnosis, the wind wailed and moaned last night. Odd to think of all the attics it ploughed through where sleepers lay awake listening for that impossible sound: the last breath. At times it seemed to be shaking out sheets, at others wringing its hands. In such costume dramas the wind is cast and identified. It seemed short-breathed, bluffing its way through the saplings. It seemed irritable, repeating itself and falling again against whatever deaf ear obstructed its meaning. In the morning after a long sleep, Margaret is determined to get up, but is so weak and dehydrated that she can hardly take more than a few paces. After this experiment, I help her back up the stairs – which she counts out one by one.

This strikes me as melodramatic, as this kind of self-torture is unnecessary. During the day the doctor and nurses come running, and slowly seem to grasp the suddenness (and the enigma) of the decline. She grows irritatingly critical and tetchy, perhaps because lying abed is associated in her mind with dying. I feel instead the pathos of the bed itself, the history of lovelessness it represents and the Alexandrian library of dreams it has steadied and poured over the edge of dawn into the tepid sunlight. The mattress knows the outlines of her body better than any living soul. Every night there lies next to her the ghost depression of her husband; they did not touch much in life but have constant intimacy in death.

The telephone rings: a bed is available at a 'high quality end of life care' hospice. Margaret accepts; arrangements are made; an ambulance arrives; the stereotypes take over. All day I play the parts expected of me, but wonder who wrote them. In sympathy with her murmurous wheezing, I lower my voice. Even connected sentences seem insensitive, as if any kind of projection into the future must be painful. But who decides hush is the appropriate signifier of sympathy? A kind of vocal couvade is performed, as if the carer will bow his head and go under the portal wordlessly with the deceased. I close up the little theatre of her life and hold her hand under the blanket as we drive to Oxford. How loving we become when the other acquiesces to our power; we wax lyrical about the meaning of life and speak as angels might of love. I hang around at Sobell House, whose name I misheard as 'sobbing house', and concentrate on overhearing. Someone whistles at the end of a corridor. Do they draw the hounds of Death away? Or, as cigarettes are banned, is it sibilant smoking? Someone blows their nose. The fire doors open and close like heavy breathing. When I called my lover, I heard her crying at the end of the telephone. I never saw or heard my mother cry. It must explain my lifelong tearlessness. (Later, the same sympathetic induction made me cry when my lover left me.) The doctor turns up with the staff nurse and we discuss in hushed and reverent tones Margaret's case notes. We could be archaeologists examining tomb hieroglyphics, mindful that our breath risks corroding them. We are double-timing her: we know it. Her life melts away: the ice thins under my feet. But we go on talking, playing our parts, as if Death will be to blame for interrupting us.

In Caitlin Thomas's *Leftover Life to Kill* (another refugee from the front room's book shelves), I read, 'There is an indefinable hostility about a house that has been left, even for a day, as though it has taken offence, and is spitefully having its own back, by giving a frigid reception, in its most aloof manner.'[30] But I have the opposite impression. After the tension of the last few weeks, the interior has relaxed and stretched out. It is as if the silences can congregate and speak again, as if the ghost breaths of departed conversations can reassemble like an atmosphere. They are not afraid of being dismissed any more. The wind breaches the window frame and the curtains stir, imitating an actor waiting in the wings. The regular *tuck-tuck* of the rayburn's gas fire is synchronized with the parade of shadows projected onto the ceiling. The sound reminds me of the butterfly fluttering against the lampshade when my father was dying; so unseasonally woken from its winter dormancy, it had to be a portent. I remember another occurrence. In 1995 researching the life of William Light, I visited his childhood village in Suffolk. Accompanied by the lover who first set me adrift from my marriage, I stepped into Theberton Church. Because of our situation, we placed a ban on photographs. But I wanted evidence of her before she ran back to her ghost existence. I asked her to prance up and down the nave, accelerando, diminuendo and pirouette, and captured the dancing clatter of her heels. Listening to the recording later, I could hear a butterfly fluttering against the glass.

In 'Light' the sound history of a life is told in breathing. Soundscapes are generated from the auscultation of troubled lungs. Auditory association decides the compositional logic as can be seen from the sound directions. For Act One, these stipulate, 'Sound of sleeping infant breathing; changes to child waking and beginning to cry.' Across this sequence there is heard the 'oceanic' inspiration and expiration, or breathing path that in various modulations is heard throughout the work. Before the crossfade to voices some in-between sounds are heard: clearing the throat, some coughing (as of actors warming up), as well as other ambiguous 'natural sounds', in particular the 'go back' call of the red wattle bird, a fragment of frogs croaking: 'All very stylized, as if approaching us from a great distance'. Following on from the various birdsongs already mentioned, Act Two's instructions explain, 'The voices of Captain Swing, the Doctor, Punch and Judy must resemble those produced by a ventriloquist. They reproduce the effect of pectoriloquism described in the script. The falsetto voice of Punch is already ventriloquial, being produced with the aid of a "swazzle".' As Light's life is recapitulated (and breathing becomes more difficult) the *râle* changes. Laennec describes different liquid sounds associated with pulmonary congestion: a gurgling noise, like that produced by the escape of water from a bottle; a slurping sound, as of liquid sliding from one cavity into another. Citing these descriptions, instructions for Act Four advise that, where the crossing of Acheron, the sea-passage from life to death, is alluded to, such sounds are also thematically appropriate.

In other words, all of the sounds of the spirit drama are inspirations of the breath – with one exception, the Theberton percussions, which, brought over to 'Light' and exclusively used in the final scene, perform a double return of the spirit. Was the auditorium of Light's childhood channelled, a sound reproduced that he might have heard? The real motivation was magical; I wanted her movement form, which, like Shelley writing about Emilia Viviani, I had compared favourably with the antelope ('In the suspended impulse of its lightness … less ethereally light'), to fly over the sea and alight next to me. I fantasized her teletransportation and re-embodiment through the rhythmic hammering into shape of a musical phrase, as if the dancer could materialize out of her dance. Repetition was morphine for the memory, a drug against loss. And it was addictive: two years later, when my father was dying, I went with my new lover to Suffolk. Where the author of *The Mayor of Casterbridge* used to go with his mistress, we went. Did he ever, like Knight at Windy Beak with Elfride, say musingly to Elfride, 'I wonder if any lovers in past years ever sat here with arms locked, as we do now. Probably they have, for the place seems formed for a seat.'[31] We went where (also in 1951) Benjamin Britten realized his own *Dido and Aeneas*. It rained and was dark. We sat in the hotel's bay window, watched the seagulls labour north over a grey sea and said goodbye. But not before I insisted on visiting St Peter's Church. In the chancel, women were preparing floral tributes for a wedding. The dry slap of our footsteps, like the crackle of water expelled from a hosepipe mistaken for stubble fire, was ambiguous. A repetition, an echo, a ghosting or a putting to rest?

When Margaret left the house, my sound history ended. She died five days later but my visits and our bedside devotions lacked resonance. After she died, I tried saying 'I love you' into her hair but heard back the electro-mechanical hum of the building and its systems. Her spirit had fled or, perhaps, had never really been there. I took the bus home but found it difficult to sleep. Surrounded by my schoolboy library I pick at random a guide to minerals and rocks and learn of the existence of *vugs*: 'Normally the crystals growing at right angles to the walls [of lodes] interfere with one another, but where the fissure was not completely filled, cavities or *vugs* occurred and these are often lined with well formed crystals. The majority of the most beautiful mineral specimens have been obtained from such *vugs*.'[32] What is the music of these inner crystalline spheres? No outer space can contain her soul, but perhaps hollows exist where spirits can hear. 'The Native Informant' carries an epigraph by Emanuel Swedenborg: 'In the structure of the internal human ear alone there are amazing and unheard-of particularities which would fill a volume. And yet if anything be said concerning the spiritual world, from which all and everything in the kingdom of nature exist, scarcely anyone believes it.'[33] What spiritual sonorities might resonate in the vug's particularities! The hymn she selected for her memorial service, 'O Love that will not let me go' (448 in *The Methodist Hymnal*, sung to the tune 'St. Margaret')? The coda to Handel's *Ode for St Cecilia's Day* I chose (the Chorus sings Dryden's lines, 'The dead shall live, the living die, / And Music shall untune the sky')?[34] The robin breaking into song as we exited the crematorium?

Or the voice of my sound recordings? I am sorting out Margaret's collection of tape cassettes when there is a rat-a-tat-tat at the door. An ambulance stands there, tall and white, its flank open, and a bed like a stage set awaiting its drama. Death denounced my mother and they have come to take her away. Due to some glitch in the system they have the wrong day; I have the impression they are disappointed that she has eluded them. Should I apologize? The exposure of administrative self-absorption is oddly reflected in the tape collection. I had imagined making a selection of favourite musical items; instead, I mainly find recordings of my own voice – the other half of the audio letters exchanged in the first decade of my residency in Australia. They make discouraging listening. The desire to avoid topics that might suggest emotional departure or intellectual growth is palpable. No admission of my passions seemed permissible. The prospect of her consolation was more onerous than the pretence that 'nothing was wrong' and nothing would ever change (even though it did). Change was treated Platonically, as an illusion to be laughed off. The Underworld did not exist, and now she is gone there. Suppressing new attachments that must signify another abandonment, I perpetuated the *folie à deux* she played with her parents. Her preservation of these recordings is ambiguous, then: when many personal items were destroyed, why do these remain? They confront me posthumously with the Arcadian idylls I used to conceal my irrevocable flight. Vehicles of reassurance, they drove us further apart. There is an analogy with the ambulance that also cheerfully denies its destination.

In the days after she died I talked to her more than when she was alive. After the breakdown of our correspondence, I preached correspondences. 'In these recent winters a robin used to come to her window,' I informed family and friends attending her cremation, 'Even when it came alone, the robin never came alone. It was a messenger from another world, bringing news of a spirit community thriving just round the corner, just out of hearing. The robin's aloneness springs from the fact that it has a home. If it did not have a home elsewhere, it could not recognize the welcome Margaret gave it.' Evidently, an *apologia pro vita sua*. But I recall it here because of a posthumous mere coincidence: having pronounced these valedictory words, I became aware that a robin was singing outside: a strange case of reversed echoic mimicry, as if my words had untapped the vessel of song. I remember how, when we were preparing 'Jadi Jadian' in Georgetown, we learnt to accept the omnipresence and interest of spirits; conversation with them could be assumed and had its protocols – the rat-at-tat-tat of the healer's wand was a wake-up call and a knock at the door. To speak and act in this mode was not to speak figuratively but predictively and rather literally. Here, too, in an English midwinter, in the cheerless production of ashes, a correspondence of invocation and response was occurring that proved to me there existed spirits of place. The lifelong conversation that my mother and I *had* successfully conducted, through poetry, here found its practical apotheosis. A mere coincidence caused the hollow of her absence to quiver and resonate.

A sound history should be not disguised memoir or biography. It is not concerned with private lives. Its object is to recall what happened in another's hearing. There is room for conversational anecdote when it suggests a cultural bearing. The funeral director recalled that in former days Margaret had a reputation in the town as 'a great beauty'. In a different story this might segue to an account of laying out her jewellery, putting on her make-up or going through a wardrobe assembled for invitations that never came. But I am listening to him. It must be strange to speak solely in the past tense. I watched the doctors in Margaret's last weeks. They spoke almost entirely in the future tense. They ran in and out of the house, and from one house to another lightly avoiding pools of suffering. They narrated in no-nonsense detail the progress of rottenness as if the right story could itself be the remedy. On the other side of life's vanishing point, however, the funeral director has no bodies left to examine and only hearsay anecdotes about the departed to relate. He is peerless in lowering his voice in the face of grief. His apostrophes, the stifled cough, the skilfully disguised mobile phone consultation, the averted gaze designed to suggest his own invisibility are the gestures of one who has learnt how to pretend not to be alive. Living in their echo, such professional speech cultures shape our own; they channel society's unconscious codes; their observation, however tangential to a personal history, is relevant.

But the character of hearing is not only phenomenal; it is residual. Besides the continuous adjustment of our speech performances to the habits of others, there is the basso ostinato of breath patternings that have provided the woof and weft of a life. There are ancestral attunements entrained in our own construction of the significant, even if, like the rhapsode, we sew our song from patches, a craft or rhythm determines their arrangement, their scale; and this rhythmic Gestalt, although camouflaged in the auditory dapple of the everyday, may come closest to pronouncing the beat of our personal existence. Margaret had been scrupulous, even obsessive, in issuing directives for the afterlife of her belongings; so it came as a surprise when I excavated her bedroom to find an unfinished patchwork quilt. Whether overlooked or forgotten, it evidently possessed no value. Yet this was hard to believe. Commenced, one had to presume, in expectation of a child, its breaking-off implied a tragedy. At the end of her life, this relic of an aborted future was perhaps too painful to handle. This was speculation. What was certain, though, was that this draft quilt, here held together with running stitches, there securely lock stitched, had a sound history. Women talk while they weave, or spin, or knit, or sew: as the Penelope of her house, I imagine my mother and her sister talking together as they cut out the latest Vogue pattern and started to assemble a garment. I suppose a feedback loop between gesture and *geste*, between stitching and story. It is surmise: when I asked Margaret to write her life, she stopped in the year I was born, explaining, 'You know all the rest.' But I didn't; I wonder whether she folded up the unfinished counterpane because she no longer had anyone to yarn with.

I do recall the sound of a treadle-operated sewing machine, later replaced by an electric one. But I am not going to fetishize this automated cricket in the house. After all, as Enid Blyton informed us children, a parrot could imitate any noise 'from a sewing machine to an express train'.[35] If I could reconstruct that soundscape, it would be with a different purpose, to note the difference between the regular metre of the mechanical needle and the rubato of the stitch as the hem was fed between feed dog and presser plate, sometimes producing a bunching of stitches, sometimes causing them to spread out. I would ascribe this variation of gap not to inexperience but to a certain reflectiveness, caused by consideration of the pattern as a whole, the emerging relationship between all the parts. From a handful of leftover dress materials, Margaret was creating an evolutionary structure: whether ultimately 'closed', like a jigsaw puzzle, or 'open', allowing a kaleidoscope of correct arrangements, each new placement was a possibility created by an earlier placement and, in turn, determined the future range of possibilities. Inter-worked structures like these, however simple, model, remember and formalize the harmonious organization of human relations considered in their mobile, continuously repeated routines. The patchwork quilt was a diagram choreographing the social possibilities she desired at home. Her compulsion to control cut her off from their fullest realization; the unfinish of the abandoned quilt is, ironically, the clue that eluded her in conversations, whose almost mechanical repetitions so often lacked the rubato of human compromise.

The patchwork quilt was Margaret's starling. Its different materials, cut up into squares and arranged in a grid, were offcuts of her appearance in the world, fragments of a cosmos known through its adornment. The materials themselves – pale pinks and blues, one patterned with a floral motif, another with ducklings, a third with bolder squares – came from elsewhere, as language does; but the way she sewed the different patches was a matter of personal taste. In the aesthetic sphere, her tastes were severe: she admired the Impressionists and the Post-Impressionists and wherever we went daydreamed painting the scenes that composed for her. But the quilt escaped the imperative of good taste; its purpose licenced decoration for its own sake, giving scope for a playfulness that never left her. It stood between the frustration of painting classes never taken and the ornament of the house where form always risked lapsing back into banal function. In this unserious craft her creativity was licenced, as it was in the annual composition of the garden into a tapestry of flowers. I like to think that she possessed a power of tonal grouping and understood how two or more notes can coexist, each with its own pattern or melody. Novalis discusses this kind of synaesthesia: 'The eye expresses itself by higher and lower sounds in a similar way to the throat, the vowels by weaker and stronger gradations of light. Might not colours be the light consonants?'[36] Her garden was like this: photographs always showed it 'past its best' because they could not reproduce its patchwork massing of flowers derived from every corner of the globe. A literal florilegium, its stitching together from many pieces could not be reproduced, but the colourful language resonated.

On the sixth day the gale comes. In the old days it would have been a signal to 'batten down the hatches', as we said in our family. The sound and fury are impressive. They trigger atavistic memories of childhood, primary experiences of care or abandonment. The complacency of the house is upset; the grip of the rafters suddenly feels puny and the tiles could stream away like a kite's tail. The window is closed but the lever arm latch squeaks wretchedly on its post. Outside, telephone wires loop sideways; drying laundry suggests survivors of shipwreck, clinging to a spar. A pile of dead leaves is kicked up and flattened against the fence. Enveloping Badbury Hill, the storm is upon us. It is tempting to act as before, to savour the vain sighing of the wind under the eaves, and the first furious rain rattling like grain against the window; to persist in the pathetic fallacy that nature wishes us cricket-wise and provident; to ride it out, making the storm our pretext for, again, changing nothing. Instead I open the windows. Margaret's spirit did not, of course, go with her body in the ambulance to Oxford. It stayed here. But it has been growing weaker, fading. Without her, it is starving. It is restless to leave. I let in the gale, which behaves predictably, scattering papers, peremptorily blowing out the candle that has burned six days by her bed, collecting dead odours and whirling them away. The spirit resists, clinging to familiar surfaces; its moorings drag; no longer breathless, it renounces control. Impetuously holding out its arms, it climbs into the wind's vessel.

FOUR

Sirens

Echoes call us to the future. 'Our strange language was repeated as glibly by the rocks of Australia as if they were those of our native land,' noted the coastal surveyor John Lort Stokes, and he looked forward to seeing the hinterland of the Gulf of Carpentaria transformed into a green and pleasant land studded with church spires.[1] What is remembered resonates because it calls us to follow. It returns because it recognizes the unfinish of what has been begun. It is the sound of the gap between what is and what may be. Responding to an inchoate cry, recognizing its desire of direction, the echo originates the project of self-transcendence. We are able to understand the cosmos, physicist Paul Davies claims, because of the 'resonance between the human mind and the underlying organization of the natural world'.[2] Ultimately, what knowledge of the Real we may possess will be a concerted act of echoic mimicry. 'Our dream pictures of the Happy Place where suffering and evil are unknown are of two kinds, the Edens and the New Jerusalems,' according to W. H. Auden. 'Eden is a past world in which the contradictions of the present world have not yet arisen; New Jerusalem is a future world in which they have at last been resolved',[3] a distinction said to build on (or echo) Louis Mumford's contrast between 'utopias of escape' and 'utopias of reconstruction'.[4] In this dialectic, the Sirens are decidedly on the side of reconstruction. In the Western tradition, they are endlessly discussed and reinvented. They are the spirits of the meadow starred with flowers, oversighting my mother's garden. Transformed into birds as humble as the sparrow and the starling, their chirruping is everywhere. As Muses of the Underworld, they oversaw my plunge to Australia. The most exhaustive enquiries appear unable to discover what the Sirens sang.[5] My guess is that the sailors heard echoes, answering to their desires.

Biographically, the summons least understood may be the most influential. Throughout my growing up, Arturo Barea's name resonated in our household as a type of the free creative spirit. He had died in 1957 and my parents regularly regretted that I had never met him; as the only writer they knew, he was cast as my mentor, politically, vocationally, professionally. The final part of his autobiographical trilogy, *The Clash*, described his role as a radio broadcaster during the Siege of Madrid; it explained his exile to rural England and second career as a broadcaster with the BBC's World Service. His 1955 novel *Broken Roots*, a story of the political exile's return to his home country and his recognition that he no longer belongs there, should have been compulsory reading for one whose future already seemed likely to lie outside England. His 1944 account of Lorca as a poet of the people would have corrected at the beginning any romantic stereotype of the lonely lyricist. But I did not read any of these books until after I had reversed the direction of his flight, taking up residence in a hill town inland from Valencia. I had to read and hear Lorca, Machado and Hernandez for myself. Besides, from the point of view of Arturo's legacy, the overlap of my Spanish residency with the last days of the Franco regime placed me in a morally ambiguous position, a theme later explored in 'Scarlatti'. But we all live in the echo of other places, one person's haven another's hell. Most sentimentalize the societies where they find their own voice. From Middle Lodge on the outskirts of the Buscot estate (where a few years later I conducted my first proto-scientific bird studies), Barea turned a cold eye on the Spanish catastrophe, but his articles about English country life are, they say, romantically forgiving.

Arturo's own romance stemmed from his immunity to class prejudice. Living extraterritorially on Lord Faringdon's estate, he nevertheless commanded the de-territorialized medium of radio. And while none of us would have theorized his appeal in exactly this way – and it was also to be noted that Gavin Henderson was from a class point of view a political renegade – Arturo's outsider status gave him a social mobility his local interlocutors definitely lacked. 'No one knew where wandering men had their homes or their origin; and how was a man to be explained unless you at least knew somebody who knew his father and mother?' George Eliot writes in *Silas Marner*. From a village perspective a new arrival like Silas has an Odyssean aura: 'Even a settler, if he came from distant parts, hardly ever ceased to be viewed with a remnant of distrust … especially if he had any reputation for knowledge, or showed any skill in handicraft.'[6] But distrust cuts both ways: it can encourage a desire to fit in or in the trickster cultivation of a deeper mystery. Exilic writing preserves only the essentials: take Barea's account of La Barraca, the travelling theatre Lorca established in 1931 to bring the great classics of Spanish theatre to the ordinary people of Spain. Why, he asks, did the working people of provincial Spain recognize in Lope de Vega's *Fuenteovejuna* a mirror to their condition? These were the same people who had supported the overthrow of the monarchist dictatorship, and the advent of a democratic Republic, and, Barea emphasizes, 'the same folk who flocked to the local bars to listen to the recently installed radio'.⁷ The experience of exile sharpens this clarity: when the same radio was censored, Barea broadcast messages of hope from London under the pseudonym Juan de Castilla.

Despite not tuning into his frequency when I lived in England, I can see, looking back, that I absorbed from anecdotes about him a model of community. Able through the power of speech to infiltrate the enclosures of our culture, he brought to Faringdon the echo of Lope de Vega's insurrectionary village. Whether through the broadcast voice or the written word, Arturo's practice seeded conversation; it opened the private and the unspoken outwards, towards a plot that could be communally cultivated. It allotted to itself a realm of gathering and sociability in-between the domestic space and the shrivelled public realm (largely in private hands). Now allotments cannot be dictated; they are an expression of political consciousness. As Joseph Arch said on his re-election to Parliament in 1892, 'Neither the Tory Allotments Act (1887) nor the Small Holdings Acts are of much, if any, good to the labourers ... we must have Parish Councils'. Further, these councils 'must have the power to compulsorily acquire as much land for the labourer as he wants at the same rent as land is letting in the district'. Above all, to advance this, 'meetings must be held in the evenings, so that the men could attend, otherwise parson and squire would manipulate them'.⁸ The transition advocated here finds an exact echo in Lorca's insistence that La Barraca took Lope de Vega not to the *pueblo* but to the public, to *un publico popular* perfectly able to appreciate Ravel or Debussy, if given half a chance.⁹

No genealogy is implied here, rather certain cultural resonances, which have, perhaps, a historical significance in mapping the prehistory of radio. I attribute no autobiographical significance to the likelihood (as far as scant records go) that many male forebears on my father's side are described as 'labourers' or 'tenant farmers'. But I note that the distinction is a post-Enclosure one: before Enclosure, a peasant might be *both* labourer and tenant farmer. He might also be independent, a smallholder who subsisted on what he could earn from his open field and from what he could graze on the commons. This pre-Enclosure figure merges with the yeoman, whose economic definition is so hard to give. This class of villagers was then defined not by the property they owned but by the social relation they enjoyed with their neighbours (and the land). And what is the distribution of voice in this ancient society if not radiophonic? 'The village', Walter Rose writes in *Good Neighbours*, 'was a cluster of ancient homesteads, formerly the habitations of yeomen who had farmed the scattered acres of the parish.' This may be romanticized, but Rose's impression of the discursive terrain is suggestive: 'The whole irregular group told the unwritten story of a developing community, a group of pastoral folk, holding their lands under the ancient system of manorial tenure, a village that had continued to grow as long as a narrowing belt of forest remained from which acres could be taken for tillage by those who had no land.'[10] How is this broadcast 'unwritten story' different from the public radio that Brecht recommends 'should step out of the supply business and organize its listener as suppliers', adding that 'any attempt by the radio to give a truly public character to public occasions is a step in the right direction'.[11]

Like a swarm of bees that forms when a scouted location has been selected, the constellation of themes associated with Arturo's memory propelled me to a positive evaluation of the creative rewards of living in another country, and his own origin influenced my eventual drift to Spain. But the character of the echo researched there remained to be discovered. Was there a distinctive style associated with exile? Did relative remoteness from the detail of the home culture act as a riddle sieving out impurities to leave only the soil essential for seeding? Ralph Kirkpatrick described Domenico Scarlatti's music as *imitating* 'the melody of tunes sung by carriers, muleteers, and common people',[12] but this is not strictly correct; the sonatas use purely musical principles of melodic development and key relation to produce their effect. It is more accurate to say that they echo the sounds of the street, like the Sirens, purifying the music of the tribe, making it new. In any case, Kirkpatrick speculated that it was a distinctively exilic style: 'Over the abysses of despair and melancholia he seems to have danced with unprecedented animation and sensibility, at times with the agility of a tightrope walker.'[13] The Italian Scarlatti found Spain a creative 'stimulant'.[14] In that case, it would be good to know what he danced, and how this was reconcilable with walking the tightrope: how was it possible to be a flamenco dancer and a bullfighter at the same time?

The twinning of remembering and forgetting is axiomatic in any art of living in another country. I mentioned how 'Scarlatti' originated in the forgetting of a name: as the echo of that forgetting, written (significantly) in yet another country, 'Scarlatti' remembers another past that points towards the future, its fiction turning out to be a faithful prophecy, for the radiophonic recapitulation of a friendship as explicitly gay, while untruthful to the autobiographical facts, correctly divined the path that would be taken; and when in an open letter released on the web in 2001 the figure on whom Paco is based, declared, 'My friend passed away because of AIDS after fifteen years,' he fulfilled a prediction.[15] But there is no end of these unconscious wish-fulfilments. When I started to think about writing the second half of 'Scarlatti', I returned to the passage in Kirkpatrick's biography that had inspired the choice of Scarlatti's sonata in B Flat Major (K 488) as our key musical reference. 'Everyday as I have been writing this chapter I have been hearing from a nearby *casern* of carabinieri a bugle call that never fails to remind me of the opening of Sonata 488.'[16] I was surprised. The entire acoustic *battaglia* of the scene where Paco and the Englishman meet, in which the radio playing Scarlatti is drowned out by police sirens, was a misremembering (a mishearing?). Although much of Kirkpatrick's research was conducted in Spain, he visited libraries and archives throughout Italy.[17] But there is more to it: for the imitation Kirkpatrick hears in the call coming from the caserma (curiously rendered in French) is adventitious, a riff of his own invention. While it suggests the biographer's susceptibility to extra-musical source material, any connection to Scarlatti's sound universe is rhetorical.

My misremembering puzzled me until by chance I took down my copy of Barea's *Lorca and His People*. It had accompanied me from my parents' bookshelf in England to mine in Spain, and had remained many years unconsulted among my Australian books. Perhaps for a time it had been lost, as, even during the period of writing 'Scarlatti', I find no reference to it. Lorca was not a 'political man', Barea explained; his enduring political significance arose from the context in which he wrote.[18] His ballads of popular life celebrated a vitality, a gaiety that, in reality, was everywhere oppressed, by the Church, by the Courts and, above all in everyday practice by the Guardia Civil, which, Barea explained, was not simply a police force but the dark face of a ubiquitous violence whose sole object was the oppression of the people in the interests of the powerful. It may be hard for a non-Spaniard, Barea wrote in 1944, to understand how far under the Nationalists the Guardia Civil had become the symbol of an oppressive and abhorred regime.[19] Gypsies might be a marginalized and ostracized group but Spanish readers of Lorca's *Ballad of the Guardia Civil* immediately identified with their oppression. In the context of institutionalized terror, a seemingly guileless account of a gypsy fair being broken up comes to stand for a struggle between organized violence and human freedom. Lorca's song may lack explicit political content but inevitably embodies a revolutionary spirit.[20] Hence, in 'Scarlatti', a repressed association surfaced: evoking a relationship forbidden under the law, I remembered the bugle as a siren.

In his essay, Barea made an additional point. As the appointed enforcers of Monarchist/Francoist politico-cultural ideology, the Guardia Civil was also the guardian of kitsch, represented by the deracination of regional cultures and their incorporation into medleys of the arts designed to showcase a national spirit or *Geist*. To speak generally, in Spain the aesthetics of tradition has oscillated between fascism and anti-fascism, between nationalism and republicanism; and the entire political pendulum swings on a beam defined by imperial nostalgia or melancholia. This is illustrated by the vicissitudes of flamenco under the Franco regime – witness the initiative of the cantaro Antonio Mairena who built a chain of flamenco clubs aimed at subverting 'the clichéd image of Flamenco that Franco's regime had built up'.[21] In that period the copla and the associated autonomous logic of the flamenco form were in a continuous contest with the regime's determination to incorporate them into the State's organicist ideology. To combat this, it was necessary to deepen and expose the anti-rational logic of the copla, to show its creative capacity to incorporate new material and to generate new artistic forms.[22]

A kind of cultural syncopation had to be developed to wrong-foot the call to conform. A determination to withstand Franco's historicist myth expressed itself as a different sense of time. Didi-Huberman explains that mastery of *el templar*, that instant of timeless stillness when the matador manoeuvres the bull to make the pass – and which he compares to Lorca's conception of *el duende* – involves a submission to a larger sense of fateful timing. This sense of a time that is doubled and interlocked, involving a subtle and continuous alteration and adjustment, can be called *acoplamiento*. Transposed to the popular four-line *copla*, it suggests the disruption of the four line's clockwork return in the final line, an *enjambement* where one line runs over into the next, as if sidestepping the pronouncements of fate.[23]

In this matter of poetics is contained an entire politics. Jacqueline Ogeil has persuasively argued that the ten years (1719–29) Scarlatti spent as *mestre de capela* to João V in Lisbon were important in his creative development.[24] Little, though, is known about what he composed there. Accepting the imitative thesis, one might surmise that his capacity for expressing 'melancholia and despair' was stimulated by exposure to Portugal's national song of exile, the *fado* – much as his 'animation' and 'agility' are easily explained as a response to the 'twinkling feet, stamping heels [and] the inevitable castanets' of Seville's flamenco. Two assumptions are implicit here: an anonymous popular culture and the intellectual freedom of the artist. They have in common the view that music-making is apolitical, its composition and reception a matter of personal taste. Is this view audible in the way Kirkpatrick performs Scarlatti's sonatas? George Steiner notes that his teacher, Nadia Boulanger, delighted in Kirkpatrick's '*virtuosismo barroco*'.[25] Boulanger held right wing views, supporting the Action Française, whose politics were anti-semitic. Boulanger did not believe in a 'method': 'I do not believe an aesthetic can be taught except through a personal exchange.' Instead, she believed in, and taught, the art of memory: 'You are enriched by all the music you know by heart; it becomes a part of you.' It is the antithesis of the passivity induced by a recording culture[26] and of the consumerist radio culture deplored by Brecht. Yet, Vico said, memory, imagination and invention are a trinity; composition may be a 'free' activity, although remembering one musical tradition may un-train the ear to appreciate others;[27] however, interpreting another's music is an ethical, as well as aesthetic, task.

Baroque virtuosity might be ideological calumny, a profusion of ornament reinforcing a myth of national exceptionalism. Hence, 'Impressionistic accounts of *fado*'s origins often invoke maritime existence and the dynamics of solitude, absence and *saudade* as the emotional conditions responsible for *fado*'s emergence'.[28] As historical claims, these have little value: '*Fado* appeared as a song form in Lisbon no earlier than 1820.' As a poetic myth they *might* usefully inform a style of performance. *Saudade* is 'characterized by its contradictory duplicity: it is a pain of absence and a pleasure of presence through memory. It is being in two times and two places at the same time.'²⁹ Evidently, it is the defining emotion of exile, but its doubling has to be actively inhabited, not passively recruited to escapist nostalgia. Scarlatti has to *learn* how to dance over the abysses. Like Baggio, improvising a new style of 'fantastication' out of the echo of the other, the composer's baroque style may, in its irony, be *anti-baroque*. In 'Pothos: The nostalgia for the Puer Eternus', James Hillman uses the analogy of maritime existence to explain psychic development.³⁰ The *puer eternus* archetype – the 'eternally youthful component of each human psyche' – is creatively drawn to 'high flights and verticality', but knows 'failure, destruction and collapse (*la chute*)'.³¹ To achieve integration, a process compared to the protection against shipwreck sailors achieved when initiated into the mysteries of the *megaloi theoi* of Samothrace, is to understand 'contradictory duplicity' as a doubling in which the solitary Ego yields to the implicated, dyadically conceived and sustained individual, a figure who can transform life's abysses into a terrain of gaps, whose cross-section is, perhaps, not unlike a musical score.³²

To claim '*fado* vocality' depends on a special relationship to saudade may lack a secure historical basis,³³ but what matters is the effect of nationalist ideology on performance. Kimberley DaCosta Holton writes that '*fado*'s diverse array of expressions not only replicated the song/dance vocabularies developed in Brazil, but also initiated a new local offshoot called *fado batido*.' This 'featured an intense balance game where one standing dancer remained grounded with feet firmly planted while the other tried to unbalance him through tricky advances and retreats and kicks or *pernadas* ideally resulting in the standing player's spectacular fall'.³⁴ The memory of a dyadically conceived movement form, pitched between abyss and tightrope, goes missing, and so also does tempo:

> In this production of society through song, the characters and heroes with which one establishes a narrative stand out as simultaneously historical and atemporal, projecting an effect of anachrony which accompanies the discourse on *fado* (essence, sentiment, or national soul, *saudade*, etc.) and the concrete experience one has of *fado*, induced by the erasure of time's contours in the nocturnal space in which fado is performed.³⁵

Mastery of the *fado batido* and reclamation of the 'concrete experience' may be the same thing; reinstating tempo over time, they are two modes of wrong footing. The fact that 'Scarlatti's music is rich in syncopations that sometimes play the role of cross-accents and sometimes frankly represent displacements of pulse'³⁶ suggests a similar exilic self-awareness.

Two baroques are in play here, two constructions of popular culture. Kirkpatrick's ahistorical assimilation of Scarlatti to a national spirit has its counterpart in Sacheverell Sitwell's characterization of the Manueline style of late sixteenth-century Portugal, which concealed its weakness behind 'a mad profusion of ornament, the carving being of an Indian closeness and intricacy of design'. He links this baroque regression to the spirit of the people:

> There is in this complexity that trait of the southern character which a traveller always notices in the entreaties and arguments of the railway porters or the cabmen when he arrives in the south of Europe. Out of this architecture almost every trace of the classical lines laid down by the Renaissance has gone out, so far has it travelled from the centre, and it is difficult to see a pillar because it is so honeycombed with carving or to recognise an arch through the web of flowering lines with which it is overlaid.[37]

A national stereotype is invoked, one designed for consumption by the foreign tourist. Its origin lies not in the sixteenth century but in the new market for travel literature, where the existence of a premodern peasantry is essential to the discovery of the exotic. The means of production are presumed to lie with the owner of the means of reproduction, a phenomenon also found in 'the development of the radio and record industry in the 1930s', which 'contributed to *fado*'s rising position as Portugal's "national song", and the concomitant diffusion of saudade as a core concept of national identity, one which was readily articulated through the expressive medium of vocal music and registered by a growing public of radio listeners and consumers of recorded music'.[38]

An alternative, *anti-baroque* baroque does not internalize exile as a national characteristic but keeps in play a migrant sense of irony. At the beginning of the Second World War, the French-Jewish intellectual Vladimir Jankélévitch fled Paris for the south of France. In Toulouse he lived a second time the exile that informed his 1936 publication *L'Ironie*, correctly associating a suspension of identity with wrong footing: '*J'ai un peu perdu l'usage de la liberté. Je ne sais plus marcher au milieu du trottoir. J'ai perdu l'habitude de mon propre nom.*'[39] Lovely Toulouse symbolized in his wartime correspondence the weariness of the West, the bankruptcy of its representations. By the end it felt like a paper-city, a baroque illusion, bereft of honour or virtue. He crept through it, a stranger to himself.[40] The musical figures of this passage were Scarlatti and Ravel. The characteristic musical tempo of irony, Jankélévitch argued in 1936, was *staccato*.[41] The pointillism of the harpsichord, the effusion of lightly touched notes, captures the double movement inherent in irony. Even as irony makes its mark, it detaches itself. Unlike the about-to-be invented piano, fortified with the sustaining pedal, holding a note until it acquires through its longevity a kind of melancholy authority, the harpsichord, free of this resonant nostalgia, is the genius of letting go.[42] In this sense, like every profoundly baroque artist, Scarlatti was *anti-baroque*, his *'jonglerie acrobatique'*, *'Le jeu de l'écho dans la musique "baroque" (un* piano *repondant a un* forte*) fait pendant au thème du miroir'*.⁴³

In 1955, writing about Ravel – the composer who, as he wrote, 'so often speaks the language of Scarlatti'⁴⁴ – Jankélévitch was dismantling a national cultural hero, and, again, discovering the figure of exile. Ravel is at least as 'dissonant' as Beckett:

> Ravel is first of all a past-master in the art of becoming someone other than himself, and he uses the real world in order to conceal the truth within himself; the knowledge of the exterior, the contemplation of the universe through intelligence, are therefore with him forms of modesty: in fact he speaks of things in order to avoid speaking of himself.⁴⁵

This self-masking is essential to the reclamation of popular culture as a site of cultural production. Personifying Ravel's 'passion for speed', Jankélévitch imagines Scarbo: 'The electric dwarf with his golden bell and his wicked laugh, as nimble as an acrobat and as impassive as a demi-god. He gallops on the wings of the wind among fantastic glissandos and flashes of steel, he whirls round on the spot in reiterated notes, he flies like a firebird from one octave to another.'⁴⁶ Jankélévitch discerns in Ravel's 'acrobatics' a way of negotiating social relations. The agility of the fingers is by extension a tactic for avoiding arrest: manual dexterity maps to a facility in threading public space, which is analogous to the subversive spatial tactics promoted by the Situationists.

In Villar del Arzobispo in 1975, the view of the *huerta* was wide but my perspective on the politics and poetics of exile narrow. Geographical and linguistic distance fostered at first a *distacco* from the implications of my position. I cultivated an aesthetic exile, reading purely for clues to the development of a style. With a dictionary, I stumbled through Antonio and Manuel Machado's play *La Lola se va a los puertos*, a non-exoticist attempt to transpose Andalucia's popular culture of flamenco, *cante jondo* and *copla* to the Madrid stage. But I was ignorant of ideological struggles simmering beneath the surface of the fraternal collaboration, and indifferent to the politics of representation – whose importance was illustrated in 1947, when a film version of the play was made, '*dentro de un estetica folclorica propia de los gustos del Regimen Franquista*'.⁴⁷ When I made a return visit to Villar in 2002, the women were still hard at work at Los Manantiales, washing, rinsing, and hanging out clothes to dry. Don Vicente Llatas⁴⁸ – the village historian who had one day shown us a ruined church, first bombarded by the Nationalists, then torched by the Republicans – had died and been replaced by a bronze bust. My tiny house survived. Although it took days to bring its details back into focus, at last I stood in front of my former self.

I felt the grain of the desk, the scope of the stone, the convergence of the rafters pointing to the *huerta* and my own desire. A great fifteenth-century poet who lived at the edge of my view once wrote, 'I am this man who is called Ausiàs March,' defining identity in exile as a journey.[49] Looking back, 'my year of hope' fitted into a poetic series; even the great Domenico had to climb into his name by going abroad.

A poetic bifurcation began there. A project to remake a popular language dried up; and a new orientation towards going with the flow emerged. After Oxford, I retreated to a rectory under Cross Fell, read *The White Goddess* and created out of Thomas Wright's *Dialect Dictionary*, a pan-dialectal lexicon of forgotten beliefs, objects and practices that would purify the language of the tribe, not by the expulsion of the vernacular but by its reanimation. In Villar, imitating Menéndez Pidal's selection of ballads, and their creative reinvention in Lorca and Machado, I wrote reams of *coplas* that experimented with the return of the third line; subsequently, in Tuscany, I made the relationship between Dante and Cavalcanti my theme, at that time favouring the latter (as appears in 'On The Still Air', where Ezra Pound's judgement is quoted). These were Edenic interludes. In Faringdon I had experienced the disintegration of community at first hand. In this sense, 'escaping' to these pre-industrial landscapes, I staged a return to the imagined community in whose mythopoetic transformation could be revived Machado's authenticity of sensation when he writes *'mi verso brota de manantial sereno'*.[50] Mine was a false serenity; its soledades courted solipsism and invited the fate that overtakes *El Sireno* in Gloria Fuertes's poem: 'Being a word that doesn't technically exist in masculine form in Spanish and a definitely unique creature, the serene siren swims off to Venice where he can be "what he is" – legless and sexless.'[51]

The predominantly four-line stanzas of my 'One Hundred and One Songs', started in northern England and conscientiously finished in Spain, were a hybrid of Percy's *Reliques* and Baudelaire's *correspondences*, presented in the cuirass of metres of Edward Thomas or Ivor Gurney. Purporting to displace the phanopoeic orthodoxy of Imagists ancient and modern, they merely proved that formal competence and emotional incompetence may coexist. Had the poems of John Shaw Neilson been available then, my grasp of the essential relationship between matter and metre would have been far more secure. Unhampered by Graves's portrait of the bard as a Symbolist on steroids, Neilson did not ask what his symbols meant. He progressively described a definite sensation. Lines like 'your foot touches lightly / The red ground, / Your lips have strange sweetness / Dying in sound' integrate the traditional ballad form and a synesthetic sensibility reminiscent of Mallarmé.[52] But they presuppose the development of an idea and its resolution – when the poetics of the echo may demand something else, a cyclic structure, in which the return does not have the character of a mathematical proof but serves instead to destabilize the original proposition or image. In the three-line copla, a central image is placed before another image and then repeated; as in a Scarlatti sonata, it is not a recapitulation but an ironic restatement. The focus is on Becoming not Being; fixed and unfixed, the image is not visualized in a mirror but as if reflected in running water, where its temporal identity will be a function of the rhythm.

An alternative, diary-like impression of passing time, a weaving together of overlooked aspects of the everyday, offered relief from the pedantic craft of proving analogies. The conjunction of different phenomena was explained by the mere coincidence in time and space; or, where the convergence between two ideas left much to the imagination, the gap itself held emotional significance. Instead of cultivating a singular voice, these verses assembled random sensations, as if two or three people were overheard bearing witness to an event. The poem proceeded like Paul Klee's line 'going for a walk' and the line zigzagged, like the echo of a refrain. Going with the flow in this way did not preclude the kind of event that occurs when a diver plunges into the waves, but those rewards of resonance between human feeling and worldly pattern were always absorbed back into the larger movement form of the poem. Later, in Italy, I encountered a popular anticipation of this in a variety of folk song called the *stornello*. 'The name is variously derived. Some take it as merely short for *ritornillo*; others derive it from a *storno*, to sing against each other, because the peasants sing them at their work, and as one ends a song, another caps it with a fresh one, and so on.' These *stornelli* invert the order of the *copla*: they

> consist of three lines. The first usually contains the name of a flower, which sets the rhyme, and is five syllables long. Then the love theme is told in two lines of eleven syllables each, agreeing by rhyme, assonance, or repetition with the first. The first line may be looked upon as a burden set at the beginning instead of, as is more familiar to us, at the end.[53]

I was intrigued that in Italian the starling is also known as *stornello*.

Embarking on this revised path, my teacher was Ausiàs March. I spent more than a year intermittently working on a translation of his poems, but his importance was ethical not metrical. He emerged at the right time to teach the essential irony of exile, the necessity of self-examination. Some of my writings suggest an inductive temper, a wealth of historical data compiled to support a cultural generalization, but my temperamental affinity is with the deductive method. The entire world of *Baroque Memories* was based on a *minimum visibile*, the physical sensation of the streets of Lecce gained during a forty-eight-hour visit. The migrant has to extrapolate from circumstantial details an entire provisional cosmos; but the perilously narrow foundations on which the new order is built are thrilling – anyone who successfully navigates their abysses will have to be tightrope walker. So it is with bringing Ausiàs March over into English: in a language I could hardly sound out, far less understand, his poems represented a *mimimum audibile*. It was as if meeting a stranger in the road I had to take my bearings from whatever sound-alike expressions we had in common. Entering the obscure paths of his Valencian Catalan (my guides the ever-reliable Provençal scholar Amadeu Pagès and the largely unreliable Portuguese-born Castilian Jorge de Montemayor), I was forced to mark my own footsteps, every step of the way into the thickets of his meaning signposting my reason for being there.

As I gained a better impression of the lie of the land, I found March already living there in internal exile ('I seem to live in another world, and my own actions seem strange to me').[54] He was

> like the child who, for his age, can walk well enough along his own street, but if, by chance, he finds himself among rocks is afraid (he does not know where to set his feet) to go ahead, because he sees no track; he will not and cannot use a level path, he cannot go back since someone else brought him (there), for he would not have come such a way by himself.[55]

I compared this with my solitary walks into the hills behind Villar, whose purpose might be to find one track among many. The identity of the one who had brought us there was ambiguous. Between Eros and Agape, sexual desire and spiritual love, March states in the first Song of Death, 'another desire, which makes its way between the two, is found to have no sure road'. What is this third way? March's comment, 'It thinks it will find haven on a deserted beach and poison seems to be its remedy,' suggests a desire drawn to the Siren.[56] The Siren is the lure of the imaginal. The poetic path to truth is through the image. The image attracts desire but desire also projects onto the image *in its own image*. The Dream of the Siren in *Purgatorio* 19 illustrates this: 'Dante's image of Beatrice draws him on to the beatific vision, yet the Dream points out the danger of trusting in images and man's tendency to construct Sirens, false images, out of his desires. How, then, can Dante know that Beatrice is a true image and not a Siren?'[57]

The ambiguity of desire's image emerges in 'Scarlatti' when Paco explains why *he* wanted to translate March into English: '"Weeping ... tears ... for ... the ... joy ... he ... can't ... restrain ... ." Remember? Translating. "Open ... armed ... he ... comes ... down ... the ... road ... ." Waiting for the bus. "Friend," comma, "abandon your strange abode." Well, I wanted to learn English, didn't I?'[58] The friend welcoming the poet with open arms is *la mort falaguera*, flattering Death, a male version of *la intrusa* who in Siren-guise beckons men to self-immolation. When I came to draft 'Siren Sonata', 'Scarlatti's second part, this passage, scarcely touched upon in the 1988 script, seemed to have resonated with the future. In the 2001 open letter the view had been expressed that 'in the AIDS age, Death, when it is not incumbent upon you' – by which he appeared to mean when not a function of normal mortality – 'is the greatest teacher of all and will make you aware of a wiser and healthier code to live by'. In a comment worthy of March, he added, 'Man must learn to define his own personal and moral conditions and then live up to them.'[59]

The debate about the exhumation of Franco's bones suggested a bitter pastiche of Lorca's 'Llanto por la muerte de Ignacio Sánchez Mejías'. Franco might have been a man of his times, but he lacked the bullfighter's tempo: '*Durante años y años la misma mecanica funcional, los mismos dias, a las mismas horas, con las mismas costumbres.*' '*Se tiene que mantener identico ritmo de trabajo.*' '*Ha llenado mucho tiempo de Espana; Franco ha muerto.*' '*A las cinco de la madrugada.*' '*Momento fatal.*' '*A las cinco de la madrugada.*' '*Consternación.*' It suggested linking March's *mort falaguera* to Falangist murder – 'Why the siren? Is there an accident?' 'Who is that, open-armed, running into the road, his arms hostaged to the sky?' 'Hateful!' 'The house! Get out while you can!' 'I come because you have not called me. The volunteers I ignore.' 'Friend, leave his strange house.' 'Why are his screams melodious?' 'La Mort falaguera! Odious!' 'Death, who has tied your hands?' 'Accident es Amor.' 'Love, who points the gun. No substance.'[60]

I say 'draft "Siren Sonata"' because it soon became apparent that any attempt to restate and complete the old material would be, as T. S. Eliot said, 'to get the better of words / For the thing one no longer has to say, or the way in which / One is no longer disposed to say it.'[61] Consider the analogy of tracks: in the *ritornello* to Villar in 2002, I walked paths closed to me a quarter of a century earlier. In 1975 the Guardia Civil had me directly under their protection; any trespass risked forfeiting this. To walk by stone walls towards the *huerta*, or climb past the kaolin mines to where the goatherds wandered, was to push into blind country and test tribal fidelities. When I came back, the old litigious spirits of place had fled: the Common Agricultural Policy had seen to that. I had no camera when I had been here previously in 1975; now, like any other tourist, I treated the fields as a photo opportunity, documenting the intricate handiwork of stone on stone, and the sentinel antique olive trees, surrounded by recently lopped-off branches, bleeding a silver light. I wondered why in the former time I had not noticed the *torrente* attracting to itself troops of plumy reeds, like spear-bearing soldiers defiling through the countryside. And what of the pomegranate trees, their exploded artillery hanging on every branch, and along the cavernous cliffs shimmering poplars? Had the lost poems of that year described these?

The old tracker had been a song path maker; becoming a regular between Calle de Cantereros and Plaza de las Flores, he carved out a territory for himself; it was recognized and he was accepted, the outsider who listened. Linking voices, doorways and chance encounters to the line of his breathing, his solitary walk became a social phenomenon, a measure of the *pueblo*. The philosopher Immanuel Kant had a regular walk:

> Near his house ran a narrow street, just a half-mile long. ... At exactly half-past three he came out of his lodging, wearing his cocked hat and long, snuff-coloured coat, and walked. The neighbours used to set their clocks by him. He walked and breathed with closed mouth, and no one dare accost him or walk with him. The hour was sacred and must not be broken in upon: it was his holy time – his time of breathing.[62]

But the song path maker times his walk to the tempo of the street, to the altering lines of shadow, to the local hauntings of the nightingale's song, or to the visitations of the wandering knife-grinder, who announces his trade on a papageno flute. His inspiration is the coexistence of consciousness and sound, and the walking art is to sew these together into an incipient rhythm. It is not an athletic skill; its Muse was lame Carmen across the alley, dragging her clubfoot behind her to the balcony, and singing as she shook out the carpets. A philosophy of finitude incorporates limping. 'To limp is to gain traction where the ground offers no rest or relief. It is the figure of desire, dramatizing the suffering involved in any approach to the other. To limp is to mark the hollow in the instant between two strides.'[63]

One day this harmony was torn apart, the coexistence of multiple lines of existence censored. Anonymous rifles shot down the fledgling swallows perched on the wire outside my window, and they dropped into the dark, their corpses smaller than mice. It was as if one of Canetti's 'voices' had been savagely removed, weakening the 'overall effect'. The mood this produced angered me; I felt once again out of key with the time, as if Death refused to licence tightrope walking over the abyss, and dragged me back to '"the sublime" / In the old sense'. It was as if time, which I cultivated as tempo, had taken the side of Death. What village *llanto* was expected, what minor expression of feeling? Sentiment was expected but it only masked Death's destructive character. True resistance is an art of passing the time, incorporating stumbling into its very stride. Later, in 'Scarlatti', I compared the 'scarlet-throated swallows' to the 'flying notes' of the harpsichord, their death to the advent of the pianoforte, whose sustaining pedal suffocated the old equilibrium of emotions with false sentiment – 'At the depression of a pedal to prolong the pain ... This decadence destroys feeling.' Playing Scarlatti on the pianoforte was 'wrong from the start'. The reign of Ferdinand VI, like the period of the Franco regime, was one giant syncope, a period of unconsciousness when time was suspended. Within and against that time tyranny, Scarlatti instituted counter-syncopations that wrong-footed the clockwork time of oppression – remember Franco's *identico ritmo de trabajo* and his equally ominous boast, 'While God gives me life, I shall be with you.'[64] The sustaining pedal of the piano, with its 'blurred fillings-out of harmony ... thick washes of colour', drugged and cancelled out a phrasing based on syncopation and substituted for it 'a confused uniformity'.[65]

If 'Siren Sonata' was a restatement of 'Scarlatti' that proceeded deductively, developing the implications of what is hinted at in the original script, then the role played by radio under the Franco regime was worth following up. 'Scarlatti' may claim to be a 'romance for radio' located in Valencia but, beyond the general idea of a radio of resistance, it is short on historical context.[66] During the Civil War the public radio to which Barea's La Barraca audience had listened changed its tune, becoming a warning siren: radio was the medium used to alert the population about dangerous bombings or to explain government actions. It was also the medium of spying: the assassination of Lorca was coordinated by radio. Naturally, it produced a radio of resistance: given the openly gay nature of the relationship in 'Scarlatti', Ian Gibson's observation that the free radio of Martinez Nadal and Arturo Barea involved a resistance to 'the claim against homosexuals' ties together the personal and the public history of 'Scarlatti'. Under the dictatorship, however, censorship was general: between September 1960 and October 1977 (the date of Franco's death and the period when 'Scarlatti' is set) some 4,343 songs were banned as 'no radiables'.[67] The cumulative impact of banning other voices emerged when Franco died. In the guise of Jaime de Andrade, through the medium of films such as *Raza*, Franco had consolidated a mythical genealogy for himself as the father of the national family. His successive returns to the film aimed to erase fissures emerging between its myth and historical events.[68] While implicitly opposed to all forms of Sebastianismo, this attempt to command the historical real and eliminate alternative accounts inclined Spaniards to attribute a messianic power to Franco. As a result, when he died, it is said that the national radio suffered an attack of nerves. It did not know what to say: 'The *generalísimo* had operated like a ventriloquist, manipulating a battalion of dummies who, now suddenly left without his direction, were unable to speak.'[69]

A sound history of this kind confirms the intuitive accuracy of the original 'Scarlatti': when the radio announces that Franco has died, Paco asks, 'What about us?' and the Englishman responds, 'What are you going to do now?' Now the pressure on their relationship has been released, they may fall apart. When I discussed this project with a prospective producer in the Australian Broadcasting Corporation's radio arts unit, he suggested that the first step must be to follow up what happened to the characters afterwards. The ending of 'Scarlatti' coincides with Franco's death; any sequel presumably stages a return to the material through its resonances into the future. In the new romance for radio, the characters would still be constructed echoically in dialogue with radio. The first step might be to help radio speak again. But was this possible? I thought of the personae cultivated by Fernando Pessoa, the different heteronymies or masks: As Zenith puts it, 'Pessoa described his artistic enterprise as "a drama divided into people instead of acts." He created, in other words, a series of characters but no play for them to act in.'[70] In this sense Pessoa inverted the relationship between Franco and the state-controlled media, placing himself in the position of ventriloquist: 'Each of my dreams, as soon as I start dreaming it, is immediately incarnated in another person, who is then the one dreaming it, and not I. To create, I've destroyed myself. I've so externalised myself on the inside that I don't exist there except externally. I'm the empty stage where various actors act out various plays.'[71]

In theory, this is actors' radio, but how much freedom is really ceded? Geoffrey Pingree comments,

> At least on the official surface, Spain's public discourse under Franco was a conversation with just one genuine voice, a drama whose many roles were performed by just one actor; in the public sphere created and monitored by the regime, all apparently pluralistic discourse was staged, all ostensible recognition of multiple perspectives fabricated. All opposing voices were effectively silenced, often by force.[72]

Ironically, speculating about the afterlives of Paco and the Englishman would theatricalize their history in exactly the same way. It would fictionalize history. Perhaps a disagreement about the destiny of the culture lay at the heart of the impasse between 'E' and 'Paco': while the future 'E', writing from the Underworld of Australia, might have become a postcolonial advocate of cross-cultural métissage discursively and erotically, Paco might have moved to the right, living out the dictates of Courtly Love in the contemporary gay scene, the temptation of a transcendental (and perhaps self-immolating) love tied to continuing state-sponsored anti-gay prejudice. Of course, the breaking of the thirty-five-year silence in 2011, and the subsequent discovery online of Toni's Testament, modified these mythic projections. But it did not solve the problem of radio: the challenge of a 'romance' that did not make radio the voice piece of private sentiment but allowed it to fulfil its public destiny.

The inhibition to realizing 'Siren Sonata' was neither thematic nor even ideological – but, instead, technological. A couple of years later, in 2013, I was in Portugal when my producer friend emailed me the news that the Australian Broadcasting Corporation was closing down its radio drama unit (after eighty years, the newspaper protested). Causes included cultural obsolescence, commercial competition, fragmentation of audience across new social media platforms. Similar closures had occurred in Germany (the other outlet for our style of Neue Hörspiel). For us, the evaporation of an imagined listenership represented a lost poetic (and therefore political) opportunity. We had no interest in a protected workshop for BBC-style, psychologically based narrative drama. We regretted the passing of radio as a medium of change. Inside the future of radiophonic art, we could resist the argument that radio had failed to fulfil its cultural and political potential. But, in what had become a post-radiophonic age, the argument seemed irrefutable, nor was there a ready alternative. According to Lander, the natural successor to radio – the internet – which, perhaps, radio always imagined, risks the same kind of suppression: 'As the radio apparatus increases its range continually, through the development of new technologies such as the cellular telephone, satellite transmissions and so on, the "primitive extension of our central nervous system, the vernacular tongue", remains suppressed.'[73]

A sequel to 'Scarlatti' could not be a general manifesto for a post-radio fulfilment of radio's unfulfilled potential. 'Siren Sonata' might take on board the techno-cultural revolution associated with the new information technologies, but would need to define the vernacular in its own way. I am not sure what Lander envisages when he writes (back in 1994) that 'the development of an all [too] often inaudible host of vernaculars into an expanding transmission of variable and multi-dimensional cultural expressions will come to radio via a fluid and transgressive noise, filtered through the minds and the bodies of those unafraid to speak in the face of mediated taboos'.[74] However, if 'Siren Sonata' is a 'romance for radio', the 'fluid and transgressive noise' is likely to be derived from the operation of radio as such, perhaps from the creative manipulation of atmospherics. In 'Scarlatti Exploded', a 2014 workshop in Germany, participants interpreted 'Scarlatti' as 'a radio romance' in a post-radio environment.[75] The fifteen participants, drawn from a range of performance and design backgrounds, were presented with a scrapbook containing parts of the 'Scarlatti' script, passages drafted for 'Siren Sonata', and a visual inventory of radio sets I had photographed in Berlin, Mannheim and Barcelona. They were equipped with largely obsolete (but working) transistor radios sourced from local flea markets, and with a simple recording studio, supplemented by their own mobile phones. Those taking part spoke a wide variety of vernaculars but generally identified as belonging to a post-radio generation, linking radio to such childhood milieux as the annual holiday car journey or grandma's kitchen. Contemporary radio in public space signified a place to avoid. Perhaps for these reasons, they found it unproblematic to treat the radios as instruments, and themselves as 'suppliers', not perhaps exactly in Brecht's sense, but nevertheless in pursuit of giving radio 'a truly public character'.[76]

The 'through line' for the workshop was an examination of 'feedback', the ways in which radio 'produces' the audiences it serves. In a post-radio epoch this phenomenon becomes visible: no longer secreted in domestic spaces, radio sets have to be collected, assembled and consciously turned on. When this happens, the materiality of the radio medium intersects with the dramaturgy of everyday life. Accordingly, on the first day, the workshop culminated in the improvisation of 'Six poses for radio'. The 'poses' produced three groups and three styles of interpretation. On the second day, these were consolidated through two exercises involving the 'exploded' 'Scarlatti' script. The first group eventually convened a speaking group around centrally placed radios, and using microphones took turns to talk over through and across one another. The second group produced an offstage *Hörspiel*-style existential monologue into dialogue, ending in the physical appearance of the three performers. The third group improvised verbal fragments around a new generation of sound effects, one of them 'DJ-ed' in real time. These exercises focused questions about the construction of meaning, the identification of audience and the politics of noise. It placed radio composition firmly in the domain of dramaturgy. These implications were picked up on the final day where, for the first time, participants heard a recording of the sonata in B Flat Major (K488), the item of unheard vernacular, around which the whole event whirled. They recorded the broadcast recording on their mobile phones, took these secondary recordings into the street and there played their own recordings and recorded each other's. Layering 'prepared' radio interference, three generations of recorded music and feedback were created when these recordings were played back through microphones and, again, amplified and broadcast produced a twelve-minute hybrid noise performance installation, evoking the auditory unconscious of a public space.

I began to realize that 'Siren Sonata' was not a project; it was the collective name for an open-ended set of exercises. 'Scarlatti Exploded' had re-territorialized 'Scarlatti', but beyond the crowd sound event there was nothing enduring to show for it. It did not represent anything. During the 2013 visit to sites associated with Scarlatti, I made street recordings and wrote poems. I recorded the poems and arranged them with the found sound; as the poems were real time improvisations echoically engaged with features of the immediate sound environment (*fado*, flamenco), post-production reconstructed a dialogue between human voice and sound environment. In another exercise, I took my cue from the fictional event that triggers the action in 'Scarlatti' and sought out No. 35, Calle de Leganitos, where Scarlatti resided in Madrid. I was sound sleuthing a double legacy. Scarlatti's biographer had been keen to track down memorials to the composer. I was following up two sets of tracks. Kirkpatrick was misinformed about Scarlatti's burial place – it appears to have been San Martín (where Calle de la Luna and Calle del Desengaño meet). As for Scarlatti's house 'with the handsome baroque doorway … now occupied by auction rooms for second hand furniture'[77] – it has been knocked down. Outside the three-storey commercial building that replaces it, I set down my sound equipment, left it playing the sonata in B Flat Major and waited to see what would happen.

Publishing his *Essercizi per gravi cembalo*, a collection of thirty harpsichord sonatas, Scarlatti advised his public not to expect any profound intention in these compositions, but rather 'an ingenious jesting with art by means of which you may attain freedom in harpsichord playing'.⁷⁸ My own exercises should be in the same spirit. Even the activity of Paco and the Englishman consisted largely of an 'ingenious jesting' with a radio set. Vico defined ingenuity as 'the faculty that connects disparate and diverse things'. It is the mastery of metaphor: 'An acute wit penetrates more quickly and unites diverse things, just as two lines are conjoined at the point of an angle below 90 degrees.'⁷⁹ On this basis, the best imitation of Scarlatti's spirit might be to cultivate connections between the remotest things: the mere coincidence that the current business at No. 35 Calle de Leganitos is a retailer of domestic alarms, for instance! Going back, I wonder what passed through the auction rooms on this site. Reading the inventory of belongings made after Domenico's death (also featured prominently in 'Scarlatti'), with its cabinets, coaches, mirrors, jewels and (interestingly juxtaposed) clavichords and chiming clocks, Luisa Morales concludes, 'Scarlatti is a clear exponent of the "consumer revolution" of the eighteenth century, where post mortem inventories – other than those of the aristocracy – show a marked improvement in the variety and quality of household furnishings and "luxury" items'.⁸⁰ Somehow, given the hard-edged glitter of his music, it seems appropriate that 'the largest percentage of Scarlatti's possessions is made up of jewellery, the most expensive item being the *venera* (arms of Santiago badge) richly adorned with diamonds, sapphires and rubies'.⁸¹ The *venera* is a scallop shell, named for Venus Anadyomene, who came ashore in one.

In the same jesting spirit I considered a revival or restatement of Lope de Vega's *Fuenteovejuna*. In Barea's view, the portrait of a *pueblo* that learnt to speak with one, anonymous voice (*Todos por uno*) anticipated the public to whom Lorca addressed the productions of La Barraca. The through line in this genealogy was radio: if there could be a pre-radio sensibility expressed in a sixteenth-century play, why not a post-radio version? After all, even the name of the play is suggestive: from the *fuente* of bubbling water springs a sound that is pre-symbolic, a fleshly noise that moistens the lips of the Muses. Graves evokes Mnemosyne as the Muse of Memory, but the Muses (and Memory) are etymologically and mythologically associated with the moist, the flowing. And what is heard might easily be like the atmospherics of a radio left on. Besides, as we have already noted, those who return to the fount of wisdom, the source whence they have sung, do not find a 'precedent' for the present: 'They leave the temporal frame and listen to the bubbling, which has gone on without interruption since the beginning.'⁸² So much for a poetics of going with the flow, but what of the politics implied by the association with sheep? When, in the third act of the play, Laurencia angrily reprimands her fellow villagers for failing to protect her honour, she says, 'You're sheep, the name / of *Fuenteovejuna* tells it plainly.'⁸³

But this is a sheepish view of sheep. In a recent community-driven flash mob project, called *Transhumance*, Porto-based computer scientist and artist Pedro Lopes and his collaborators question it. Noting how 'society, often, makes us follow trends or mobs ... superimposing a global attitude on the sole individual', they made an audio piece from 'sounds of sheep in various contexts (inside the corral, with the flock, individually)', then 'explored the possibilities offered by digital-processing of sheep sounds with noisy effects (reducing quality, forcing distortion and feedbacks), metaphorically liberating the human-communities of these social pressures'.[84] But this metaphor is obtuse, in Vico's sense, not based on any real discovery of shared characteristics, and, insofar as it continues to treat bleating as noise, anthropocentric. 'Sheep bleat when they are isolated and some more domesticated breeds bleat at feeding times, but the most common use of vocalisation is between ewes and their lambs.'[85] Ewes and lambs answer each other's bleats; in this way they keep track of each other. Bleating signifies that a territory is precarious, and calls the other to remain close by.

It has been pointed out that the sheep derivation of *Fuenteovejuna*, or Sheep Well, is mistaken. The town is known for its production of honey, 'and the true etymology of its name is from *abeja*, so that it originally meant Bee Spring'. But if, in a posthuman, biocentric world, we take the call of the animals seriously, this merely restates the same challenge. We need to listen far more carefully. Bee shamans exist. Indirectly, bees supply the food of the Sirens, as Homer's 'euphony of *melainei* (black) and *meligerun* (honey-sweet) in the Sirens' song' recognizes. It 'recalls the honey-sweet wax (*meliedea*) that protects the crew of the ship from its effects – that *melos* (melody) implied in all three words but never actually described except for the phrase *liguren aoiden*, clear-sounding or shrill song'.[86] But, to speak biomimetically, a bee well is not like a pueblo with the *pozo* at the centre: it resembles a floral archipelago whose colours occupy different wavelengths. It communicates like a radio that can be tuned to many frequencies. Here, another corrected translation is relevant. '*Fuente* is not a dug well in the sense of *pozo*. A synonym of *manantial*, it means a natural wellspring or natural fountain.'[87] The noun '*manantial*' means 'flowing water'. The related verb '*manar*' means 'to run or flow from'. The English word 'emanate' comes from the same Latin root.[88] In his *Epistles*, Horace describes the poet as one who distils poetic honey, '*fidis enim manare poëtica mella / Te solum*'.[89] But in the hive, honey does not flow from Solus, the poet: it is crowd sourced and, if it could be heard, would sound like the echoic Sirenland of radio.

Would these reflections pave the way for Barea's utopian project, a free radio of resistance that, in the post-radio environment, reclaims the streets? Attacking the 'provincial and chauvinistic anthropocentrism' of most recent cultural theory, where Nature is either cast aside as in-significant or deemed a cultural projection',[90] Christoph Cox advocates a new 'sonic materialism' beyond representation and signification. He reviews favourably Cage's conception of sound as 'anonymous flux', 'a ceaseless production of heterogeneous sonic matter, the components of which move at different speeds and with different intensities, and involve complex relationships of simultaneity, interference, conflict, concord, and parallelism. This flux precedes and exceeds individual listeners and, indeed, composers, whom Cage came to conceive less as *creators* than as *curators* of this sonic flux.'[91] He recognizes the value of Pierre Schaeffer's sonorous object enjoying 'a peculiar existence distinct from both its source and the listening subject', but he thinks it falls short:

> For sounds are peculiarly temporal and durational, tied to the qualities they exhibit over time. This temporal quality is not incidental but definitive, distinguishing, for example, the call of the cardinal from that of the robin, or the spoken words 'proton' and 'protein'. If sounds are particulars or individuals, then, they are so not as static *objects* but as temporal *events*.[92]

The upshot of this useful summary of recent theorizing around sound art is a 'realist conception of sound' as 'thoroughly immanent, differential, and ever in flux'.[93]

The 'aesthetic production and reception via a materialist model of force, flow, and capture'[94] that Cox recommends is the *manantial* as sound art, but it begs the question of territorialization. Following Friedrich Kittler's idea that 'the phonograph ... registers acoustic events as such', and that 'audio recording registers the messy, asignifying noise of the world that, for Kittler ... corresponds to "the real" – the perceptible plenitude of matter that obstinately resists the symbolic and imaginary orders', Cox's 'sounds' are, collectively, noise. But noise, at least as generated by radio, is territorial; atmospherics, as 'Scarlatti Exploded' showed, is tactical. In the context of Barea's public flocking to the local bars to listen to the recently installed radio, noise is the sound of public space. As Jacques Attali writes, 'Equivalent to the articulation of a space [noise] indicates the limits of a territory and the way to make oneself heard within it, how to survive by drawing one's sustenance from it.'[95] In short, post-radio radio is *tactical*, one of the 'cheap "do it yourself" media, made possible by the revolution in consumer electronics and expanded forms of distribution (from public access cable to the internet) [that] are exploited by groups and individuals who feel aggrieved or excluded from the wider culture'.[96] 'Those who lack power cannot claim a territory or define a space of their own, but they can operate in the spaces defined by others.' 'Such operations are always temporary incursions. ... They operate on the terrain of strategic power, fragmentarily, without taking over its entirety. Whatever tactics win, they cannot keep. Therefore, tactics are inevitably nomadic.'[97]

But not entirely – 'mobility and precariousness' are a function of localization; the 'terrain' of power cannot be taken over, but it can be re-territorialized. This is what happens when bees swarm or starlings gather. Such formations are occasional; they have a tempo, a group sensibility that keeps them together. Transposed to the practice of media tactics, 'Siren Sonata' would be the radio dramaturgy of a public space.[98] The function of the radio would be to bring about the kind of interactive rhythmic entrainment, which sociologist Randall Collins calls the 'hum of solidarity',[99] raised to an urban scale – 'As in an empassioned conversation, the synchrony of our rhythms forms an inaudibly low frequency, we hum. If you record such a conversation and speed up the recording a few octaves, you will hear it. Laughing together is a perfect example, often showing rapid synchronization.'[100] While these 'local interactions [can] combine, giving rise to another form at a "higher" degree, a social emotion',[101] it need not follow that a chain reaction is inevitable, whose outcome is Canetti's headless conformist crowd. Indeed, our very susceptibility to rhythmic identification may hold the key to a public space practice of resistance. In his 'Suggestions for an Un-Creative Collectivity', for example, Kai van Eikels defines a 'self-induced synchronisation', that is 'synchronising with things or parts of things others do', that 'never leads to perfect uniformity but always retains a difference between rhythms'. These subtle differences, he argues, make any 'collective *rhythmoi*, spatio-temporal forms achieved through synchronization ... transitory and reversible'.[102]

Van Eikels's notion of a synchronization that is imperfect is comparable to the phenomenon of entrainment in radio broadcasting. In the days when radios had oscillators, a stronger signal could 'pull' a weaker signal towards it. The resonance effect of one frequency on the other caused a rhythmic synchronization to occur. From the point of view of the broadcaster such interference was a cause of performance degradation. Taken, though, as the radio expression of a negotiation between source and neighbourhood, it has a different character. A key feature of entrainment is its oscillation between synchronization and non-synchronization: in radio this is, or was, audible. A dramaturgy of public space based on resonance – whose achievement is an effect of echoic mimicry – will likely demonstrate 'participatory dissonance', to borrow Charles Keil's term. 'According to Keil, discrepancies – particularly in timing – are what create "groove", or an activation of positive feel in the music.' 'Keil and Feld define "groove" in numerous ways that relate to entrainment, for instance, as the experience of "being together and tuning up to somebody else's sense of time".' In an ethno-musicological context, the key point is that 'musicking involves a sense of *participation* ... and that participation is founded not on exact synchronisation but on appropriate degrees of being "out-of-time".'[103] Scarlatti's music, 'rich in syncopations' is similarly 'out-of-time'.

Atmospherics in a double sense characterize a revivified public space: interference patterns territorialize and group the general public into regional agents for passing the time differently. But an atmospherics also attaches to the anonymous performance or overall effect produced in this way. The performances associated with the 'sound turn' in contemporary ethnography, for example, entrain: they negotiate a new border between self and crowd.

One expression of this is the return that does not quite return – that, like Scarlatti's sonatas, avoids recapitulation. A return of this kind may be said to territorialize sound. The difficulty with the optimistic thought that the internet can realize the emancipatory potential of 'radio art' is that it identifies the post-radio of resistance with the globalization (or de-territorialization) of communication. The potential of a post-radio radio art to bring about new 'Public occasions' which directly address the relationship between listening and action, 'forcing participation in real space and concrete, responsive thought rather than illusionary space and thought'[104] – the project attributed to the artists associated with the 'Sound/Art Foundation' – depends on its articulation of a feedback or echoic region. 'When the infosphere is too dense and too fast for a conscious processing of the information, people tend to conform to shared behaviour,' Kevin Kelly writes.[105] Discussing the problem of resistance to this, Franco Berardi introduces Guattari's discussion of the *ritournelle* as the magical device mediating between the rhythm of singularity and the chaosmotic rhythm of the world. In our context, though, the *ritournelle* is simply a special case of entrainment, a characteristic expression of 'rhythm', which is 'singular and collective'.[106]

The walking patterns of life in Villar were *staccato*; an awareness of treading on foreign ground fostered a lightness of touch. The *andante* was always at risk of becoming *agitato*. Territories as small as the width of a house, uneven street corners demanding an intuitive give and take; and the obscure divisions of the *paseo*, implying a particular serpentine, advance and withdrawal: these solicited an agility not uninflected with anxiety. One day I recognized these village manoeuvres formalized and transformed in an exhibition of flamenco dancing; an extraterritorial performance in Valencia, possibly aimed primarily at tourists, it nevertheless hooked me. Somewhere between these extremes of marking there must exist a dramaturgy of public space. The street I lived in, Calle de Cantareros, or Street of the Potters, was associated in my mind with singing, with the *cantar* (song or epic poem); and together I imagined the song flowing from a deep and serene source. There exists in Spanish a word '*cantarada*' that refers to the liquid that fits into the *cántara*, a narrow-mouthed pitcher; and when the pitcher is up-ended the *cantarada* disgorges at its own pace, like the pulse of the heart. When I first came to Villar, I was given a *botijo*, the sweating jar that keeps its contents cool because of a continuous slow transpiration across its inner and outer surfaces. I thought of Silas Marner, the misanthropic weaver who breaks his water jar because his shrunken life is as 'the rivulet that has sunk far down from the grassy fringe of its old breath into a little shivering thread, that cuts a groove for itself in the barren sand'.[107] And I realized that the jar was amphibious life, filled and poured, refilled and tilted, a still dancer accompanying me and measuring my days.

Whatever the rhythm that mediates between the One and the Many, it will flow from an untroubled source; its depths will be transparent, its *profondeurs* will be light. Of his *Exercises* Scarlatti wrote, they are not to be scrutinized for 'profound intentions': they have a modest object – the person who learns to play his sonatas will acquire 'freedom' in the art of playing the harpsichord. They will assist her to step beyond technical competence and find a path of her own, one free of the false sentiment of nostalgia. Like March's lover, who 'cannot go back since someone else brought him (there), for he would not have come such a way by himself', she experiences 'freedom' as a struggle. Her guide in that realm where tempo yields to timing might be *duende*, which, Lorca says, is 'a force not a labour, a struggle not a thought'. In the same way, Georges Didi-Huberman claims, the spirit of the *cante jondo* declares itself in Israel Galvan's dancing as an effect of the surface – it is deep because light.[108] The 'commotion' his dance creates opens a space of flight – you could say a fissure in reality. The instant between two strides is materialized as a movement form, one that is labyrinthine. There is opened up in the imperial space of the spectacle an intriguing maze of contingency, fateful, dangerous and – in confronting the real – liberating. These exercises emancipate the exile from going back – 'it's impossible for it ever to repeat itself, and it's important to underscore this. The *duende* never repeats itself, any more than the waves of the sea do in a storm.'[109]

A good example of how the spirit of a work migrates across technological and historical borders is offered by Barcelona-based Eugeni Bonet's 'Throw Your Watch to the Water, Variations on an Intuited Cinegraphy by José Val del Omar, 2003–4'. When he invited the FMOL Trio to create a soundtrack, they had to admit, '*We* knew nothing at all about Val del Omar.' Nor did I, until I came across a retrospective of his work in the Museo Reina Sofía in Madrid in 2013 – a scandal, really, given the extent to which the first part of his Triptico Elemental de España, *Aguaespejo granadino* (1953–5), achieved the synthesis of dance step and flowing water that I was imagining in 'Siren Sonata'. The central trope of *Aguaespejo granadino*, the correspondence between the fountain's animation and the rhythm of flamenco, depends on a technological and phenomenological 'deepening' that is not at all descriptive but aims instead at discovering a kind of audiovisual *cante jondo*.[110] A lyrical representation is predicated on a polyphonic reconstruction. The new lyricism is dramatic but its drama is shorn of all naturalism. This is clear in the language used to frame the presentation of the work at the Berlin Film Festival in 1956:

> Spain proposes a mecha-mystic action in opposition to Neorealism. It is not content with reality, or willing to stop at the frontiers of the mystery. In opposition to the electric light beam of the television with its wide-ranging reporting of the present, this director wants the magic lantern of the cinema to go beyond the panoramic screen and situate the horizon of its great collective visual field at the central point of the infinite.[111]

The physical trope that makes sense of this is the vortex, a figure of a creative turbulence that endlessly revolves around a centre that does not exist except as the figure of change. The vortex is a rotatory version of the *templar*, a bullfighting term, which is, broadly, the technique of maintaining the space between the toreador and the bull through the manipulation of the coat. An extreme potential approach or rapidity of movement is balanced with a virtual immobility (comparable with the way the flamenco dancer revolves on the spot). To be 'contemporary' in this traditional context is to make sense of the passage, the arrest, the turn and the departure.[112] It is to attend to notation correspondences across technologically different movement forms. On this basis Didi-Huberman finds that the work of Pedro G. Romero, photographer, video artist and artistic director with Israel Galvan, is consistent with Galvan's aesthetic. Romero produces 'types of montage', samples electronic music clips and appreciates flamenco burlesque. According to Didi-Huberman, Romero poses the question of what it means to be contemporary in Seville, responding with '*une forme errante d'archaeologie urbaine*'.[113] He recovers other times in the present moment, to build '*une archive poetique et politique*'.[114]

Because of its constructivist aesthetic and praxis, Val del Omar's work can continue to be revisited, and made contemporary. Hence, when he came to explore Val del Omar's intentions by way of the notes and images he left behind, Eugeni Bonet intuited a way of giving new life to the images, techniques and abandoned projects of a great artist who produced only a small body of work. At the same time he avoided the impossible role of medium, recalling Dziga Vertov's generative precept about films that stimulate and generate other films; that said, the ideas of the Russian filmmaker present numerous affinities with those of Val del Omar. 'The film', Bonet notes, 'is a kind of remix of a work that never had a first mix.' Nevertheless,

> a number of sequences and elements are very close to the original material. This is true, for example, of the 'whirlwind of ecstasy' that breaks out near the middle of the film and goes on for more than twenty vertiginous minutes, containing several grafts of and various variations on the picto-luminic experiences of the late and always surprising Val del Omar taking as a source some Super-8 footage that provided documental testimony to those experiences.[115]

Bonet showed the FMOL Trio 'some twenty minutes or so of the first part of the film, already edited, and explained his idea about the general structure, approximate times, key themes and the kind of montage he intended to apply to the different parts'. The group established 'an open general structure', aware that 'the music would be generated in the actual moment of recording'. They 'sampled sound material ranging from Falla to Japanese improvised music; we manipulated discs by everyone from Miles Davis to Enrique Morente; we recorded acousmatic material, water, voices, and so on, and we had sound recordings of Val del Omar himself, previously selected by Eugeni'. They 'wrote motives and graphic scores for our improvisations', and then, translating Ausiàs March's ethical dilemma – not seeing where to set their feet, not ready to lay down tracks – into a compositional practice, they got into the studio.

> After an appalling first take, it all came together at once. We recorded intensively over three whole days. As a novelty, in addition to our usual instruments, Pelayo brought along a violin, Sergi a set of glasses and a toy piano, and Cristina her Tibetan bell and some cans that we incorporated spontaneously. Òscar Celma played Spanish guitar on several themes and Jan Schacher provided some samples and additional sounds for the theme 'Granada'. The recording technician was José Lozano, who also helped us with the final mastering in a highly creative way. ... In June 2003 we gave Eugeni some first mixes to use as a support in his visual montage. About a year later Eugeni sent us a workprint that astounded us by its visual richness and the tremendous subtlety and empathy of the interaction between images and music. On the basis of this version, during May and June 2004, we produced the final master in 5:1.[116]

Exercises of this kind are *contemporary* in the strong sense of living in the same time as another. They take their cue from what is to hand. They interact and remain unfinished, like a swerve or any passage between two banks. They model the turbulence of flowing watersides, punctuated with vortices. In the Rua Augusta, Lisbon, I recorded the wandering Austrian musician Ananda Krishna playing hang and didgeridoo and singing. I have observed many hang performances but Ananda Krishna's intuition of the pace of the street, and his uncanny ability to channel its potential flows, stood out. In an interview reported in the *Havana Times* (26 December 2012), he described his resistance to studio projects: 'I have to be on the street, because life, children, impulses, all of that is in the street.' Street music and street presence are the setting of a performance practice that 'let[s] me play the music that comes out of me, which is my natural music'.[117] This attitude also expresses the aspirations of the Integral Hang's creators. In 'Letter from the Hangbauhaus', (2009) Felix Rohner and Sabina Schärer explain,

> Our concepts, developments and implementations are far from the musical norms of modern times which require study, practice and performance. Playing with this Hang can lead to a form of freedom, an intimate conversation that can only unfold without pressure and coercion. If individuals are aware of this concept, they will be strengthened by this Hang. Thoughtless use can weaken a person.[118]

To be in the flow, responsive to impulses, is to describe a spectrum of human movement, ranging from the leisurely urban *paseo* to flamenco's *paso*. It describes a disposition to entrain, to get into step with a polyrhythmic environment. Bonet characterizes his synthetic video as '*una pasarela, un lugar de paso, un agente intermediario entre disciplinas o direcciones diversas. Y que su identidad reside ahí, en esa cualidad borrosa o vaporosa.*'[119] The freedom playing the Hang brings is similar to the freedom Scarlatti envisages flowing from the ingenious jesting of his *Essercizi*. In another exercise for 'Siren Sonata', I set the poem 'Fadisto with Megaphone', written in Lisbon in 2013, to street sounds recorded during the poem's composition. Although the sound-mix was in the tradition of the auditory picturesque, featuring police sirens and clocks striking the hour, the integration of the found sounds was rhythmic, timbral and harmonic. Through the fourth stanza, Ananda's music is heard; a clear entrainment of recitation and performance occurs. Both elements listen to a hollow, awaiting an echo from another. This other realizes the potentiality of the hollow, and exists psychologically and musically like one half of a dialogue whose expression lies outside language and whose desire of reciprocity is communicated mimetically. Inside the grain of my voice, across the breath's extension of a phrase, comes the pulse of the hang, and a double syncopation is heard as the two strains pass in and out of phase, a phenomenon that corresponds to the endless parallax of passers-by composing the movement form of the street.[120]

Moved by a sense that 'Scarlatti' was unfinished, I embarked on a sequel called 'Siren Sonata'. It seemed logical to go back over previously trodden ground, to locate the autobiographical and aesthetic wellsprings from which 'Scarlatti' had flowed. This preparatory act of recollection would bring to light forgotten or neglected influences, and probably identify undeveloped themes that could (twenty years on and with the advantage of new materials to hand) form the basis of the *restatement*. But, instead, what I have found, as I have tried to order the Spanish current percolating through my life, is not a well-marked path of conscious recall, but an anti-history of unconscious propulsions towards a source that always lay round the corner in the future. As a sound history, it consists of calls only heeded in the delayed recognition of their echoes: it is one written into the future not the past. I imagine a visitor to the Delphic Oracle, accosted by the inscription 'Know Thyself', might have experienced a similar sensation: the source of that knowledge may lie within, but only through the echo of the other is awareness of it possible – and in this sense it will always lie up ahead. To come upon the source – to hear and heed the Sirens – is to hear the echo of one's own desire to be called. Tiresias is said to have been blinded for taking things literally: he lifted the skirts of Pallas Athene expecting to see a woman, when the 'being' of the goddess is 'in the multiple appearances she assumes'.[121] A comparable reward awaits the literal hearer who supposes the Sirens can be tracked down through their echoes. In fact, their being is always a coming into being at this place and time; in this they resemble public space, listened to for its resonances.

Consider in this context one last *periotographical* shadow: the poet Luis Cernuda. When, in 1947, Arturo Barea moved into Middle Lodge, his host already had a long association with the Republican cause. In 1937 Gavin Henderson, the 2nd Lord Faringdon, had accommodated a group of *niños de la Guerra*, Basque children confiscated from their families. Among the teachers of this 'stolen generation' was Cernuda, himself in post-Franco flight. Cernuda never returned to Spain, embracing exile (ultimately in Mexico) as his fate. Now the points of resemblance are many. They are biographical, thematic and even stylistic. Biographically, Cernuda inverted my journey south. As the chiaroscuro of the English countryside brought a new depth to his poetry, my residence in the hills overlooking the Valencian *huerta* produced a sequestration of poetic ornament. Thematically, it would be presumptuous to suggest any comparable articulation; but, to take one humble coincidence, the feelings Cernuda describes on hearing an English blackbird sing are those of my father. There is the same experience of sharing a moment out of time '*todo abstraído en una pausa / De silencio y quietud*', of occupying '*El instante perfecto*'. There is a comparable awareness that the blackbird's labour falls outside the human system of needs and rewards: '*Tan so lo un mirlo / Estremece con el canto la tarde. / Su destino es ma s puro que el del hombre / Que para el hombre canta, pretendiendo / Ser voz significante de la grey, / La conciencia insistente en esa huida / De las almas.*'[122]

Stylistically, there is the development of a shared taste for writing poems in which the phrase does not coincide with the formal line; instead a phrasing is preferred that continues from one line to the next. In this way the basic rhythm of the line and the ordinary pattern of accentuation of the phrase interfere and overlap to form a kind of double rhythm. This rhythm can be compared to what happens when one person walking is joined by another. The French word for a phrasing that continues from line to line is '*enjambement*', a striding over, rather as if the rhythm of another flows into the line. A counterpointing occurs that introduces eddies into the metrical formality of the line; the line of thought is bipedal, and flexibly responsive to the emotional weight of the meaning; the poem conveys an impression of continuous self-adjustment, of a movement responsive to what comes along in the course of composition. Threading one's way through a crowd, or even adjusting one's pace to another's, produces analogous alternations of glide and pause, aside and renewal. Cernuda associates this kind of syncopation with exile. Hearing the conventions of the *copla* through the differently phrased and accented literature of the English and German Romantics suggested a choice – between nostalgia for the old forms and the path of innovation.[123]

I took Cernuda's poems to Spain more as a talisman than as a text; I kept them as Petrarch kept Homer in his pocket, their grip on my imagination in inverse proportion to my understanding of them. Cernuda wrote that '*el cambio repetido de lugar, de paìs, de circunstancias, con la adaptaciòn necesaria a los mismos, y la diferencia que el cambio me traìa, sirvio de estìmulo, y de alimento, a la mutaciòn*'.[124] In a way, I read his poems as a key to my own mutation. Slow to work out their meaning, I was quick to practise reading them aloud. The impulse to hear the auditory imagination informing them embodied my desire to step out of myself, to walk in the path of new breath patternings. But it would be autobiographical fiction to suggest that I found in his mapping of political, sexual and geographical margins a discursive mirror held up to my condition. In those days I was unaware of his local connection. These and other points of resemblance emerged later by what might be called a process of poetic consilience. The Victorian historian of science William Whewell called the process by which the raw materials of natural science yielded progressively more accurate hypotheses *consilience*, or a 'jumping together'.[125] So, perhaps, with the clarification of elemental themes – it may take a lifetime to identify their sources accurately. In the meantime, there is an illusion of travelling solo; until a deep listening is developed, able to tune into these subtle accompaniments from the rhythmic environment, one may feel haunted by echoes but rarely on a path where they converge.

In the end, the auditory anamnesis 'Siren Sonata' really documented emerged through a process I have encountered before: a mishearing, or misremembering, turned out to provide the *via regia* to the sound that was sought – the fascinating sound that is feared, that the other sounds merely mask – as good names are used to placate the Furies. Among the Basque children exiled to Buscot were survivors of the Fascist–Nazi bombardments. One of them later recalled,

> The droning of aeroplane engines overhead, shrieking of warning sirens, bombs explode. ... The murderous aviators destroy the life of a loyal town. There is crying in the streets. Women run with their children clasped to their breasts. Like wild beasts, mothers crouch to defend the lives of their young. 'The black birds have come to exterminate us.' The droning ceases. Once again the hoarse siren tells us that the black birds are gone. The daily air raid is over. Left are the bodies of age-worn men, little children, harmless women, all victims of the tragic flight. We cannot believe the stillness. Our ears still ring with the sound of bombs, and we think that somewhere in the distance another harmless town is suffering as we did.[126]

Reading this, I remembered Cernuda's experience in 1936 and 1937 when at night in Madrid to the sound of shelling he read Leopardi and composed *Las nubes*.[127] I also remembered his description of the *sang froid* of hotel guests in Liverpool during a Luftwaffe attack: the air raid warning siren sounded but no one panicked. Returning to Liverpool later, he found the hotel destroyed.[128]

The siren is, as the Basque testimony emphasizes, twinned with silence. The siren sounds twice, once to announce the imminence of deafening explosions, and once to terminate the end of the 'stillness' that follows. A British government circular from 1939 explained,

> When air raids are threatened, warning will be given in towns by sirens, or hooters which will be sounded in some places by short blasts and in others by a warbling note, changing every few seconds. ... When you hear the warning take cover at once. ... Stay under cover until you hear the sirens sounding continuously for two minutes on the same note which is the signal 'Raiders Passed'.

At the same time, the siren is a summons to listen: it interrupts normal services, and placing citizens on high alert, it causes them to hear (perhaps) for the first time the normally muffled hinterland of children's cries, garden birds calling and middle-distance conversation. In Edward Thomas's poem 'Adlestrop' the cessation of the steam train's hissing suddenly brings into auditory focus the birds singing – as it seems all the birds of Oxfordshire and Gloucestershire. The air raid siren created a similar illusion. In the interval of auditory dread ahead of the raid the same kind of attunement must have occurred, the environment being perceived for the first time as a soundscape.

Firefighters in the Blitz experienced a supplementary schizophonia:

> There are the noises. But there is something here that is more terrible: the silence. I believe that in the course of great conflagrations there sometimes occurs a moment of extreme tension: the jets of water fall back; the firemen no longer mount their ladders: no one stirs. Noiselessly a black cornice thrusts itself forward overhead, and a high wall, behind which flames shoot up, leans forward, noiselessly. All stand motionless and await, with shoulders raised and brows contracted, the awful crash. The silence here is like that.[129]

The twice-sounding siren and the acoustic event it framed create a new urban *rondo*, where deafening noise and eerie silence alternated before returning to an ultimate silence. The novelty of this new ambience was its return to silence: the trauma of the air raids had produced a new kind of listening. It was as if the public was on constant alert; for the first time, they listened to the silence *ahead of the siren*. In the next generation, the role of the siren was taken over by the telephone, whose unpredictable summons popularized auditory anxiety.

In local firefighting histories, the host of the Basque children, the 2nd Lord Faringdon, receives honourable mention. During the Second World War our small town managed a continuous firewatch: a whole time complement covered the day, while the so-called 'retained' covered alternate nights: 'Alexander Gavin Henderson who was a pacifist joined the Fire service, serving at Faringdon, sleeping with the retained and serving with great courage at London, Bristol and other City Blitzes.' Likewise, my father, unrecorded in these modest annals, fought alongside other members of the Auxiliary Fire Service, including Henderson, attending London blazes as well as conflagrations in Exeter, Bristol, Coventry, Southampton and Avonmouth.[130] Perhaps it is also worth mentioning that the 'retained' used to sleep 'in a Nissan hut opposite the then Fire Station', an odd architectural coincidence given Bill's later role in founding a different auditorium for echoes, the Faringdon Dramatic Society, in another Nissan hut. If these ancestral spirits, hovering, as it were, above the crib of my future existence, shared a sound history, it was not the blackbird-loud dawn chorus of garden hedgerows, or the charming colloquies of exiled poet and incurable *borracho* Pedro Garfías and the landlord of the local pub – according to Neruda: 'Long after closing time the two soul mates would pour out their feelings, each in a language unintelligible to the other, and yet communicating with the heart.'[131] It was the siren's wail and the echo of silence heard in it.

At the time of writing 'Scarlatti' I lived opposite a fire station. With care for a three-year-old son, I was sensitive to sounds that disturbed his sleep. Sleep-deprived myself, I had the impression that the fire brigade was mainly called out at night. Preparations oddly resembled a fire taking hold. The sound of a telephone ringing, amplified through the station, issued a preliminary alert. The duty officer answered the call, and after a tense interval, during which the fate of our rest was decided, either the station would quieten down, letting back in the background a hush of cars driving home, or, as was more likely, a second broadcast would be heard. As information about the incident crackled through our walls, the great red doors of the station could be heard squeaking and screeching on their rollers as, like theatre curtains, they folded back left and right. A growing pandemonium ensued, voices were raised, doors beyond the number imaginable slammed, engines revved and, from inside the cave of the station, the sirens began to wail. The sirens were a source of dispute: to lend weight to our complaint, for it seemed that a simple objection to the agonizing noise fell on deaf ears, we observed that, sounded in this way, the siren was inaudible to traffic in the street. Only our house, directly opposite, heard the warning. We drew their attention to the fact that, far from pulling out of the station heedless of traffic conditions, the fire engine driver waited until the road was clear. The siren served no purpose.

I wish I could have cried out, echoing Heidegger, towards where does the out-turning turn out:

> for the siren sounded and the red revolving lights flashed, a *blitz* that did not retrieve that which it caught sight of but signified 'the injurious neglect of the thing' where 'we do not hear, we whose hearing and seeing are perishing through radio and film under the rule of technology'.[132]

For the news the siren announced was directed to the firemen. Like the flaming firebrand of Karl Kraus's *Der Fackel*, its news ignited in separate individuals a burning desire to act as one: to rise up and destroy. In the same months I was intermittently attending to 'Scarlatti', our son was giving sounds to things: the plum tree, for example, he designated the 'a-toot-toot'. He copied my telephone voice and even improvised the answering murmur audible through the receiver. I wonder whether it is this sonic *intrahistoria* that explains the unfinish of 'Scarlatti' and the sense I had of an unanswered echo. Miguel Unamuno, another of Barea's subjects, derived his notion of *intrahistoria* from an earlier notion of the *intracosciente*, which is a kind of the unconscious that exists within the ordinary content of consciousness (rather than buried underneath it).[133] Evidence of this intra-consciousness was to be found in language. If the *intracosciente* 'consists of the structural deposits or equivalents of psychic activities which were repeated innumerable times in the life of our ancestors',[134] then *intrahistoria* was the cultural counterpart of this, 'a repository for ancestral patterns of apperception or behaviour'. It is at once minor history – 'the millions of small details of human life that lie between the schematized generalisations found in books' and, as the expression of a cosmopolitan, eternal tradition, something rather like Heidegger's 'Presence', 'the ancient name of Being'.[135]

FIVE

Echoes

Charged to look after the literary archive of the poet Martin Harrison, who died in 2014, I was surprised, in the course of going through his materials, to come across a proposal we had jointly written back in 1987 to produce a '90 minute stereophonic sound-piece' called 'Spring Song'. In 1899 and 1903 eight wax cylinder recordings were made of 'the last Tasmanian', Mrs Fanny Cochrane Smith, singing and speaking both in Tasmanian and English. She gives her own biography (her father was a sealer), sings three versions of a 'corroboree' or 'dance' song, two versions of 'Spring Song' and 'improvises a hymn'. On other cylinders, the members of the Royal Society congratulate themselves on being present at the making of what will be a permanent record 'in future years when this and the remaining representatives of the native race have passed away'.[1] We emphasized that we had no intention of producing a 'tragic' re-enactment of the dispossession of black Australians: our aim was 'to begin a history of Australian sound', 'through various kinds of studio-mixing and treatment, to offer a new, celebratory version of these songs and their singer'. 'Spring Song' would be a 'sound history', a 'genuinely radiophonic and sonic history rather than a written, literary history "on air"'.[2]

The conversations of that period served as a creative solution. 'Spring Song' was one of a number of possible crystallizations. 'Memory as Desire' and 'What Is Your Name' (which Harrison produced in the previous year) were other expressions of our desire to create a new Australian sound vernacular. We pursued this goal by many means. I worked as a poetry reviewer for Martin's program 'Books and Writing'; when I became editor of *The Age Monthly Review*, he contributed a pseudonymous column satirizing contemporary Australian literary life. Although 'Spring Song' did not go ahead, its materials catalysed other operations. The extraordinary photograph of Fanny Cochrane Smith 'recording Aboriginal songs' in the rooms of the Royal Tasmanian Society (dated 5 August 1899) formed the frontispiece of my 1992 publication *The Sound In-Between* because of its relevance to the book's opening items: 'Spirits of the Dead: a sound history of "Cooee"' and "Cooee Song' a performance work for two actors and their voices'.[3] Meanwhile, Harrison worked on the poems later published in the brilliantly titled collection *The Distribution of Voice* (1993).[4]

If a fundamental lexeme existed for a history of Australian sound, a kind of *Urwort* from which an original *Ursprache* might have sprung, it would be, I suggested, that distinctive sign of Australian sociability, 'cooee'. My advocacy of a word sounds associated with making contact reflected my interest in foregrounding the migrant experience of reorientation and affiliation connected with living in a new country. Our history of the Australian sound vernacular would use the everyday experience of the contemporary migrant to expose a neglected history of colonial cross-cultural communication. 'Cooee' might be a fundamental word sounds, suspended between vocalization and conceptualization, but it was also the least fixed and most playful of expressions. Where people meet without either language or interest in common, efforts at finding common ground are mimetic. In the absence of anything to say, their communication is echoic. They seize on mere sonic (and gestural) coincidences. They improvise a new language of exchange that is peculiar to the situation they are in, and which afterwards has the potential to become formalized as a multimodal lexicon of sociability negotiated at that place and time. The sound in-between, imagined here as the relationship between the two vowel sounds 'oo' and 'ee', is simultaneously a historical phenomenon, a psychoacoustic experience of everyday migrant life – and a signature feature of the 'open work', where (as in 'Cooee Song') the actors are free to interpret the performance instructions in a number of different ways.[5]

Later, I generalized these performances (colonial encounters, migrant miscommunication and the construction of the persona in post-dramatic practice) as types of 'listening as a cultural practice'. 'Ambiguous Traces, Mishearing, and Auditory Space' (2004) considered 'the *locus classicus* of cross-cultural collision', arguing that Columbus's encounter with Taino people in October 1492 largely revolved around an ambiguously signifying word sounds 'ca'. Echoically generated from this core were apparently significant words 'Guanahani', 'canibe' and 'caribe', and their variants 'canoa', 'Canarias' and 'canna'.[6] The same spirit of 'invention, improvisation, innovation, and fantastication' was detectable in R. A. Baggio's echoic mimicry of his own name's mispronunciation.[7] Whether guileless or ironic in intent, echoic mimicry opens up a space in-between where power relations remain for a time fluid. In a related context, David Tomas's studies of cross-cultural encounters discover 'the existence and dynamics of a transient, sometimes humorous, often dangerous, and periodically cruel intercultural space – generated in situations governed by misrepresentation or representational excess'.[8] Such transcultural spaces are 'predicated on chance events, unforeseen and fleeting meetings, or confrontations that randomly direct activity originating from either side of geographic or territorial, natural or artificially perceived divides that separate and distinguish peoples with different constellations of customs, manners, and language'.[9] As an instance of the actor spoken through by another, and thereby achieving a doubled identity, I offered my experience of listening to *myself* in 'The Native Informant'.

The transposition of radiophonic practice into post-ethnographic speculation is, I suspect, unusual. I tried something similar when I extrapolated from the experience of writing 'The Calling to Come' for the Museum of Sydney in 1995 a far-reaching historical claim, the existence of a distinctively erotic intercultural space in the early days of white settlement at Sydney Cove. The historical sources for the male and female voices heard in the Entrance Cube to the Museum of Sydney were, respectively, William Dawes, a First Fleet officer, and Patyegarang, a young Eora woman. Their intimate bantering dialogues, recorded in Dawes's notebooks, bore witness, I claimed, to 'the gestural, pantomimic and flirtatious aspects of communication which the ethnolinguistic (and historical) record habitually neglects'. They 'sketched an erotic zone where power relations were negotiated differently', and the basis of this was echoic mimicry: 'Like the "coo-ee", which becomes a way-finding call only when it is returned, the discourse they developed to keep in touch was imitative, a matter of doubling up phonic sound-alikes and behavioural mimes.'[10] But echoic mimicry is too slippery to furnish a method. Close cousin to irony, it causes serious people discomfort. They await the day when it becomes the victim of its own quick tongue, and like the saltimbanque, makes an ill-fated leap: 'You'll roll on the ground with your head split open, / and that tragic stunt, distressing and speechless, / will be your most memorable and applauded stunt.'[11]

In my experience, sound has always been a calling. Listening has been educative, soliciting a response. Less a signal than an environment, sound invites incorporation into the surroundings. An acoustic environment is a twinkling sea of sound events on all sides. Within its endless rise and fall, the rapid onset and dying away of individual events and the discernible rhythmic alternation and counterpointing of spatio-temporally coincident auditory impressions, the listener is a swimmer at sea: listening seems inseparable from orientation, and both from the dispersed life pulse of indiscriminate noise, understood here as something like the undying 'S' of ocean. In my thought, there has always been an intuitive connection between the impressions that the sound environment makes on me and the impression that I make in return. This relationship is prefigural, analogous to the sensation of buoyancy. It helps to explain the flights of the sonorous body described earlier. According to Massimo Cacciari, the divine message that the angel brings is communicated in the passage itself. He does not transmit information: it is his transmission that informs. The angel is an icon of the instant, an impression of movement or, more formally, the *typon* of *rhuthmos*.¹² Just as it is impossible to separate the wind in the trees from the fluttering of the leaves, so there is no operational distinction between breathing and pattern-making. Beyond sound there is always an environment listening. In certain places this is audible: in the mountains, for example, primary echoes assemble. Generally, though, places attentive to our passing speak too softly to be heard. They gather to listen, as perspectival space gathering doorways and gardens in Fra Angelico's *Annunciation* does, but their overhearing remains discreet.

This volumetric sense of sound as pacing, as a rhythmic architecture of intervals, may explain the transition from sound installation to public typography in the later 1990s. Apart from 'Out of Their Feeling', a sound sculpture created with Hossein and Angela Valamanesh for *An Morta Gor*, the Irish Famine Memorial at Hyde Park Barracks (1999), the soundscapes for the Museum of Sydney, including the epic-scale 'Lost Subjects', were my last forays into institutional sound designs. The text-based public art commissions for, respectively, the Sydney Olympics and Federation Square, Melbourne, seem to mark an abrupt break with that which had gone before. We later created the polyphonic works 'Relay for Radio' and 'Nearamnewspeak', both broadcast on public radio.¹³ But these dramatic adaptations were secondary: the ground writing was primarily composed and designed to create a field of *typoi*, characters or letters whose interpretation demanded physical movement and orientation: walking beside or across them, we said, integrated reading and treading. 'In his essay "Typography", Lacoue-Labarthe shows that the relationship between foot and footprint, or stamp and impression, isn't simply a physical one. It implies a stance or attitude. The relationship is ethical. This association is present in the word "character", which denominates both the form of a letter and a human disposition.'¹⁴ Looking back at this passage, I would now add that the relationship becomes ethical when the traveller-readers overhear themselves reading, integrating movement and desire, passage and inclination. This seems close to Brandon LaBelle's idea of an 'ear ethic', where building a place 'between you and me' depends on a 'third body or third ear, the one sitting over there, in the wings, off-screen, out of frame; the one, that is, who *overhears*'.¹⁵

As regards writing a history of Australian sound, the migration from predominantly indoors sound art to entirely outdoors ground writing reconnected me to a geographical line of enquiry largely absent from the 'Spring Song' proposal. There we had described 'an interest in Australian sound both as technology and as event'; we recognized a material history bound up with the way that 'certain cultural sounds become fixed and foregrounded in more permanent, recorded form'; and we appreciated the social life of sound as 'a history of transactions – of how one group of Australians have chosen to represent another, or how we have chosen to represent ourselves to ourselves'. These are the themes illustrated by the

Cochrane Smith recording sessions: in one photograph Fanny seems to peer into the horn of the phonograph; it is difficult to tell whether she is making a recording or listening to the playback.[16] In another, Fanny sits stiffly but complacently, hands clasped in her lap; six men of science listen spellbound. The only dissonant note in the composition is Fanny's nephew: ignoring the strange ghostly wail of the horn, he stares past the photographer into the dark. Here is technology as event *and* transaction; the immobility of the group is suggestive. Something is being fixed forever, recorded and put to rest. The philological fantasies of Hermann Ritz are a logical corollary: in 1908, offering a 'translation' of Fanny Smith's 'Corroboree Song', based on his theory of 'essential powers', Ritz expressed the pious hope that his 'sympathetic efforts' would enable members of the Royal Society of Tasmania 'to hear once more "The sound of a voice that is still"'.[17]

The 'histories' we identified as proper to a history of Australian sound were dangerously like *cultural* studies of sound. True, we intended to create a work of acoustic art whose sound structures and technological mediation (radio) would offer a transcript (somewhat in Berio's sense) of the original materials and their situation of production. But, if the object of acoustic art is to use 'the technical possibilities of radio productively, not reproductively', that is, 'not to construct or imitate a non-acoustical event', how could the interpersonal dynamics of those attending the 'Spring Song' recordings be conveyed?[18] How is the auditory ambience of the event to be portrayed? Despite his essential powers theory, Ritz entertained the possibility that 'Spring Song' was 'an imitation, not of a Highland bagpipe, as [James] Bonwick opined, but of a melody of the native magpie, which most unmelodiously the zoologists call a "piping crow"'. Distinguished ethnomusicologist Alice Moyle thought this suggestion should be taken seriously, 'especially if the trills and other "coloratura" effects which distinguish this melody are considered'.[19] Besides the environmental melopoeia lurking in this music, what of the audience? Fanny was 'a very popular person in the district and using her barn as an improvised concert hall, she would entertain by singing her native songs'. Horace Watson 'regularly gave concerts with his Edison phonograph'.[20] The phonograph recordings functioned like pre-recording-era transcriptions for piano. They were neither solitarily authentic nor silently final. They documented and promoted new listening cultures.

Both of us wanted to take sound out of the technological series defined by the recording studio; we wanted to repatriate it on country, but to a country that was, we suspected, as yet unheard of. The auditory moment we cultivated had not pre-decided what would matter; hearing and overhearing blended. The noise spoke. The transient character of sound, its blending, appealed to migrant writers in search of a vernacular, perhaps via an echoic acoustics. As Harrison argued in *Ancient Noise*, the auditorium of the studio characterized sound against a background of silence: there is always more in the listening space than what is attended to – 'the specific properties of the room in which the sound is heard, the distorting effect of the passage-way, the watery influence of the lack of carpets, the mist of bird noise outside'.[21] To focus on

> the writer's, the recordist's or the engineer's text as the source of the sounds is simply to ensure that this moment of acoustically transformative blending is given an edge or boundary across which a listener cannot travel because it is a limit across which a hearing which exclusively listens to 'things' cannot pass. ... It is a boundary past which there is always only a nothing, past which there is an unnameable absence.[22]

To capture the 'auditory moment', rather than 'to silence the world', is to extract the listening experience from time and place and to understand it as a remaking of the present, which Harrison compares to the composition of a flock of birds: 'The metaphor's introduction is not about discovering where the birds go to, nor from where they have come. For a non-specific temporally based audition of language shifts all definitional parameters – including indeed the parameter which asserts that utterance is ... situated within a larger field of audition.'[23]

In this discussion of a 'primary moment of audition', Harrison referred to the passages in 'Memory as Desire', where the explorer Sturt followed the direction of birds in flight but ignored the organization of the flock. His persistent mistake on sighting birds in arid country was 'to "read" them as signs which produced specific meanings within a supposedly given text of dryness and refreshment'.[24] If our desire is to 'comprehend the totality of the auditory moment and the situatedness of utterance within it',[25] then the explorer's curiosity about where the birds come from and go is *not* natural. A natural stance would not subordinate the ear to the eye and recruit both the senses to the horizon hunger of colonization. Journeys remained possible but explored a sound geography, mapping 'the unfolding of sounds over time and across space, the contingencies of speaking and hearing and the complex environmental interference patterns that characterise our auditory experience'.[26] As a sound explorer, I conducted a couple of expeditions of my own, one into western Victoria, the other to Lake Eyre. I discovered many 'in-between noises, sub-vocalic phenomena that hover on the border between human and non-human sounds', and which leave their imprint in names (Babel Island, James Voce, the sealer, who answered the call of a drowning man, the Aboriginal ironist, Tommy-Came Last, whose imitation of a Scotswoman was 'ridiculously true, through all the modulations of that particular accent, although, strange to say, without the pronunciation of a single intelligible word'[27]). At temporary Lake Eyre I was in the country Sturt longed to reach. Avoiding the impatience of a diary, I recorded any parrots I saw or heard in an occasional notebook.[28]

For Harrison, an Australian sound history bringing the auditory realm back to life[29] held the key to a distinctively Australian poetics: if in common experience 'words, phrases, intonations, repeats and interruptions and the entire sound-structure of language "disappear" as soon as expressed'[30] because they are now *understood*, then 'the here-and-now temporal shape of what is understood as meaning' needs to be reformulated.[31] Primary audition is not bracketed off from the past and the future; no Husserlian *epoche* frames the sound event and makes it available for study. It is *historical* when history is understood as spatio-temporal marking, when time and space turn into rhythmic geography or ground writing. It is Aboriginal when 'the cultural location of the auditory moment requires that we can *see through* the (outlining of) fully formed conceptual pictures of the land and still hear its living multitude of voices'. A good place to do this is in one of 'those thousands of "sites" where the original cave, the ancestral escarpment, is now a tourist look-out'.[32] Our idea at that time was that the history of sound took the form of vibrational diagrams, scores physically etched into bark or body; an embodied listening was synaesthetic, mediated through energy patterns. In the great poem 'Red Marine' a yacht on the estuary furls its sails, challenging the poet to find 'the meaning of that movement'. 'It was as if, just then, a river shone, / as if, behind that wave, lost voices spoke – / voices heard after they had gone away.' In Harrison's imagination this auditory memory coexists with the actualisation of space as movement:

And yet you see that movement as it is,
crossing like tide itself, through mobile space:
on the sea edge a sail topples, a red
tulip-flame twists in wind.

You see it like this because it defies focus:

    The bright sea's
glitter, with people bobbing in it, swallows it up
like interference blizzarding a screen.

The impression develops like an old negative: 'Half-noticed things remain as glints, / leaving behind them latency in time.'[33]

A new sounding of country happened when our son began to speak. I studied his utterances for evidence of a primary echo; I recorded his imitations and improvisations, as if they were a unique and holy clue to the labyrinth of the country. His utterances would be at once original and received, a cry echoing down the track and coming back as a dispersed multitude. So when at the weekend we drove up to our bush block, a lightly forested shoulder of land near Mount Buninyong in mid-Victoria, I listened for resonances. The children's primary babble merged with the measured cadence of the treecreeper or the scattering trill of the bronze cuckoo. To hear the world learning to speak its name in the child finding his way into speech was no doubt naive. The voice of the poet might, as Bachelard said, be the voice of the world, but the parrot babblings of a child surely conjure up a different, less settled environment. According to Bachelard, 'A word dreamer recognises in a man's word applied to a thing of the world a sort of oneiric etymology. If there are "gorges" in the mountains, isn't it because the wind, long ago, spoke there?'[34] But what is the etymology of babble, of protolanguage hanging expectantly on the air? What volume corresponds to the baby's unsupported howl? Without eloquence, Bachelard's 'great natural words' collapse into the pulsion theory of Hermann Ritz, imaginatively amplified by Mary LeCron Foster when she hypothesizes that language may have originated in a process 'by means of which states and movements in space [were] translated into spatiosonant, articulatory counterparts'. Instead of listening and repeating the numbers lisped by nature, the first speakers, according to Foster, *physically imitated* the valleys, paths, groves, rocks and grottos. Shaping mouth, lips, tongue and vocal tract in imitation of external states and movements, they produced the vocal equivalents of those places.[35]

The Road to Botany Bay arose from my sensation that the Australian landscape defied the descriptive terms available in English: it lacked valleys and groves; its so-called lakes were nothing like lakes; its mountains bore no resemblance to mountains. This was not only a topographical dilemma: it thwarted emotional identifications. If no words could be found, no communication could be initiated. Repeatedly, the glib echo of 'our strange language' reported by John Lort Stokes turned out to be ironic: nothing except projections of 'our native land' came back. Visually unclassifiable, the country was auditorily inarticulate. Defying picturesque enclosure, the country resisted rhetorical capture; the derogatory place names of the colonial map record this frustration. Violent colonization thwarts the hope that 'human and cosmic tonalities reinforce each other'.[36] Aboriginal communities are immune to this accursed cartography; settler descendants have mixed feelings about belonging. A migrant, keen not to recapitulate colonialist violence, has, whatever else they may do, to remain open to 'the movement as it is'. The politics of dispossession and repossession are inseparable from the poetics of listening. To hear the 'voices' is to shape oneself into a place where they can echo, to investigate the proper names of things that the English language and the European sensibility have failed to notice or connect to sense, to mouth these, inhabit their cadences and guess at the resonances they score. In denser country where the architecture of the sound environment is obscured, it involves venturing new utterances of one's own: the guttural that imitates what the wind says when it is *not* confined to gorges sounds more like a *didjeridu*.

One manifestation of 'the movement as it is' was the parallax effect experienced walking through remnant bush. In the corners of mid-Victorian paddocks, along uncleared gullies and clustered around the edges of country towns were patches of messmate, peppermint or manna gum, saplings some of them, overarched by trees that carried their bark like cloaks on their arms, and whose beginning (I imagined then) predated white settlement. Freestanding as stars, their verticals construed my passage as a constellation of eclipses, one trunk sliding before or behind another, unpathed coulisses opening on either hand and closing again like fans. I did not share the colonial disparagement of these arboreal associations as untidy, jumbled or confused; their free scribbling always compared unfavourably with the copperplate handwriting of the grove. I had the impression that they preserved a movement form, a flock-like consciousness of the Many as One sacrificed in the quincunx and the line. I recalled Descartes desiring to find 'the cause of the position of each fixed star', who did not doubt that, 'although they seem very irregularly distributed in various places in the heavens. ... There is a natural order among them which is regular and determinate.'[37] To discover this would be to walk in the right path as, one supposes, the ancestors did. To find this way, 'the motion of which is constant and uniform', as Leibniz wrote in a related context, 'according to a certain rule such that the line passes through all the points',[38] would be to fall into step with the metre of the place. The breath pattern of the walker keeps pace with the inspiration of clearings: pneumatic correspondences of this kind would make even the child's vocalizations descriptive – much as the sonogram of a bird call often looks like a graphite sketch of tall timber.

In this new country the poet was an explorer, the explorer a poet – as I proposed in *The Lie of the Land*, taking the magisterial Paul Valéry to task for supposing that advancing through difficult country was prosaic.[39] Earlier, I had had an inkling of the poetic explorer in the toponymy of Spencer's Gulf: Australia's circumnavigator Matthew Flinders had named the Sir Joseph Banks Group of islands after villages in his native Lincolnshire. I was intrigued that Flinders had transposed not only the names but also their spatial arrangement, as if a strange topological harmonization were intended.[40] Flinders's Group was not a new Cyclades; Dalby Island, the insignificant centre of the Group, could not compete with Apollonian Delos. However, a poetic geography was in play, and, to me, the *approximate* fit of one geographical constellation into another increased its interest. Flinders's poetic geography was, I felt, a heuristic tool, a template for ordering complex arrangements without sacrificing their irregular distribution. In the midst of this enquiry I learnt by chance of the poet Paul Verlaine's stay in Stickney – one of the villages whose name now graced Flinders's chart. Toponymy is for me a principle of geographical transmutation, and I fantasized reading *La Sagesse* to the sea lions. I imagined speaking there '*L'échelonnement des haies*', dated 'Stickney [18]75', which, in the perception of 'a movement as it is', effects a strange metamorphosis of enclosure into infinity, as if England's closed horizons could yield an ocean: 'Row upon row of hedges / billow into the distance, / like a pale sea in the clear mist. ... A moment ago, like a scroll unfurling, / a wave came rolling and breaking, a wave / of flute-like bells / in the milk-white sky.'[41] Verlaine's 'scroll unfurling' and Harrison's toppling sail are two slopes of one *turning to attend* that transcends the 'locative fiction' of hearing as placement:[42] the synaesthetic twinning of seeing and hearing produces a new figure, a curl, like the involute of a shell.

The Stickney fantasy inaugurated a habit of reading foreign poets in the forest. How did poems in a 'strange language' sound in places where they had never been read before? I avoided English. Transposing Leopardi's 'L'infinito' to a moon-striped plain in the Mallee or reciting Hernandez's *Elegía* where coincidentally shell drifts broke through the surface, I did not risk the solipsism of rocks answering back in my native tongue. Beside the River Murray at sunset, I recited Lamartine – *Au sommet de ces monts couronnés de bois sombres,/ Le crépuscule encor jette un dernier rayon*.[43] Cockatoos circled overhead: as if their raucous calls could be metrical markers, I fitted the lines into the intervals between calls. Other, subtler calls could be counterpointed with syllables. I was listening for shared cadences or overlapped rhythms, as if these musical lines could operate as echoic scores, discovering an absolute rhythm. The kind of sound in-between these exercises explored did not fit into Roman Jakobson's classification: neither sense nor sound defined the meaning of musically inflected, foreign sounding cadences. They were intended as echoic patternings. The poem is a rhythmically organized region of sound. Imagined as a flock of syllables like the cockatoos alighting in the red river gums, it aligns its outline with the environment. Dissolved into the sonic solution of a forest, it might *crystallize* a distinctive echoic ecology, not consciously composed but collectively constitutional of the movement as it is. These solitudes were too subjective to have a binding value: substituting bird calls for fixed stars, rhythmic choreography for the calculus, may simply have deferred facing the fact that the ground not given was political as well as poetic. Yet they provided a provisional, migratory attunement, sensitizing me to a different kind of cultural auscultation.

Later, in the *humming* of the Mallee poet John Shaw Neilson I found a principled exemplar of what I was trying to achieve. Poetry came to him as a kind of response to what could not be heard:

> I have not a very good ear for music and I have no voice. Riding along slowly on a quiet hack I would try to hum the tunes I knew. I would become dissatisfied with these. I would try to hum tunes of my own. This vanity I believe has been a great help to me. In a quarter of an hour or so I would find out that I was quite powerless to compose a tune of my own. Then as a sort of consolation to my wounded pride I would start to make a rhyme.[44]

Practised here, I wrote, was a kind of rhythmic symbolizing:

> Over the *basso ostinato* of the humming Neilson layers the contrapuntal rhythm of the horse's hooves. There emerged from this a rhythmic region whose filaments were pitch relations or scraps of melody strung between the 'points' or beats of the hum. At a certain point a musical skein is woven, a metrically indicated region that starts to attract to it words and phrases. In the first instance these words are selected on a principle of echoic mimicry: they do not represent anything but stand in for melodic phrases whose contours they imitate. Gradually, as the phrases thicken and extend filaments of sense there emerges a significant pattern woven into the hummed ground. After that there may be a good deal of revision to strengthen the meaning, but the essential form, the extended breath pattern, the metrical region, has been fixed.[45]

Revisiting here our project of an Australian sound history, launched in the mid-1980s, and until now largely forgotten, I can see how it shaped this interpretation (some twenty years later) of Neilson.

Neilson was the poetic conscience of Australia's arable and pastoral transformation, an itinerant worker, sometime smallholder, who by night sang back to life the trees he felled by day. Other poetic interpreters bypassed white history altogether. Bringing Sappho's poems to Broome, for example, I imagined a resemblance between the coastal landscapes of Lesbos and the Dampier Archipelago. In comparison with Melbourne, Broome is close to the Mediterranean. It has its own middle sea to the north and, like Mytilene, it faces Asia. I thought of the sponge divers of the Greek Islands, and of their Aboriginal and Japanese brothers here. The half moon hangs over the port; the Pleiades steer through an ocean sky so limpid they can be enumerated. And, I found, they had their place not only across the Creation Myths of the Yawuru people but suspended in the metrical bunting of Sappho's restored poems. A Sea Breeze Dreaming in Marri Ammu language runs, in English, 'Oh, brother Sea Breeze, he is eternally manifesting himself here and now.' The words translated in this way mean that 'he makes himself active', 'he lies' or 'he has done it forever' and 'right here and now'.[46] Sappho embodies the same apprehension in her worship of the goddess of love. Ruskin claimed, 'There is no Venus-worship among the Greeks in the great times',[47] but Sappho lived in great times and many of her 'poems' are nothing more than prayers addressed to high Aphrodite, the Cyprian, the shimmering-throned, the weaver of wiles. When she calls on the daughter of Zeus, to harness swift birds to her chariot to descend, 'fluttering over the dark earth, from heaven through mid-space',[48] her poem purports, like the Sea Breeze Dreaming, to make manifest the eternal in the here and now.

Common to both wisdoms is a conception of poetic Logos, consciousness as an act of attention, a calling to come solicited and heeded. Sappho's

Aphrodite is summoned up, named in the only way she can be named, as a figure of the poet's self-consciousness. She does not predate invocation but is coeval with it. She is immortal in the here and now of language poetically deployed. She is conceived as the archetypal name of language – language as Logos, as 'the source of all law and order and meaning both within the universe and within the human community' and anciently identified with 'the Godhead'.[49] In the living word Logos is incarnated: To be poetically wise is to embrace Logos as the principle of life. To embrace Logos is to embrace consciousness, an act that is unthinkable except in language. That self-awakening brings eternal life, but only on condition that the experience of one's death is also accepted – 'If we take eternity to be not infinite duration but timelessness, then eternal life belongs to those who live in the present.'[50] Hence, Logos is inherently creative, it is the ground of all things coming into being, constantly, without end, and always in the here and now of consciousness. In this case, Logos, as the ground of consciousness, dissolves the old dualism between world and mind, and between self and other. Through Logos all individuals participate in the whole. Likewise, because the Sea Breeze is elsewhere, it is here: grammatically articulated in the phrase 'he makes himself active' is a 'self-manifesting and eternally active nature' that corresponds to Heraclitus's Logos. Unlike the poetic enigmas of the Dark Riddler, though, Maurice Tjakurl Ngulkar of the Daly region is heir to a poetic tradition that without disguise sings the Logos repeatedly into being.

The project of an Australian sound history was a product of migrant self-consciousness. It reflected a determination to find a third way of belonging, neither autochthonous nor ruthlessly colonialist. As a poetic project, it demanded that we go back to the beginnings of utterance and, listening to the voices of the environment, remake language. Harrison's extraordinary poetic oeuvre is one triumph of its meditation. My own experiments took a different path, one that proved equally fateful. A fascination with colonial toponymy, sketch maps for a sound geography and solitary marriages of strange languages to strange places – these might seem to be subjectivist and wilful and, except for the serendipitous discovery of a poetic master in John Shaw Neilson, largely wasted divagations. Yet, as the other aspect of utterance, audition is kinaesthetic, irremediably entangled in other senses of encounter, exchange and passage. If sound is not to be confused with the temporally infinitesimal, its power to spread out and relate must be embraced: the horizon of the present must fuse with the infinite. In other words, as Harrison observed, it is hard to detach the auditory experience from a sense of going somewhere. I think this apprehension arises from a primary inclination or attention, from a desire to be called where the act of making sense is understood creatively and relationally (or echoically). If it is true that 'we produce the lived present, not as a synthesis of temporal points, but as the self-orientation of erotic striving', a condition, Stanley Rosen states, that is only possible 'through "being by or next to"',[51] it is also the case that desire is 'openness to otherness', and this, like the Sea Breeze, means being open to 'change' – 'Becoming both is what it is, and yet is not fully what it is because it exists only in process towards its own realisation.'[52]

An exfoliation of something richer from the parsimony of the present is every migrant's hope: from the shallowest beginnings to go deeper. In an inversion of genealogically derived authority, the migrant insists on extrapolating the greatest dreams from the most minimal of grounds. Substituting echoic mimicry for the sluggish complacency of local parentage, migrants amplify the ambiguities audible in everyday speech, improvising out of them in-between states of becoming, fluid, provisional, ironically inflected. In this way a new territory of names emerges, tied to the places of their first utterance, and the poet-stranger begins to acquire a conscience – an awareness of responsibilities that long predate his arrival and in whose web his own pathways vibrate either responsively or destructively. For instance, in this sense, the stone letters of *Nearamnew*, a public artwork I devised at Federation Square in Melbourne, which surely gave physical expression to Jean-Luc Nancy's claim that the work of the poet in building the *chora* is simply 'a certain way of covering a territory of words', are a distinctively migrant labour: as 'The Migrant's Vision' states, 'Be what you alighted here to be ... warmly touched, these sleeping worms are shapes of time awaiting our awakening.'[53] In this case (although nowhere else in the 'federal visions' composed for 'Nearamnew'), the migrant appears to pull rank, asserting a special relationship with Aboriginal people that the white settler stock, sullied by a colonial past, cannot share – the 'sleeping worms' underneath the surface of Federation Square allude to a local foundation story told by the Woiwurrung people.[54]

But such identifications cannot escape the metaphysical and physical erasure of ancient presences, the elimination of local affiliations and the entrenched aesthetic appropriation of what was ethically offered.[55] And the poets also know this: shortly after coming back to Broome for the second time (in September 2006), getting my regional bearings with a map, I noticed the name Mount Jowlaenga. I had to catch my breath: the name was as familiar to me as if it had been my own country. From Mount Jowlaenga came the coloured sandstone used in *Nearamnew*: The name 'Mount Jowlaenga' is engraved into the surface of Federation Square. There came back to me the story of the Italian migrant, road maker and builder, Dario di Biasi, who discovered the extraordinary rainbow-hued stone (and whose story is also written into the ground).[56] There also came back my bad conscience. The Mount Jowlaenga stone had been quarried without the permission of the traditional owners. While the Kimberley Land Council had apparently advised the architects that Indigenous people with a direct interest in the place could not be identified, I had always felt that the absence of the traditional owners from the opening of Federation Square was a bad omen and that further efforts could, and should, have been made. Now, I thought, was my opportunity to visit the place and begin the process of reparation, and with a local quarryman and a four-wheel drive I made the trek to Mount Jowlaenga: along the Euclideanized escarpment I found the quadrated negatives of the blocks that we had cut into tiles and cobbles, but not the local community that might have made the stones speak.

It appears that *'jowlaenga'* is a version of a Gija word, *'juwurliny'*, *'juwurlinyji'* or *'juwurlinybe'*, meaning 'boulder, big round stone'.[57] The community nearest to the site could say nothing about this. But, I realized, the poet could; for Sappho had foretold my situation. In her fragmentary way – Delphic in the context of wanting to know what to do – Sappho had spoken on behalf of the stone. There is the description 'shot with innumerable hues'.[58] There is the artist's prayer to a stone composed of ancient sea shells, and addressed as a Tenth Muse: 'Come, O divine shell, yield thy resonances to me.'[59] And when the stones speak *(saxa loquuntur)*, enjoining him to speak, there is his exclamation: 'And this I feel myself.'[60] But more telling is the fact that his own engraving, while eloquent, forgets the circumstances in which those resonances were born, and the theft that followed: 'Thou forgettest me,'[61] a rebuke illustrated by the inaccurate information about the location of the mountain which I printed in my commentary on *Nearamnew*[62] in *Mythform*. Finally, made sceptical by the artist's faithlessness, there comes the mountain's belated admonition: 'Stir not the pebbles.'[63] These utterances were, for me, touchstones: evoking universal correspondences at a particular place, they instantiated the Logos ethically. Bringing care to consciousness, they mapped a distinctive region of responsibility. If Logos is 'the hidden harmony', the unity or order in chaos,[64] as well as the 'substratum', 'the unshakeable ground of all our knowing, the "fixed stars" about which the human enterprise and all of consciousness revolve and are given meaning and point',[65] then to know a name is to know its place. This place is nothing other than the primary moment of audition when, in the collapse of visualist coordinates, there is realized, like a sea breeze, the movement as it is. The generalization of this way of entering, passing through and inhabiting the country is *poetic geography*, and its 'fixed stars' are composed of a multitude of voices, human and non-human.

I mentioned before the almost superstitious hope I placed in our son's initiation into language: in this place, at this time, a voice was being drawn into the mystery of utterance, a self-listening was happening that both echoed what was heard all around and improvised its own auditory cues. I fantasized in this conjunction of origin and generation a new communication with the voices of the place. The relationship between vocalization and the development of concepts is informally followed in the notebooks I kept through the mid-1980s, but, glancing back at these notes, their salient feature is my fascination with what appeared to me his free improvisations. First through gesture, then through the symbolic medium of sound, Edmund twirled about himself an atmosphere that lacked external object or communicational destination. Reading at this time the newly revised translation of *Thought and Language*, I found that Lev Vygotsky had anticipated this observation. He referred to the child's 'gesture-in-itself', which only becomes a 'meaningful communicative act' when others, usually the mother, 'interpret the child's grasping movement as an indicatory gesture'.[66] I could also immediately see how the sense Edmund gave to word sounds (imitated words or word sound-alikes) conformed to Vygotsky's description of inner speech. At this stage, a word rarely functioned as a generalized sign of socialized discourse: 'The sum of all the psychological events aroused in a person's consciousness by the word,' it was 'dynamic, complex, fluid' and had 'several zones of unequal stability', changing its sense in different contexts.[67] Importantly, Vygotsky had considered that the interface between inner dialogue and external communication provided 'a psychological mechanism for creating new symbols and word senses capable of eventually being incorporated into the cultural stock'.[68]

To illustrate how these reflections informed a radiophonic and sound installation practice, let me take the word sound 'Yarra'. On 6 July 1989 I recorded (under the heading 'Mirror States') that for two or three days Edmund had been calling out 'á-rrr-a, á-rrr-a', and I wondered whether he was imitating an external sound or simply celebrating a new-found power of articulation. On the same notebook page, under the heading 'Susannah Chorus', I then wrote the following variations: 'árra-ara; are-harry; goanna-anya; aria-gaia; galla-dutigulla'. Finally, under these variations there is a note on Newton Garver's preface to the English translation of Derrida's *Voice and Phenomena*. The logical status of 'private language' is under discussion, in particular, Wittgenstein's critique of 'some conception of private understanding or inner speech such that it is possible for linguistic expressions to have meaning for us in "private mental life," quite independent of any reference to private objects or external circumstances'. Typically, I resist this private–public divide and suggest that the question of reference appears differently to the migrant, 'where speech implies presence – a code of communicating presence – *in the absence* of a shared sign system'. It was, after all, a strange feature of those language discussions that they never considered the performative, kinaesthetic character of multilingual environments.

These three entries on the same page illustrate the involuted nature of any creative process. They are not causally tied together but through their contiguity a mere phonic or conceptual coincidence occurs that begins to resonate. Edmund's 'á-rrr-a' strangely echoed the name of our local river, the Yarra, across whose liquid expanse 'Mirror States' (mentioned in an earlier chapter) intended to stir up an acoustic whirlwind in the spirit of the Second Creation described in Woiwurrung myth. Yarra was not really a name, a generalized sign for a distinctive geographical feature. Applied to flowing water widely throughout Victoria, it also meant hair ('they say the White Woman of Gippsland was nothing more than my palomino's long tail combed out when we waded McAllister's Creek').[69] Yarra was a strange poetic attractor, allowing near sounding words from different cultures to mingle with environmental sounds and noises, producing a new emotional entrainment. I recalled the Aboriginal woman camped on the south bank of the Yarra who borrowed Susannah, the name of the Aboriginal Protector's wife, for her own daughter. Then, in Parachilna Gorge in the Flinders Ranges, on the way back from Lake Eyre, we staged our own 'timely utterance', as Edmund's lustily chanted 'y-arr –a' echoed through the mountains. Prefixing the palatal 'y' to the vocal gesture in itself of the trilled 'r', he produced an indicative gesture. Coming up from the sonorous 'world of original words', where the gorges are, says Bachelard, speaking throats, he stood on the threshold of creating a new word sense. Only the poet who 'exalts' in passing from human vocabulary to the vocabulary of things was in our case a small boy. The beneficiary of this, where 'human and cosmic tonalities reinforce each other',[70] was his father: the Parachilna recordings were a leitmotif in 'Mirror States'.

Reflecting on these experimental sound histories, I find myself in agreement with P. Christopher Smith, when he insists on 'an earlier, pre-metaphysical dimension of speech that remains inaccessible to Derrida's deconstruction and Heidegger's *Destruktion*': 'The origin and starting place here is the lyrical-poetic event of *onomatopoiein*, naming according to the sound of the thing that we hear.' When the poet taps into 'the genesis of vocal naming in the primary reverberations of acoustical, musical experience',[71] she realizes the power of speech to name what cannot be put into 'voiceless' text: 'My problem (in writing verse, and my reader's problem in understanding it)', writes Marina Tsvetayeva, 'consists in the impossibility of the task: for example to express the sigh a-a-a- with word (that is meaning). With words/meanings to say the sound such that all that remains in the ear is a-a-a.'[72] It is one of the oddities of *sighing* (in English) that it accommodates the pre-rhotic heaving groan of the cancerous breast as well as a variety of near-fricative sibilances emitted across the spectrum of waking and sleeping. The different positions of lips and tongue cause the air from the vocal tract to emerge as an 'er', 'ooo' or 'agh' or as an 'sss': either way, one hears the breathly origins of the voice. Tracking this oscillation of sigh types in Margaret's last weeks, I had a practical demonstration of Logos still embedded in *pathos*, 'speech still embedded in a "feel" or affective tone'.[73] In Faringdon, in my mother's house, 'the temporal rising and falling tone of *phonê* and *melos* that we hear, feel and undergo physically', blended with the noises, the refrigerator's shiverings, the strangled columns of water in the tap, the tortured hinge, and the 'anarchic, unfathomable, bottomless' rumour of the street.[74]

Between the sound portals of 'aagh' and 'sss' the voice withdrew into the underworld of death. But under different circumstances, the same 'lyrical-poetic event of *onomatopoiein*, naming according to the sound of the thing that we hear', brings about an Orphic return and the birth of a new word sense. This was the case in 'Underworlds of Jean du Chas' where the name of the eponymous hero was treated as the outcome of primary reverberations:

> HOLLOW is asleep, his mouth is open. He breathes with some difficulty; his outbreath is a kind of cooing or whooing accompanied by a salivatory crepitation. His in-breath is a sharp hissing, also accompanied by a salivatory crepitation as well as a sibillance. HOLLOW begins to dream. The sound of his breathing slowly alters as his vocal chords begin to work. The *ooo* acquires as its prefix *d*, modulating to *do-ooo*, *doo*, eventually *du*. The hiss of the in-breath acquires a shrill *shhh* outline, and resolves itself into *shhh-aa*, eventually *sha*. During this transition the rhythm of his breathing becomes more pronounced, the 'in' and 'out' more definitely differentiated. HOLLOW repeats aloud, as deliberately as a coxswain calling *in-out: du-sha! du-sha*.[75]

As in Faringdon, so in 'Underworlds of Jean du Chas', external sighs counterpoint the internal ones. Walt Whitman described how the ocean breaking on the sand 'with slow-measured sweep, with rustle and hiss and foam, and many a thump as of low bass drums' evoked the echo of 'half-caught voices' and inspired his own 'garrulous talk'.[76] Whitman's erotic vigour, oceanic, orbic, is somewhat over the top in anything connected with Beckett: its parodic counterpart in 'Underworlds of Jean du Chas' is, indeed, Jean himself who, as a super-virile bathing attendant, copies the powerful men who in mid-Victorian England took up female bathers in their arms ('awaiting penetration into the liquid element, the feeling of suffocation, and the little cries that accompanied it all so obviously suggested copulation that Dr Le Coeur was afraid the similarity would render bathing indecent'[77]) and therapeutically dunks Nausicaa's daughters:

> Jean: Cool, turbulent! Under you go!
> Sharon (*spluttering*): Ooh! Ah!
> Jean (*immersing her again*): Just as the wave breaks. Down! Up!
> (*Many Oohs and Aahs from the girls.*)[78]

Still, transposed to the state of *acedia* appropriate to this Beckettian riff, the same principle holds. Reclining in deckchairs by the river Styx, Hollow and Didi 'alternately doze and wake up. The sound of the waves breaking at their feet is soothing. It sometimes mingles with their own noisy respiration.

Sometimes siren-voices seem to issue from it, fleeting giggles and cries, Nausicaa's girls playing ball somewhere nearby.'[79]

Environmental respiration, breathing and dreaming, sighing and waking, the birth of vocalization, the articulation of the movement as it is – this is the sound history of sound histories like 'Mirror States', which, in its opening passage, compresses poetry's entire genealogy:

(SOUND ENVIRONMENT 1: SEASHORE)

Voice 1: Susannah, Susannah, Susannah!
Voice 2: However musical …
Voice 1: You: Anya. Yes, you!
Voice 3: Yana, Yana.
Voice 1: Not Yar-Yar: Anna. Your name, my little one. Yeah! Yeah!
Voice 2: Being born beside the river yonder …
Voice 3: Run softly …
Voice 2: They gave her …
Voice 3: Yarra.
Voice 2: Its name – it is not long.
Voice 3: Or was it …
Voice 1: Your name, YOU.
Voice 3: They gave the water?
Voice 1: My daughter.
Voice 2: Water: run softly till I end my song.[80]

Alvin Curran has remarked that, as a composer who arranges found sounds, 'it doesn't matter if I use a piano or ship-horns or fireworks or recorded environmental sounds. They're instruments. The music is what you feel.'[81] A sound history operates on the opposite principle, that there is a generative rhythm, or compositional potential, within a local acoustic region. It is not a lost or compromised 'tuning' but more a product of creative attunement. So, in gathering sound materials for 'Mirror States', we recorded the 'seashore' at the mouth of the Yarra, as well as building a sound inventory of the site where the work was to be installed, listening throughout these recordings for acoustic forms or *algo-rhythms* that could be imagined, Chladni-like, coalescing into the three-dimensional towers of a city.

This led to the claim that

> a material history of the city's soundscape would show that, underlying the different dialects of the jackhammer, the telex machine, the jolt of couplings, the putter of the outboard, the police siren, and the shuddering exhaust stack of the semi-trailer changing down gear, there is a common grammar or syntax, a simple percussive phrase, a recurring digital back-and-forth that the pronunciation of the Yarra's rolled 'r' physically mimics.

Such a history would also show that this moat of sound screaming north and south across the modern river, overlays yet another babble – the east-west undercurrent of water lapping babyishly around pylons and platforms and, nearer the river's source further inland, gurgling over the sunken remains of pontoons, wooden bridges and stoved-in punts.[82]

This is crudely expressed, authentic evidence of my own sound history as I tried to make sense of phenomena that were new to me. I would not now compare the effect I was trying to describe to a 'grammar or syntax' except, perhaps, in the *antithetical* sense that Vygotsky gives these terms when he writes, 'The grammar of thought is not the same in [oral speech and written language]. One might even say that the syntax of inner speech is the exact opposite of the syntax of written speech.'[83] The 'inner speech' of the city may be a fascinating concept but it will not be detected, let alone recuperated for art or pleasure, using an approach that presupposes a 'deliberate semantics – deliberate structuring of the web of meanings'.[84]

The last line quoted above adapts a famous refrain from Edmund Spenser's virtuosic *Prothalamion*. Certainly it is a quotation, and quite possibly its selection here opens it, as Umberto Eco might say, towards new interpretative possibilities. In other respects, though, it is *not* a quotation but an unusually explicit exposure of the prosodic bedrock of much of the radiophonic writing from that period. English poetry has been overwhelmingly written in an elastic four or five stress, ten or eleven syllable line whose resilience stems from its easy assimilation of ordinary speech patterns. The older alliterative line with its heavy caesura has an echoic structure, concentrating into one line the variation and return found in the *stornello*. The rhythmic flexibility of the more familiar line of Spenser, Shakespeare or Keats allows an entrainment of breath, sound and concept whose emotional possibilities appear infinite. Any serious student of English literature must internalize its archetypal rhythmic structure; it becomes the rocking cradle of speech that seeks to stay close to the sigh; its metrical closure provides a convincing poetic surrogate for the expression of any emotion that must seek measure. Certainly, it is my metrical fallback position – remove the Voices from the passage above, and the lines could be recast as casually deformed iambic pentameters into whose series the line from the Elizabethan poet fits without strain.

The metrical subjugation of the Bush has not been much studied or, indeed, recognized. Beyond the habitual disparagement of Aboriginal singing, playing and dancing, there has been little attention paid to the failure of auditory sympathy due to the white listener's own rhythmic conventions. The Stickney fantasy was an attempt to expose this: in the difference between the sonic organization of the place and the metrical enchainments of the foreign verse read there, a historical parenthesis was improvised, a spot in time where the colonialist split between the deaf language of administrative authority and the multimodal communication of Aboriginal society (henceforward relocated in the auditory picturesque unless entirely silenced) was suspended. And here, I believe, is a partial answer to the obvious objection to those solitary soundings – that a proper assumption of historical responsibility must begin in acquaintance with the present traditional custodians of the land. For it is, or ought to be, the case that anyone entering another's country uninvited should wait to be invited; and also that, on being beckoned forward, they should come prepared with gifts, and able to say, in authentic tones, where they have come from and to give some reasonable account of the passage. To tell a story in these circumstances, one that creates new symbols and word senses capable of eventually being incorporated into the cultural stock, is desirable; but it will be a product of listening, not speaking.

Where there *is* speaking in the old, foreign way, it should be for a good and instructional reason. 'Sweete Themmes! runne softly, till I end my Song', Edmund Spenser beseeches the river in his liquid-syllabled *Prothalamion*, a poem composed in 1596 to celebrate the twin marriages of the daughters of the Earl of Worcester.[85] What can this possibly have to do with the events on the Yarra evoked in 'Mirror States'? The juxtaposition is ironic. In the commentary on the 'Federal Visions' inscribed into the floor of Federation Square, a short history of Western civilization as 'a history of dessication is given', one of whose manifestations is the draining of swamps and the confinement of rivers to fixed banks. This 'drying principle', illustrated in the engineering of the Yarra ('Poor Yarra! Once many-named, three humours mingling, tightlaced with banks ... straitjacketed, a picnic for the picturesque'[86]), extends to the training of speech, which is subjected to the deaf regulation of writing. But this subjection of places and peoples comes disguised as the promise of new fluency: this is the meaning of Spenser's eloquence – the Thames will run but only softly – only so long as it does not disturb his song. In sonorous disguise, Spenser's poem illustrates a direct relationship between imperial pretension and environmental engineering. The freedom of his poetic narrative to glide decorously past, to survey expanding territories of classical allusion and contemporary fame, depends, metaphorically, on the steady, soft running water. The smoothness of the numbers – the even keel, as it were, of the poem's rhythmic measure – symbolizes dominion over turbulence of every kind – social, political, environmental. In Spenser's rhetorical *tour de force*, language and nature celebrate their own marriage, in their consecration creating the discourse of empire.

This is not mere rhetoric. In the same year that the *Prothalamion* was composed, Spenser wrote his lengthy and informative memorandum, 'A View of the Present State of Ireland', in which he recommended the repression of the Irish language and prohibition of bardic composition.[87] Spenser's objection to the bards sounds like a projection of his own bad conscience. Instead of choosing 'the doings of good men', they set up and glorify 'the most bolde and lawless in his doings, most daungerous and desperate in all partes of disobedience and rebellious disposicon'.[88] They are political poets who defy the colonizer: well might Spenser complain that poetry savouring of 'sweete witt and good invencon', which the poets have 'sprinckled with some prettye flowers of theire own natural devise', should be 'abused, to the gracing of wickedness and vice, which woulde with good usage serve to bewtifie and adorne virtue'.[89] Spenser was keen on weddings, and in his own allegory of virtues *The Faerie Queene* describes the 'spousalls ... Betwixt the Medway and the Thames', to which, among others, all the rivers of Ireland came – 'Why should they not likewise in love agree?' the poet asks, somewhat disingenuously one feels as these lines were written on land seized after the Serwick massacre (which Spenser almost certainly witnessed).[90] In any case the rivers arrive in good order 'To doe their dueful service, as to them befell', and, after them, the inevitable Nereids, whose freshwater sorority, 'A Flocke of Nymphes ... All lovely Daughters of the Flood', is found, in the *Prothalamion*, gathering flowers in a meadow.

Obedient to the authority of Ocean, the Nereids disport themselves decoratively in the margins of Spenser's mythopoesis. From the point of view of a country whose waters have been irrigated, canalized and polluted, their marginalization has a different meaning. It refers to the silencing of the 'half caught voices' of places, spirits and interests that have only been heard when fully caught in, and subordinated to, the administrative grammars of empire. Under patriarchy, the voices half heard or left out will be predominantly female. One of the aims of 'Mirror States' is to find a way in which these subalterned voices can speak, not in the language of the State Department but according to a different echoic logic, one infiltrated with the sound of water flowing. The imitation of informal discourses when vowels and consonants briefly flow through one another, creating temporary tourbillons of sense on the surface of colonial discourse, allows an alternative culture of associations to appear, one where, say, the imaginary Dutigalla people that pastoralist John Batman said had signed over 600,000 acres of grazing land could acquire poetic existence as a word sound in a new discourse of female determination – '"Dorta" or "Dutta": their name is uncertain.'[91] Besides these speculative sound histories, poetic archaeologies of the distaff line conjured up in the constellation of sound-alike syllables, 'Mirror States' also stumbled across evidence that names had been historically fateful: I mentioned the inspiration derived from learning that an Aboriginal woman had named her daughter 'Susannah'. I also knew that William Thomas and his wife, Susannah, had named one of their own daughters Susannah. Only later, however, did I discover that on 18 November 1845, 'In the morning about 10 o'clock [Susannah] was about the house in her usual spirits, but when the family sat down to dinner she was nowhere to be found ... on the return of Mr. Thomas, in the evening, the water hole was dragged, [and] the body of the unfortunate young lady was found.'[92]

Water and water management are fateful: Susannah's drowning is, in my mind, connected to inarticulateness, to the prevalence of the wrong kind of relating to country; as if she had succumbed to the imbalance of dry thinking, the engineering habit of draining swamps and installing in their stead the harsh infrastructure of banks, steps and locks, and their devitalized counterparts, motionless canals and abyssal wells. A sound history that gave voice to the behaviour of uninhibited water that recognized the derivation of the name of the *Muses* from 'the Greek *mois*, that is, water'[93] would be turbulent in character and form. Writing about *Agua viva* by the Brazilian writer Clarice Lispector, Hélène Cixous contends that it 'does not give itself to be read without escaping, drowning, submerging, retreating'.[94] I take it that the navigator who plunges into this writing has a taste for turbulence and knows well how to swim. Then, two fluencies, of vocation and of vocalization, merge into a different humid history of country. Its environmental image is not the compliantly fluent Thames or any similarly engineered antipodean stream but the 'Waters breaking' of the Eighth Story of 'Mirror States', when the second redemptive Coming ('It was as if the river stood on its tail and climbed into the sky'[95]) coincides with the birth of a little girl. However, this opening to the other is swiftly curtailed:

> Voice 2: No, listen!
> Voice 1: A splendid howling effect!
> Voice 4: And she as yesternight says: named after 'your lubra Susannah'.
> Voice 3: From this day the natives are prohibited from entering the town.
> Voice 2: Susannah!
> Voice 1: Susannah!
> Voice 3: Susannah!
> Voice 4: Susannah!
> Voice 1: Yarra, run softly till I end my song.[96]

Water stories ran throughout the 1990s, notably in my collaborations with the Bharatanatyam dancer and choreographer Chandrabhanu, where time and again the association of female movement histories with water was unforced. After the success of 'Light' in 1996, Chandrabhanu and I decided to tell the story of William Light's mother, Martinha Rozells, whose former place of residence outside Georgetown, Penang, was next door to where Chandrabhanu grew up. At the climactic moment of 'Jadi Jadian', the way back to Martinha's house is barred, not by a river but by a bridge that has 'closed' and bound the river, and to open the way it is necessary to overpower 'the destructive character', and free the water to flow again, a task that Chandrabhanu memorably achieved by balancing his way across the abyss on the bridge's handrail.[97] In 'Old Wives' Tales', an arrangement of autobiographical stories, the protean powers of the river are embodied in the Chee Chi Wi, a fairy-tale character from Chandra's childhood, who, when the rajah summons 'that young girl up the river' to come to the palace to be his wife, gets into the *prahu* next to the girl and before the boat can complete its journey ('Over the water, down the river') transforms itself into her and takes her place – a dance for children and all people young at heart, I remember it with affection.[98] At the end of the decade the same preoccupations migrated into public writing. In *Relay*, the public artwork made for the Sydney Olympics, the struggles of female swimmers to overcome male prejudice were commemorated: 'semi-monsters, O swift and blithesome, waterborne, adroit as divers!' 'Hear, O Hera, our blood is racing'.[99]

Name borrowing was a conventional mode of affiliation or incorporation in Aboriginal society; before Aboriginal people were 'prohibited from entering the town', black and white people bearing the same name populated the streets. Institutionalized in this way, an echoic onomastics produced a double, and doubled, social identity, although not necessarily a shared understanding of the obligations this entailed. Local Aboriginal people readily took English names, but the reverse never happened. What one party understood as a conditional courtesy, the other interpreted as the admission and flattery of conquest. Conflict arose when the white settler failed to understand that the involuntary gifting of his name implied voluntary gifts to follow. In post-first wave invasion negotiations, arrangements were no doubt both more pragmatic and more nuanced. Extended coexistence, and domestic intimacy, produced mimetically marked relationships that were subtle, unconventional and enduring. For instance, Isabella, daughter of James Dawson, a Scottish settler in south-west Victoria, lent her name to the daughter of Kaawirn Kuunawarn ('Hissing Swan') (*c.* 1820–89), a shepherd and general worker for James Dawson and Patrick Mitchell on Kangatong station from about 1845, and also an Aboriginal leader, sometimes known as 'King David', chief of the Kirrae wurrung, or 'Davie'.[100] However, the situation was complex: Jan Critchett writes, 'King David had a close relationship with both Patrick Mitchell and James Dawson'. Isabella Dawson, who was widely regarded as King David's daughter, was in fact the daughter of Patrick Mitchell and an Aboriginal woman whose name was recorded as 'Mary', and Critchett speculates that King David and Mitchell 'shared' 'Mary'.[101]

Isabella was the co-author of *Australian Aborigines: The Languages and Customs of Several Tribes of Aborigines in the Western District of Victoria, Australia* (1880), although only her father is recognized on the title page. Isabella had been born at the Kangatong property outside Warrnambool and had, according to her father, an 'intimate acquaintance from infancy with the aboriginal inhabitants of that part of the colony, and with their dialects'.[102] This phrasing is probably exact: *Australian Aborigines* purports to record the vocabulary of three Aboriginal dialects; however, ethnolinguist Barry Blake classifies the Dawsons' Kuurn Kopan Noot or Small Lip and Peek Whurrong or Kelp Lip as dialects of the coastal Warrnambool language (his preferred term for Gunditjmara) and Chaap Wuurong or Broad Lip as Tjapwurrung, a dialect or language spoken further inland.[103] The intimacy was of a different kind. After the violent extirpation and dispossession associated with 'first contact', the survivors had negotiated a limited coexistence. The situation (recorded in photographs) in which Isabella sat down with farm hands and domestic servants from the Gunditjmara and Djabwurrung[104] language groups at their pastoral property can be characterized as a 'second contact' one. In contrast with the formal questionnaires on which Robert Brough Smyth based his official account of Victorian languages in *The Aborigines of Victoria* (1876), Isabella's information-getting sessions were occasional, leisurely and stretched out, perhaps, over ten years. Above all, they were primarily records of women listening and speaking.

These circumstances determined the language of communication – and the language that was communicated. Dawson indicates that lists of English words were distributed amongst the participants at the word-getting sessions. The evidence of the wordlists is that Isabella explained the meaning of unfamiliar terms, sometimes by reference to Indigenous terms for related concepts, sometimes by paraphrase or deictically or pantomimically. The resulting lexicon is far from a cold ethnolinguistic exercise. It is a genuine act of 'thirding', recording a new cross-cultural discourse improvised affectionately and consciously maintained and stabilized. My impression is that the Small and Kelp Lip word lists preserve idiolects (people speaking one language differently), and not ignorantly differentiated dialects. It shows a grateful human respect for the 'living multitude of voices', and an unwillingness to dumb down this discursive identity in the pseudo-scientific pursuit of inventing a self-referential system of generalized concepts – a surmise supported by Dawson's sincere acknowledgement of the patience and intelligence Yaruun Parpur Tarneen and Wombeet Tuulawarn, in particular, had brought to the project.[105] Above all, the vocabularies and grammars of *Australian Aborigines* bear witness to a speech not simply unrecorded in a literature overwhelmingly written by men, but (strictly speaking) unrecordable. Vygotsky speaks of 'peculiarities of grammar and syntax characteristic of inner speech',[106] but similar 'peculiarities' arise where women, operating inside patriarchal conventions of language authority, attempt to preserve the flow of speech, the echoic humour of mutual recognition and the irony of inevitable failure, errancy and opacity.

One curious expression of this feminized sound history was Dawson's unusual decision to name his second place of residence after the local word for lip, tongue or language. After leaving the Kangatong property and spending a period in Melbourne, Dawson returned to the Western District around 1865, building a house outside Camperdown, and calling it *Wuurong*, the word in the Gunditjmara language for 'lip'.[107] Using Aboriginal names for white properties was not unusual: *wuurn*, the Djabwurrung word for 'a habitation', for example, was occasionally adopted, as were terms like '*laar*' and '*mia mia*'. Why, then, did Dawson depart from this culture of toponymic kitsch? Although *we* may think that *wuurn* and *wuurong* look suspiciously similar, their forms and meanings are clearly distinguished throughout *Australian Aborigines*. The Dawsons took care to preserve different pronunciations of similarly sounding words, and their sensitivity to conversational context and the potential polysemous application of terms is inconsistent with an elementary confusion of this kind.[108] Still, 'Lip' is not an intuitively obvious choice of name for a house, and one wonders what motivated its selection.

Analogies between landscape features and body parts are habitual in Aboriginal toponymy. In the Dawson wordlists we find that the 'Aboriginal Name' for the 'northern peak of Mount Leura' is 'Lehurra', which means 'Nose'; another 'Mount' is called 'Puutch beem', translated as 'High head'.[109] Similarly, *wuurong* – the word found in Peek Whurrung (translated by the Dawsons as 'kelp lip') and in Chaap Wuurong or 'broadlip' – is also found in such place names as 'Wuurong killing' ('Lip of waterhole', with reference to a 'particular spring where a bunyip lives')[110] and Bukkar whuurong ('Middle lip', referring to a 'Bank between Lakes Bullen Merri and Gnotuk. A gap in this dividing bank is said to have been made by a bunyip, which lived at one time in Lake Bullen Merri, but, on leaving it, ploughed its way over the bank into Lake Gnotuk, and thence at Gnotuk Junction to Taylor's River, forming a channel across the country').[111] Now Wuurong was almost certainly a language house – if the photographs we have of the language sessions were taken at the *Wuurong* homestead, then perhaps a loosening of lips was being commemorated, the flow of Aboriginal speech, a moistening of country occurring when locked-up affections were released and allowed to percolate through dried-up pastures, and relate what the enclosure acts of the settlers had attempted to dam up and stop.

It is possible that *wuurong* has a more active, eventful sense. Further to the south-west lie the eel-harvesting complexes whose ingenuity and sophistication attracted admiration from the earliest days of white colonization (and whose engineering achievement is likely to be recognized by UNESCO Listing). The revisionist archaeologist Harry Lourandos demonstrated that the eel economy underpinned a dynamic regional trading culture characterized by periodic human flows into and out of the district. Basing his work partly on the evidence of complex economic patterns derivable from the excavation of particular sites, and partly on Dawson's descriptions of the 'great meetings', Lourandos argued that the focus of the eel-trap design which produced the abundance of produce underpinning the intensification of sociability witnessed at the great meetings was the *lip*, the artificially constructed overflow where outgoing eels could most easily be trapped. *Wuurong* referred not simply to a body part and its anthropomorphic analogue in the landscape but to the climactic event in the regional economy, the production of the excess that transformed a subsistence culture into a trading one. Groups of people were from this perspective, Harry Lourandos argued, not separate and fixed in their habits. They were defined relationally and were co-dependent; and the spatiality of their life worlds should be reconceptualized on the model of the great meetings 'in terms of "relationality" and "fluidity"'.[112] Lip, in this context, refers to the potential of the country to produce. It is wealth defined as overflowing.

Yet there seems to be more to it than this. As the lip of a waterhole, *wuurong* is associated with that environmental incubus of the colonial imagination: the Bunyip. What constellation of poetic associations can explain this? In *Australian Aborigines*, James Dawson mentions that

> there is a superstition, called *Wuurong*, connected with the tracking and killing of kangaroos. In hot weather a doctor, or other person possessed of supernatural powers, looks for the footprints of a large kangaroo. On finding them he follows them up, putting hot embers on them, and continues the quest for two days, or until he tracks it to a water-hole, where he spears it. He then presents portions of the body to his nearest neighbours, and takes the head home to his own *wuurn*. There seems to be no special meaning attached to this custom.[113]

What is the meaning of this story? Perhaps it makes more sense when juxtaposed with the information that

> the coal sack of the ancient mariners – that dark space in the milky way near the constellation of the Southern Cross – is called 'torong', a fabulous animal, said to live in waterholes and lakes, known by the name of bunyip, and so like a horse that the natives on first seeing a horse took it for a bunyip, and would not venture near it.

And Dawson adds, 'By some tribes the coal sack is supposed to be a waterhole; and celestial aborigines, represented by the large stars around it, are said to have come from the south end of the milky way, and to have chased the smaller stars into it, where they are now engaged in spearing them.'[114] Hence, Isabella's Vocabulary gives 'Torong' as the Kuurn kopan noot word for 'Coal-sack in Milky Way' and 'Bun'yipp' as the equivalent name in in Chaap wuurong.[115]

In this astronomical myth, the Coal Sack is *both* a water hole *and* the spirit that inhabits it. Something pours over the lip there but may also be caught there: where the kangaroo hopes to put the hunter off his scent, he is speared: passage and arrest form two aspects of a single movement form. Yet the primary manifestation of this form is different. The same Kuurn kopan noot word said to mean 'Coal-sack' is elsewhere translated as 'bark' in the sense of 'ship'.[116] In Peek whuurong, which, as I said, is usually taken to be a dialectal variation of Kuurn kopan noot, 'Torong' is translated as 'steamboat'. In both these Gunditjmara wordlists, 'Torong' also means 'trough for holding water'. As the spirit of the waterhole, the Bunyip is responsible for carrying water over from one place to another. We are talking about a movement form defined in terms of differential flows: there is a secret, unarticulated or unconscious impulse (it could be the undetected leaps of the kangaroo or the emergent seepage of water yet to form a definite stream); then there is the lip, the bank or the edge that brings the unconscious movement to consciousness – makes it articulate – and causes it to pour into a receptacle (the water hole). However, the water hole is not a closed body – it resembles the *fuente* of the Muses: creating a gap, or groove, or channel, in the surface, it provides a conduit where the bubbling water audibly flows. The term 'anastomosis' refers to 'intercommunication between two vessels, channels, or distinct branches of any kind, by a connecting cross-branch',[117] but Hillis Miller notes that it is also the name of a figure of speech, referring to the 'insertion of a qualifying word between two parts of another word'.[118] His point is that the word seems to mean different things depending on which part of the intercommunication is taken to be primary. Is the intercommunication between two closed vessels (two lakes joined by a stream) or between two open vessels (two channels joined by a branch channel)? Or, to put it another way, is the anastomosis 'an external link between two vessels or channels' or does it enter into the vessel it opens 'so that it becomes a version of the figure of container and thing contained'?[119] So with the Bunyip, is it a vaguely dreadful spirit of place or something altogether more subversive, the spirit of a communication that resists invasion, conversion and enclosure?

In short, in a history of Australian sound, the Bunyip signifies the origin of natural utterance. The water sounding of the landscape is due to it. The 'loud gurgling noise' of the Bunyip[120] vocalizes the creative energy of the environment. It is the Muse that makes flow fluent: 'Its groanings and bellowings were heard at certain times by all the people of a tribe when they encamped near a lagoon, or by deep water holes, or by the seashore.'[121] It is, indeed, the amplification of the lip. The Ngarrindjeri people, near the mouth of the Murray, describe a 'booming sound'; the Reverend George Taplin confirms that 'it resembles the boom of a distant cannon, or the explosion of a blast', adding, 'sometimes, however, it is more like the sound made by the fall of a huge body into deep water'.[122] In Victoria the Bunyip's significance is not recognized: among the Nyungar of south-west Western Australia, it is revered as Wagyl, the creator of rivers, of rain and rainbows. According to the Ionian philosophers, the coming into being of the Cosmos is a threefold process: firstly, the state of 'indistinction', defined as *pre-Chaotic* – corresponding to state of the country before it achieves a distinct topography of flows and catchments; secondly, the coming into being of Chaos, described as the 'yawning gap' between heaven and earth that makes space for the elements to separate and find their proper places – and whose auditory signature is noise, the 'loud gurgling noise' of the Bunyip, say[123]; thirdly, 'when the gap has come into being, between the sundered opposites appears the figure of Eros, a transparent personification of the mutual attraction which is to reunite them'[124] – which, in our analogy, corresponds to the birth of poetic language, where a word like *wuurong* attracts and constellates different concepts into a single overarching movement form.

'The past life of émigrés is, as we know, annulled,' wrote Theodor Adorno from the United States. To avoid this fate it is necessary to revisit that past through the mirror of migration; for what one will find in the new country is, to an extent, the fulfilment of a frustrated quest in the old country. 'The Native Informant', for example, evokes a pre-Enclosure Vale of the White Horse where dialect is a faithful index of spirit of place. But the approach to that ancient region of separating noises is via the archaeology of sound recording engineered from Australia. Going back in this way winds the past onto the future. The geography of colonization is one of territorial separation but the topology of a migrant's life might stage a reunification of distant places under the reuniting aegis of Eros. Colonial allusions to the horse-like Bunyip may be few and fragmentary, but they exceed in richness the stories about the Uffington White Horse. Brough Smyth himself speculates that the Bunyip ('voracious in its appetite for human beings') should be compared with the Dragon slain by St George,[125] whose story is associated in Berkshire, with the White Horse. The status of the White Horse in this triangular enquiry is ambiguous: Does it represent the Dragon or St George's white charger? Or something else, a kind of English Bunyip perhaps? Is St George in this scenario the silencer, the one who drove out the sounds of nature (according to Bachelard, among the sources of poetic reverie – 'One dreams in front of a spring and the imagination discovers that water is the blood of the earth'[126]), insisting instead on the arid Jakobsonian divorce between meaningless environmental sounds and meaningful speech sounds?

To establish the dialectical significance of this migrant preoccupation with Australian sounds – the fashion in which it might elucidate the repressed soundscapes of growing up – James Dawson's story of the White Lady is a good starting place, for it seems that she is not only a sister of the Irish *Cailleach* but of the holy lady of Gog, Lady Godiva, who, in the guise of the 'old lady' in the nursery rhyme, 'rode on a white horse to Banbury Cross'.[127] As an observer of Aboriginal practices and beliefs, Dawson was alive to parallels with certain beliefs and customs still maintained in the old world. Thus, he noted in passing, 'It is a remarkable coincidence with the superstition of the lower orders in Europe, that the aborigines believe every adult has a wraith, or likeness of himself'[128] and, more particularly in relation to his own place of origin, 'For pains in the joints, fresh skins of eels are wrapped round the place, flesh side inwards. The same cure is very common in Scotland for a sprained wrist.'[129] In terms of parallels between Aboriginal and Scottish/Irish cultures of place-making, though, the most suggestive passage in *Australian Aborigines* concerns a certain 'clever old witch': 'The aborigines had among them sorcerers and doctors, whom they believed to possess supernatural powers. In the Kolor tribe there was a sorceress well known in the Western district under the name of White Lady, who was the widow of the chief, and whose supernatural influence was much dreaded by all.'[130]

Dawson had difficulty in understanding her social role. On the one hand, 'this cunning woman possessed such power over the minds of her tribe that anything she fancied was at once given to her.'[131] On the other, she belonged to an odd social class:

> Witches always appear in the form of an old woman and are called kuin'gnat yambateetch, meaning 'solitary', or 'wandering by themselves'. No one knows where they come from or where they go to; and they are seldom seen unless at great meetings. ... They belong to no tribe and have no friends; and, as everyone runs away on their approach, they neither speak to anyone nor are spoken to.[132]

However, this anomalous social status soon makes sense when compared with the Irish *Cailleach Bhearra*, 'whose powers and activities have resulted in the shapes of hills, the courses of rivers, the location of islands and the presence in the landscape of numerous natural features'.[133] In relation to later Celtic patriarchal myths – and the yet-later Protestant usurpation and enclosure of Irish land – the cailleach constitutes 'an overarching female matrix of sovereignty and fertile power that is as vast and as untameable as the wild, wide landscape, and that is yet as nurturing and as intimately fruitful for human beings and for human existence as are the services of the [midwife, the wise woman and the keening woman]'.[134] And why should she be keening? Not simply for a human life but for the ebbing vitality of the land: the Old Woman of Beare boasts, 'My flood has guarded well that which was deposited with me,' but her own 'flood' ebbs: 'Today there I scarcely an abode I would recognise; what was in flood is all ebbing.'[135] Dawson's White Lady stands in relation to 'the *logos* of the Dreaming'[136] – bound up in hundreds of tales that represent 'a cosmogony, an account of the begetting of the universe, a study about creation'[137] – as the *bean feasa* or 'wise woman' does to the overarching cosmogonic *Cailleach*. She is a shaman or, as Dawson put it, 'a sorceress': on one occasion, 'having left the camp for a while on a moonlight night, she pretended, on her return, that she had been to the moon'.[138] Nothing is known about the spells brought back but I guess they concerned the blood of the earth.[139]

How the White Lady earned her epithet is unknown: the connection of the Mistress of the Animals, the *Cailleach* or Veiled One, with white is better attested. As a Celtic goddess, she could shape-shift or change into animal form: Epona, the Horse goddess who 'in her humanlike form gave sovereignty of the land to the chiefs, was either pictured riding on a white horse or as a horse with foals'. The *Cailleach* could take less elegant, more Bunyip-like forms: 'In Scotland, this powerful sorceress, also called Scotia, is depicted with the teeth of a wild bear and a boar's tusks.'[140] In any case, I like to entertain the idea that the Uffington White Horse transcends, or predates, the symbolic breaking-in of the wild represented by the stories of St George or the Invisible Blacksmith. Carved at the lip of the escarpment where the freshets rise, it is another *wuurong*: 'Half-way down the hill ... the coomb narrows down to a considerable gorge, where fifty or a hundred springs burst out.'[141] And 'lower down, where the streamlet is stronger and has worn a groove – now rushing over a floor of tiny flints, now partly buoyed up and chafing against a smooth round lump of rubble – there is a pleasant murmur audible at a short distance'.[142] Graves, as noted, thinks that the spring on Helicon named Hippocrene was originally struck not by Pegasus but by 'the moon-shaped hoof of Leucippe ('White mare'), the Mare-headed Mother herself [Demeter]'.[143] Ignorance preserved me from casting White Horse Hill as Mount Helicon but this network of associations suggests that going to drink at those freshets would have missed the point. It is not the Pegasus of vaulting imagination that is imbibed there but another outcrop of the ancient lunar or scintillant logic that preceded the triumph of the day. I do not want to be a Blue Rider, if blue is the 'typical heavenly' – and therefore spiritual – colour.[144] I do not want to be a white knight unless it is to ride a horse drawn as an archipelago of parts, a figure that expresses the movement form of *dissipatio*, not a fatal fragmentation but an assembly of like-minded portions of the Orphic body, singing together although severed, star-like.

However, in my view, this misrepresents the White Horse, which, in my life at least, is not representative of a poetic ideology – a resumption of matrilinear, lunar ascendancy, say – but corresponds to a rhythmic ground. The White Horse is less a totem than a hieroglyph, figuring forth an archetypal choreography. It scores a movement form composed of syncopated glissandos and could be imagined as the *algorithm* of the hill when the hill is conceived (as it is in Gilbert White's fantasy) as the geological instantiation of a breath pattern preserving an 'air of vegetative dilatation and expansion'.[145] In an essay published some years ago, I speculated that after all the White Horse *had* been brought over to Australia: as a ground design composed of gaps, it had supplied a different way of inhabiting a new country; as a dance-floor modelling a rhythmic geography, I carried it about with me, a potential arrangement that cast its encouraging shadow wherever I landed.[146] 'Backwards and forwards they must have leaped,' as De Quincey put it, with reference to the way consciousness involutes sensations to form vortical meaning centres, emotional structurings that discover genuine analogies between external reality and its internal ordering – much as the spiralled shell generates consistent forms out of hollowed space or the human ear, apparently passive, sifts what it receives echoically, infusing sense into sensation. These gap-leaping orderings of experience seemed to me archetypally migrant, evoking a distinctive psychic geography of active re-membering.[147]

I described an experiment in which I lay the outline of the White Horse over different ground patterns I had designed in Melbourne and Sydney. I found a shared vocabulary of gestures, arabesques and involutes that corresponded to a constellated movement form, a spatial arrangement of lines and curves that delineated a gathering arrangement – as if the dispersal associated with Chaos was being reversed under the aegis of Eros and the parts flung apart were being drawn together again to create a federation of forms, an archipelago of gaps. I associated the apprehension with 'the eido-kinetic intuition', our instinctive capacity to measure the distance of things and calculate pathways between them.[148] I speculated that in this abyssal view of the migrant's fate, the recovery of an underlying eido-kinetic history is obviously therapeutic. Giving measure to the gap, both geographically and psychically, it suggests what of the self might be carried over from one place to another. The concept of identity here is structural. It is not that an image of the White Horse was imprinted and reproduced in a new country. There was no image, only an arabesque, an eido-kinetic reduction of all the paths taken as a child whose play was unconsciously adjusted to the ground rules proposed by the figure of the White Horse; and this in-between place – not an object, not even strung between objects, but a gap always wherever I was working – was a place belonging to every place. In this sense, the movement history recovered is not marginal to the clearing of everyday life but is dotted like stepping stones throughout it, so that over and again, we make and remake its intuition of a meeting place whenever we articulate this primary relation.

And here the movement form enters a sound history; for what is rarely mentioned in descriptions of soundscapes is their beat, usually complex, contrapuntal or confusedly syncopated, which provides the diffused noisy ground and environment of communication. On one of the 1903 wax cylinder recordings, Horace Watson (the sound recordist) is heard saying, 'This record by Fanny Smith, daughter of Tangnarootoora, presents the song of the natives, when holding their corroboree, who endeavour to sing with all their might, accompanied by the beating of sticks and skins. This is a dance song by Fanny Smith.'[149] Watson's information is supported by a mid-Victorian report of women accompanying themselves 'by laying upon greasy kangaroo rugs which were rolled up in some peculiar manner so that when struck by the open hand the sound resembled that of a muffled drum', or, alternatively, 'by beating time with two short dry sticks'. But Moyle comments, 'References to the musical use of the sound-producer, i.e. to rhythm, tempo, initial and terminating percussive effects etc., are omitted from Tasmanian reports.'[150] Keeping the beat is the primary way of joining in; and finding the beat, as Neilson does when the percussive striking of hoof on stone blends into a hum, or as the listener feels when 'With idea, sound, gesture', *el duende* baptizes 'all who gaze at it with dark water',[151] is the basis of any echoic alignment with the movement as it is.

SIX

Recordings

In 'Memory as Desire' the characters stage the passing of time: polite conversation papers over the awkwardness of being cast together in a strange place, at a strange moment. Topics arise and trigger personal associations: someone overhearing them would suppose Captain Sturt and his wife and Aboriginal Protector Robinson and *his* wife were talking to each other – there is a kind of auditory wallpaper effect. But individually each is caught in the spell of personal memories and the intersection of one free associative reverie with another is almost coincidental. Except for the theme of communication itself – the languages of Australian Aborigines, the onomatopoeic suggestiveness of Australian birdcalls – which surrounds the entire encounter like the fear of an embarrassing silence, the conversational triggers lie buried in reminiscences that, it seems, are only intermittently noticed, and then largely misunderstood. At the beginning of Scene Twenty-Eight, Rose Robinson suddenly announces, 'My father was a painter,' prompting Charlotte Sturt to observe, 'How picturesque.' Rose's remark seems to have been triggered by the explorer's earlier explanation of his parrot-based theory of the inland sea – 'Now, if we were to draw a line from Fort Bourke to the west-north-west.' The word 'draw' jolts Rose out of her reverie; she does not catch the rest of the explanation. Charlotte's reply is topically at the crossroads of two lines of thought. A few scenes earlier, Sturt and Robinson had compared career notes; then the topic had lapsed. In the echo of that, Rose intends to assert her own genealogy. Perhaps Charlotte Sturt receives this assertion of independence defensively; or else she finds Rose's bohemian connections somewhat *infra dig*. Secreted in the obvious pun *picturesque* are such overtones.

Also indicated is a listening that screens out the silence. The emergence of the auditory picturesque in England is associated with the Enclosure

Acts: the congregation of song birds composing the 'dawn chorus' depends on stitch-working the commons with hedges. The romance of the middle distance is middle class: it depends on excluding other, human voices. As the aesthetic mask of colonization, the musicalized sound environment gets round the responsibility to heed what is being said: the wind conveys the sound to the distant listener but the meaning is winnowed as so much chaff. Treating vocal intonation purely as tonal, it places the unspeakable beyond response. When I was preparing materials for 'Out of Their Feeling', a sound work for An Gorta Mor, a memorial to the Great Irish Famine, installed at Hyde Park Barracks, Sydney, in 1998, I came across this description of *auditory* Ireland post-Famine, already alluded to:

> The 'land of song' was no longer tuneful; or, if human sound met the traveller's ear, it was only that of the feeble and despairing wail for the dead. This awful, unwonted silence, which, during the famine and subsequent years, almost everywhere prevailed, struck more fearfully upon their imaginations, as many Irish gentlemen informed me, and gave them a deeper feeling of the desolation with which the country had been visited, than any other circumstance which had forced itself upon their attention.[1]

This can be interpreted in a number of ways. The writer of this narrative, George Petrie, was in search of native informants. He was a song collector interested in enriching his store of nationalist, anti-colonialist protest songs, enjoying as a side benefit the frisson of a particularly fine performance on the *uillean* pipes. Perhaps his musical bias leads him to identify 'the feeble and despairing wail for the dead' with an 'awful, unwonted silence'. Or, given that lamentation had always been part of Irish mourning ritual, perhaps the special sense of 'desolation' arose from the impression that the whole of the country had become a grave.

Whatever the case, these Irish gentlemen hear the silence. It presses in upon the auditory imagination. It embodies the terror of unjustly exercised political power. Recalling Williams's description of the noise in the railway factory – 'In a moment you seem to be encompassed with an unspeakable silence – a deathlike vacuity of sound altogether' – it is the pressure of modernity whose other expression is the creation of artificial silences for recording purposes. The silence that haunts the characters of 'Memory as Desire' is the absence of Aboriginal voices, and it is the haunting that is the drama. Meeting by chance in a French seaside hotel, the two couples speak from other places: not only England but also colonial Australia. They are personae or acoustic masks, echoic mimics of accents that betray both geographical and social origin – and something else, a memory of the 'Underworld' where they communicated differently. This is explicit; early colonial descriptions of the 'chattering in the camps' or the 'unique concert' of frogs are used in 'Memory as Desire'. When the Sturts rhapsodize

> Char: Summer sun and cloudless skies are better than gaslights.
> Capt: Paganini himself need not be regretted, when you can listen to the music of rural sounds.

they copy another South Australian sentimentalist.[2] They find amusement in the bird calls too but there is also a sense of unease, a feeling that something important has been missed:

> Capt: But for the birds …
> Char: Their melodies.
> Capt: Their words.
> Rob: The chick-o-wee.
> Char: Harbinger of hay-fever.
> Capt: The ricocheting stick.
> Rob: The Do-Not-Whip-Me.
> Char: The lyrebird's cymbals.
> Capt: And even-paced glissandos.
> Rob: There is a history in it.
> Capt: The ventriloquist.[3]

A call has been issued but its import eludes them. Perhaps *glissando* provides a clue to its identity. It conjures up Australian composer Percy Grainger's free music machines – 'Free to roam thru tonal space as a painter is free to draw & paint free lines, free curves, create free shapes.' In a curious anticipation of poet Martin Harrison's desire to 'see that movement as it is', Grainger considered that free music 'entirely inspired (heard in the inner ear)' was 'the only type of music that tallies our [*sic*] modern scientific conception of life (our longing to know life AS IT IS, not merely in symbolistic interpretation)'.[4] Grainger's 'Hills and Dales' machine suggests a comparison with the lie of the land known by walking it, a poetic geography defined rhythmically. Its antithesis is the determination of colonial explorers with literary pretensions to arrest this movement form and subordinate the ordinary appearance of things to a taste for the picturesque. In effect, exploration became a search for picturesque views, rare occasions when the prospect ahead momentarily mimicked a painting composed like a park by Repton or a painting by Salvator Rosa. The corollary was the disappearance of the non-picturesque environment at the very moment of its first encounter. Excised was the glissando of passage, the timeliness of footsteps adjusted to the irregularities of the earth's surface. When timeliness is taken into account in the management (or duration) of the film shot, film director Andrei Tarkovsky writes, 'You sense something significant, truthful, going on beyond the events on the screen; when you realise, quite consciously, that what you see in the frame is not limited to its visual depiction, but is a pointer to something stretching out beyond the frame and to infinity.'[5] It is the dim awareness of this unconscious environment that the visual picturesque seeks to screen out which explains perhaps the defensiveness of much explorer writing, the aggressive disparagement of Australia's frameless stretch.

'Memory as Desire' is a prehistory of the recording studio. It imagines the auditory culture that made the split between noise and silence, between the frameless complication of the world's ceaseless rumour and the anechoic technology of permanent sound impressions, possible, even desirable. As a radiophonic work, it continually questions its own acoustic limitations: the evident alienation of the characters from the world of sound is echoed in the radio production itself. As a pre-recorded radio drama, programmed according to a public broadcasting radio schedule determined by untested sociological convention, 'Memory as Desire' is picturesquely framed – 'symbolistic' – in advance of its reception. Repeatedly, the characters reach out beyond the frame towards infinity, but their efforts are stymied by lack of the right lexicon and, perhaps, insufficient self-knowledge. Following the flight paths of budgerigars, Sturt's historical expedition into Central Australia managed to reach the borders of Pitjantjatjara country. Beyond the expedition's caged starling whistling the tune of 'Bonny Doon', he heard nothing of interest. How different his map of the region might have been if he had understood the way in which Pitjantjatjara people classify sound, using sound, as we saw, as a form of echoic mimicry (onomatopoeic), or naming, through which the spiritual essence of things is reproduced. As Catherine Ellis emphasises, in this performative mode many ideas may be conveyed simultaneously – melodic sections evoke body designs, dance movement and via the interlock of these 'the totemic ancestor who is being recreated at that performance.'[6] Hence, again, song and song lines are not projected into the future but gather the three extensions of time together. At the heart of this active recall is rhythm: different places referred to in the song may have their own distinctive rhythm, a genuine sound geography.[7] Finally, in reflecting on the hunger the early white explorers experienced in Australia's inland, it is instructive to learn that 'the melody to which these rhythmic patterns are set conveys information about the "taste", the essence of the totemic ancestor.'[8] In compensation, as Ellis notes elsewhere, there is no evidence that Sturt (in contrast with his contemporaries) ever shed 'native blood.'[9]

The colonization of the ear is subtle and pervasive. Charlotte complains, apropos some half-overheard anecdote, 'I do find young people today so needlessly obscure', as if, like my mother, she thinks the meaning will be plain if the informant simply *speaks up*. This perennial impatience with mumbling paves the way to the recording studio, where the articulate modulation of the voice is de rigueur. Raising or lowering the voice is a power game, a colonizing technique. 'It is peculiar to English-speaking travellers that they endeavour to impress their meaning on the "foreign" natives by speaking very loudly and distinctly, and by using what has often been called "jingalese" syntax ... which consists in uttering a series of names of things and actions without any attempt at connecting them.' Without irony, the Tasmanian Hermann Ritz goes on to claim that, in Tasmania, 'this is precisely the style of the Aboriginal speech'[10] – this predictably from the ethnolinguist who assisted in the production of Fanny Cochrane Smith's studio recordings. People who speak up are said to like the sound of their own voices. They certainly drown out the sound of other voices, but the motivation may be something else: by hearing themselves speak, to fashion an identity of sorts. 'Here, it strikes me, every man seems to have two voices, both of which he uses, as if with no other desire than to hear himself speak,' the Captain reflects.[11] *Per suonare*: persona: personality: through sounding, this etymological sequence suggests, one acquires identity. In colonial cultures, though, it is hollow, the sound of one's own voice amplified. Its prostheses are the megaphone and the ear trumpet.

In Australia I learnt that auditory colonization was not simply geographical. My reflections on the communicational tactics of empire put my own sound history in perspective. *My* native informant was not a Tasmanian Aboriginal woman or one of those singers sacrificed to the Great Famine: it was the sound recordings that I had made in 1966, used in 'The Native Informant'. As I said before, remastering these, I might have expected a sound diary, recordings of our domestic interior, perhaps an improvised drama of my own, or simply the experimental chatter of fourteen-year-old boys. I was not entirely wrong: there were, as I say, two attempts to recorded environmental soundscapes. A combination of inadequate recording technology and unpropitious recording conditions frustrated both. But mainly I mimicked his master's voice, represented here by the subordination of my voice to efforts to learn prestigious languages, alive and dead – and by the near magical status of the tape recorder itself, which seemed to demand exceptional material, rather than a report of the everyday. '*Luo*', that staple of Greek verb conjugation, means 'to loose'. Ironically, my tape recorder had not loosened my tongue; I heard myself performing as if tongue-tied. My imagined interlocutors were my teachers who, in another circumstance, could have been curious white scientists visiting my camp. Perhaps the search for a voice of one's own was a hangover from the old days when the sovereignty of the self was fetishized. Perhaps this nostalgia for an authentic voice – one that provided a *via regia* to Lockean self-consciousness – was an effect of exile. Undoubtedly it was a product of cultural chance: but for the investment of the Australian

Broadcasting Authority in radiophonic compositions of this kind, my home thoughts from abroad would have remained unaired.

In those days I imagined loosening the tongue mimetically; I still believe that, in principle, new and just social arrangements are negotiated performatively. However, there are limits to what can be said and sounded, and after every human effort has been made to tune the world, the communications of the earth remain obdurately indifferent to our quest for meaning. Our soundings fall on deaf ears. Looking back on that period of radiophonic production, and its associated search for a way to find a voice of my own living in a new country, I identify three significant extrapolations or projections by means of which I thought to be at home. One stratagem was etymological, crossed in Australia with descriptions of Aboriginal philosophies of place. The result of this cross-fertilization is a poetic phenomenology. The discussion of the word *wuurong* in an earlier chapter exemplifies this approach. A virtue is made of the uncertainty of the information; the impossibility of checking any interpretation licences a self-absorbed reverie, which yields beguiling poetic insights but whose historical value remains a moot point. The feeling that an obscure lexical item from colonial Australia could help a migrant articulate a new sense of place presumed the enormous, even constitutive power of the imagination. According to Bachelard, the poet undertakes two journeys. The first is inward:

> While metaphors are often only displacements of thought, occurring because of a desire to express oneself better or differently, the image, the true image, that is, when it is the very wellspring of imagination's life, will leave the real world behind for the imagined, and imaginary world. Through the imagined image, we come to know that absolute of reverie that is poetic reverie.[12]

Then, as already noted, 'Poetic reverie revives the world of original words,' and in a second, outward journey produces 'The words of the world' – a double movement that, Bachelard stresses, depends on the fact that 'human and cosmic tonalities reinforce each other'; therefore, in returning to this world, the poet recreates it.[13] An ontological claim of this kind may be plausible in post-Lascaux Europe – in Australia it perpetuates colonial deafness. To exalt and universalize the poet's voice masks a lack of historical imagination, a failure to notice, let alone listen to, what the local custodians of the imagination say.

An antithetical approach to achieving inwardness with the place is as already discussed mimetic. Instead of attributing primary resonances to words, an original 'spatiosonant' identity between utterance and environment, an onomatopoeic theory of meaning production imagines the origin of meaning cryptically communicated in the sounds all around us. Some of my radiophonic works make the most of this. Besides the echoic translations in 'Memory as Desire', alphabetical transcriptions of birdsong abound in 'The Native Informant'. Has a historical study of the rendering of bird calls ever been made? In my ornithological days, the doyen of scientific bird vocalization studies was Terry Gompertz; looking back, it is striking to find how deeply her experimental techniques are indebted to recording technology: 'Contrived experiments, such as playbacks, are needed to confirm critical conclusions or to solve puzzles that cannot be elucidated naturally, such as the testing of response to songs from a different area or from a geographically-remote species.' For 'hearing quiet sounds, seeing small movements, and for controlling changes in the birds' environment', 'some form of captivity is essential.'[14] Of course, the older writers were solely concerned with identification in the field, and by the late 1960s bird-sound recordings had rendered their quaint onomatopoeic transcriptions and picturesque analogies largely obsolete. So when, in 'The Native Informant', I derived certain bird calls from T. A. Coward's *Birds of the British Isles and Their Eggs*, first published in 1920, I was consciously cultivating a scientific anachronism consistent with the nostalgic cast of the piece as a whole. The voice that evokes 'the inner speech of the country' as 'Tschizzik, tschizzik, / stip stip-stititiititipp, swee-ee, / tit-tit-tit, churr' faithfully reproduces Coward's versions of (respectively) the pied wagtail, black redstart, garden warbler, yellowhammer and tree sparrow.[15] If they are mixed up, sometimes hard to differentiate, it is because, post-Enclosure, flung together along the littoral of England's inner empire, they speak a new, mimetically based 'jargon'.

A mimetic theory of communication does not locate the origin of language in Bachelard's primordially sounding gorges but in the jungle of everyday noise. Writing 'Columbus Echo', I came to regard this counter-ontology as distinctively Fourth World, or American in a non-Anglophone sense. The totem of this new world of communication was the parrot, 'America's ambassador,' whose speech was not merely 'tangled' but, as long as the Bahamas were confused with the outskirts of Cathay, possibly polyglot - a medieval Arabic *periplus* reports that in Java there are white, yellow and red parrots 'that speak every language.'[16] In *Columbus Echo*, they even translate the nightingale – SPAN[ISH] 1: El ruisenyor PID[GIN] 1 (imitative, cantabile, exploratory): Tu, teu, teeewu, ti you, chi you, tre-ou.' How mimetic communication arises and may be sustained is suggested by Asturias in *Cuculcan*, an experimental 'ethnographic play', based on the *Popol Vuh* and the books of Chilam Balam. The play enacts a series of encounters between the supreme serpent god Cuculcan and the Guacamayo, or sacred macaw, 'who as a parrot by definition speaks with quotation marks, that is, with the words of others, [and] embodies the persistent foreignness of language'.[17] In the Guacamayo's presence, 'Cuculcan's orderly journey through the days' is turned into 'a symphonic babble orchestrated by the poetic parrot's linguistic play', as indicated in the stage directions: '*dog barks, chicken cacklings, tempest thunderclaps, serpent hissings, troupial, guardabarranca, and mockingbird warblings, are heard as the Guacamayo names them, just as the cry of children, the laughter of women and to close the commotion and chatter of a multitude that passes.*'[18] Guacamayo's poetic genius consists of channelling the sounds around him, and creating his discourse mimetically from these sounds. His word does not emerge from silence but from noise. It is multiple, not one, born of incessant, motiveless movement. It sets the exchange rate that makes translation possible. 'A market is like a great Guacamayo,' Chinchibirin, the warrior, says, 'everybody talks, everybody offers coloured things, everybody deceives.'[19]

The idea that meanings – equivalences between sounds and concepts – are the provisional profit of social exchange is, of course, attractive to anyone who does not belong; anyone whose foothold is precarious, dependent on a prayer being heard and answered. But, like the myth of first contact, it is utopian. In reality, as I found in Lecce, the noises of the street are sampled, entrained, ironized and manipulated associatively; the hearer/speaker's auditory imagination, shaped by an acquired acoustic lexicon, triggers sound associations; and these, heard echoically in the jungle of noise, elicit the most immediate reaction. As in any exchange, the interlocutors hear their own interests first. Ya Lur, the parrot whose ventriloquial genius was mentioned earlier, must come as close as is possible to a speaking in tongues that suspends self-interest; but her authority still depends on mimicking the interests of the local Tzutujil Mayan people of Santiago Atitlán. A scene in 'Columbus Echo' illustrates this inevitable auditory prejudice. It riffs on onomatopoeic bird names recorded in Frederick Cassidy's *Jamaica Talk*. Columbus and his crew are in the Caribbean but they *want* to be in the vicinity of Cathay. The calls/names of (respectively) the nighthawk, the long-tailed hummingbird and the short-mouthed quit become entangled in the 'imperialismo of signs':

Pid[gin] 1: Gi-me-me-bit, gi-me-a-bit.
Ital[ian] 1: Cuba, Coalbha, Alba, nella mattina.
Ital[ian] 2: Creo que deve ser Cipango.
Pid[gin] 2: Chi-chi-bud O, Chi-chi-bud O.
Ital[ian] 1: Can Grande, Can I Bale, Cabo de Isleo.
Pid[gin] 3: Cho-cho quit, cho-cho quit.[20]

Here, the potential intelligibility depends on word sounds being ambiguous. At this time, I hoped the ambiguity of cross-cultural improvisation might hold the key to decolonizing the violent mishearings of colonization. Silverstein's classic paper, 'Goodbye Columbus' had set out to establish 'the very great range of semantic ambiguity of jargon sentences vis-à-vis sentences of any of the primary languages of its speakers',[21] arguing that 'the essential "ambiguity" of the surface forms ... lies in abandoning a search for a single grammar for Jargon, and recognizing that each speaker produces surface forms convergent with those of other speakers by use of a modified form of his [sic] own grammar'.[22] The performative utterances that correspond to this situation are the product of echoic mimicry; rather than communicate a stable, translatable intention into words, they bring into being a new situation; they perform communication. In the precarious environment of invasion or refused entry, 'a dialogue cultivated in this way works if it produces nothing else than a tradition of such meetings'.[23] Perhaps it was a good thing not to understand one other. If, as Drechsel thought, pidgin speakers really engaged in a game of double-bluff, then the commitment to making sense resided purely in the performance as such. It is the echo with interest that brings people together. The business of greeting, let alone setting exchange rates, could become interminable: Pid[gin] 1: Elo. Pid[gin] 2: Aloe. Eng[lish] 1: Hello! Pid[gin] 2: Hoi. Ital[ian] 1: Pesce. Span[ish] 1: Hoy. Eng[lish] 1: Oy, you. Come 'ere.' And so on: truly, an economy without endings.[24]

Here were antithetical approaches to the challenge of finding a voice of one's own. One asked you to descend into the cave of original sounds and undergo shamanic initiation. The other took the opposite route and recommended the would-be poet haunt market-places and carnivals. Ordinary, unconscious speech has no need to explore the limits of the communication; ordinary speech is the servant of powerful conventions it agrees not to question. But anyone who stands aside, even momentarily, to notice the strangeness of any language, printed or pronounced, hears the loneliness of the sky pouring in. It is the first association of wind or clouds. Among people it is the christening experience of childhood, marked by those exploratory formations and mishearings parents like to remember; it is the habitual experience of the stranger thrown back on sound symbolism to make sense of the unfamiliar languages spoken all around. Unable to decide between one route or another, the dumbstruck refugee might make the impossibility of choice into a drama and, instead of striving to command a language, perform the contingency of communication. The actors of Kantor's 'Impossible Theatre' came to mind: 'They do not imitate anything, they do not represent anybody, they do not express anything but themselves, human shells, exhibitionists, con artists.'[25] They are no longer actors reproducing the ambiguities of reality. They are players whose own performance is ambiguous. By this device they bring into being a new, in-between place: 'Playing is identified in the theatre with the concept of performance. One says, "To play a part." "Playing," however, means neither reproduction nor reality itself. It means something "inbetween" illusion and reality.'[26] And, as Kantor stresses, redefining the poetics of performance alters the 'situation' of the audience-member, who is no longer passively in 'a state of hearing', but 'a potential player'.[27]

The poet of impossibility abandons theatrical illusionism, treating the ordinary contingencies of communication as opportunities for dramaturgy. Discussing pidgin Delaware, which evolved as a means of communication between the Indians and the Dutch, Silverstein makes the intriguing point that this language 'seems to have been the contact medium used by the interpreters, rather than by the general populace', an argument supported by the intrusion of non-Delaware Natick and New England Algonquian forms into the lexicon.[28] Apart from its historical suggestiveness, the attribution of a creative role to the 'interpreter' is relevant to the approach I took in compiling the materials of 'Columbus Echo'. The goal was not to re-enact historically possible cross-cultural encounters but through a concomitant mode of production to identify the emergence of emergency languages as events in their own right. The introduction of new sound-associative elements, the juxtaposition of phonically convergent materials from different sources and the sub-Joycean reworking of these to evoke the post-Babel history of human kind through the long European preoccupation with reaching China was an act of literary dramaturgy rather than authorship. If 'Dramaturgy is building bridges, it is being responsible for the whole ... a constant movement. Inside and outside',[29] my function was to be an interpreter in a double sense: through the sound composition to create a 'contact medium' that aquarium visitors foreign to the immediate dramatic *mise en scène* could follow and by the physical distribution and orchestration of the soundscapes throughout the aquarium interior to suggest a continuous coastline of conversation whose navigation was an end in itself.

As 'sailing directions' for the actors, the scripts of 'Columbus Echo' are poorly represented by quotation. Still, the following may give some idea of their prelingual or gestural quality. In this passage from Soundscape 4 (Putting to Sea), each line consisting of one word or a phrase: '*Al Genuvese a i voi dunej la vitta / La, la / Mentres tu esperes el retorn magnific / Pigliono quanti ne vogliono / la e la e la e la / A sol ponent / Mi ghe dirò ch'l 'e 'na donna crudele / E se va per lo mare oceano / E la, e la, e la / De la primera barca / Un Munferrin e l'altru Genuvese / Que sortira del mar / Della della / Ora, ora / The shore, the shore / Tota olorosa / ora, ora / The mouth, mouth / Per molte diverse isole / Tota dorada / La, la / A un paese molto abondante di pesci / Che si chiama Isola Lamori / La la / Amore, amore / E oro*'. This is glossed as: 'To the man from Genoa I will give the prize (Genoese dialect) / La, la (unspecified or There and there etc (Italian)) / While you await the wonderful return (Catalan) / They catch as many as they want (Italian) / La e la etc (unspecified, etc)) / At sunset (Catalan) / I will say to them that it is a heartless woman (Genoese dialect) / And they go sailing the ocean (Italian) / La e la (unspecified, etc) / Of the first boat (Catalan) / One from Monferrato, the other from Genoa (Genoese) / Which will issue from the sea (Catalan) / Of the / From there etc (Italian) / Now, shore, mouth, hour (Italian, pseudo Mediterranean pidgin) / The shore etc (English) / All perfumed (Catalan) / Now, now (Italian) / The mouth etc (English) / Past many and various islands (Italian) / All golden (Spanish) / To a country abundant in fishes (Italian) / Which is called the island of Love (Italian) / Love, love (Italian) / And gold (Italian).'[30]

An archetypal voyage is evoked here. Through the sound spell cast by the lap of the oars (e.g. la e la), a local fishing trip out of Genoa evolves into the universal quest for gold. This semantic metamorphosis is mediated through a sound evolution that is entirely echoic: *ora – dorada* (also contemporary Italian *orate*, a popular Mediterranean fish) – *oro*. When language is stripped out in this way, it almost inevitably becomes echoic. In one way, an elemental reduction of this kind conforms to Bachelard's reverie of original words. In another sense, it parodies it. It is, perhaps, impossible to decide. Inside the sequence *ora-dorada-oro*, for example, a reasonably literate actor familiar with Romance languages might hear the entire word family clustering around Latin *orare*, 'to pray', a term broadly linked to ideas of eloquence. To descant on such semantically relevant near-homophones – to be eloquent about the impossibility of eloquence – would be entirely in the spirit of a writing conceived as 'sailing directions' for verbal free association. Perhaps anti-literature of this kind is authentically American (in a non-anglophone sense): it seems to illustrate Andrés Ajen's call for a writing that, 'not erasing or even recombining the differences between traditions of inscription but, by bringing them face to face and exposing them, [makes] way for an encounter between different cultural provenances and languages', perhaps producing a '"poem" memorious … and unprecedented'.[31] Such an achievement may be the death of 'literature',[32] the beginning of a different kind of recording – a possibility grasped in 'The 7448', the sequel to 'Columbus Echo', where the minimal phonemic combination, 'la', acquires a distinctive rhythm and stable semantic field.

Whatever the future of such writing, it revolutionizes the idea of recording. The script no longer represents language. It no longer attempts to reproduce anything that has ever been said. While 'The Native Informant' illustrated the ethnographic claim that the native informant might be none other than the ethnographer talking to himself or herself, there remained embedded in this solipsism a desire to represent something. In this new iteration, all desire of representation is abandoned: the interpreter constructs a medium of communication that transcends the problem of translation. Likewise, the new writer is a translator who adopts the paradoxical position of not believing in languages. Is this feral intransigence, perhaps through wearing the mask of the oppressed, convincing, emancipatory or patronizing? A recent study of writer-ethnographer Hubert Fichte's polyvocal world, 'in which no authorial vision is empowered to impose a unifying vision and gloss over contradictions and counterhegemonic struggles',[33] suggests that the dialogical presentation of material 'risks eliding the fact (the labour and the control) of textualisation through the evocation of unmediated speech'.[34] Sharing textual space and authority with local informants (or, for that matter, a cast of multilingual actors) 'obscures the author's orchestration of such dialogues within the agenda and the parameters set by him'[35] and, in the case of public soundscapes, the hosting institution. Fichte's response is to refuse to grant the language of the victors' victory, to keep ethnographic notes 'of errors, false conclusions, rash actions' and to state famously, 'fundamentally different language existed', 'in which the movement of changing and contradicting opinions, the dilemma of sensitivity and conformity, despair and practice, could be made clear, I should use it'.[36] It does not exist but, if it did, its sensitivity would, Klaus Neumann suggests, be 'a state of being open to magic, and open, too, to the liberating propensities of language'.[37] It is no accident that Fichte's radiophonic works were part of the Neue Hörspiel movement and no doubt helped create a receptive environment for the German version of 'The 7448', '7448, Eine Kolumbische Phantasie'.

Another work that went into German, 'What Is Your Name' ('Wie Ist Ihr Name'), illustrates the temptations of 'orchestration'. The various stratagems of etymological reverie, echoic mimicry and a kind of performance that seeks to solve 'the dilemma of autonomy and representation, with ease', are all easily recruited to the exercise of power. Anyone who does not share the utopian ambition to reground language in the primary sounding of the world can easily exploit these improvisations. Speech that is stripped of the grammar and syntax lends itself to misinterpretation and exploitation. Stepping out of one prison house (Western thinking's violent imposition of unifying and generalizing categories), one risks a new vulnerability. It is all very well to promote a pure power of saying that does not convey a general form of knowledge or law:

> It acts in its own weakness. … That this potentiality finds its *telos* in weakness means that it does not simply remain suspended in infinite deferral; rather, turning back to itself, it fulfils and deactivates the very excess of signification over every signified, it extinguishes languages. … In this way, it bears witness to what, unexpressed and insignificant, remains in use forever near the word.[38]

But what happens when someone armed with violent intent disagrees? You bear witness at the risk of your life. The rule of the carnivalesque may replace the Master's Voice of dictatorship, but turns into a new tyranny. Mikhail

Bakhtin's mighty force of communal, popular experience becomes a new dictatorship of the proletariat – in capitalist society, the monopolistic voice of the media. The immersive 'translinguistics' of the carnivalesque (where 'language is not a neutral medium that passes freely and easily into the private property of the speaker's intentions', but 'is populated – overpopulated – with the intentions of others') all too easily produces nonsense.[39] On the horizon of 'boundless polyphony' loom 'silence and insanity'.[40]

I found a quaint local instance of this easy alliance between masquerade and murder. 'The Native Informant' was indebted to *A Glossary of Berkshire Words and Phrases* (1888) for its dialectal versions of common verb forms. However, the author of this volume also related the following anecdote:

> Amongst country folk the notes or calls of many birds are given their equivalents in phrases. I remembered an old shepherd at Hampstead Norreys, 'Shepherd Savoury', who seemed to have words or phrases for all birds. As an instance, he one morning said he had been walking down a lane with his gun (a recent conversion from a flint arrangement), and found there a small flock of sparrows flying along the hedge in front of him. When these birds saw someone coming, they began to argue as to his identity; some said, ''tis he, 'tis he', to which others replied, 't'yent, t'yent'.

This discussion went on until the birds fell a-fighting over it, and all flew close together in their struggle as their manner is. 'Then,' said the Old Man, 'I thate the time had come vor to show um "'tis I," an' zo I let vly an' killed a dozen on um.'[41] In this case, the complacent shepherd behaves like a cruel puppeteer. Personification does not foster sympathetic identification. Instead, it is a power trip. Having projected his own insecurities onto the birds, he interprets their response as rebellious insubordination. Caught in his fantastic interpellation, they are made responsible for their own deaths. So much for the apparently harmless practice of imagining the birds speak: their descant will only be as good as the character of the hearer. Freed into speech, they may be deprived of their liberty.

'What Is Your Name' opens with a scene of overheard torture:

1. No.
2. Ah! Ah! No-ah.
3. Here, Noah, here, Noah.
1. Who are you calling Noah?
2. That is your name.
1. My name? All right.
3. Everybody has one.[42]

From 1's unheard exclamation of pain (an 'Ah!' which, we must imagine, immediately precedes the action), and from his verbalized refusal to yield

('No'), the interrogator derives a name. Here is an act of echoic mimicry, certainly, but no 'dilemma of sensitivity' is exposed. While 1 has no alternative but to accept the rules of the game (everyone must have a name), accepting a language 'overpopulated' 'with the intentions of others' does nothing to alleviate the precariousness of his position. The interrogator's sole object is to seduce the subject into colluding in his own disappearance, as emerges in a later scene:

1. Noah.
3. Noah? We've got a hundred Noahs.
2. Absalom.
3. Achilles.
2. Alexander Luck.
3. Sahib, Samson, Solomon Grundy.
2. Jacob, Peter, Stephen Duck.
3. But no *nobodies*.
2. Mr McGuinness and myself, we'd suggest.
3. You know.
2. Nobody.
3. To distinguish you.
2. Just for the record.[43]

The irony is that this system of naming eliminates the colonial subject from the record; only, in exceptional situations, the disappearance survives – hence the epigraph for 'What Is Your Name': "Mexico City, 6 Dec. 1985 - four bodies found in the ruins of the building that contained the Mexico City prosecutor's office were not like the thousands of others dug out after the devastating 19 September earthquake. For one thing, they were bound and gagged . . ." Reflecting on *Antigone* and the 'politics of the witness' in Greek tragedy, Marc Nichanian writes that, in the wake of catastrophes that leave no one alive to bear witness (the Mexico earthquake anecdote is an instance), 'there is no possible location, no possible stage, on which she could appear and present the limit she represents.' The logical corollary is that 'theatre is not theatre anymore (I mean theatre as the other politics that it used to be in ancient times)'.[44] The names in the quotation come mainly from nineteenth-century Australian mission station records. They are sardonic, satirical, even comic human mnemonics improvised in the absence of Aboriginal cooperation (or comprehension?). Special 'Naming Days' were set aside for these rituals of administrative initiation.[45] Nichanian's *Not Theatre* imagines a performance practice not predicated on the eye-witness illusion; it recognizes that eye-witness history may collude in the terror it purports to represent. The 'criterion of verifiability', as Nichanian calls it, not only captures the victim 'in the logic of the executioner': it causes the 'immediate transformation of memory into archive for the sole purpose of providing proof, that is, of dispossessing the victim of his own memory'.[46]

To *stage* the passing of time aestheticizes labour. It anaestheticizes the history of passion (passing as suffering), embalming the human production of space and time in a coffin that represents time going on as before – time immortalized. Recording technology theatricalizes time and puts to sleep memory as desire. Imperial history colludes in this silencing – Carlo Ginzberg refers to 'history's practice of putting past events "in perspective"'.[47] A sound history reanimates time. Following the suggestion that Greek theatre housed the politics that found no expression in the agora, Ginzberg champions a return to the public space cognate with the emergence and maintenance of democracy. The performances of everyday life witnessed *here* are perspectiveless. They are 'not a series of facts to be contemplated at a distance, but a series of situations into which one could somehow be existentially drawn'. These situations, while not entirely novel, cannot be old: like memories, they are 'reactualisations'.[48] Looked at in this way, the oscillation between English and Australian soundscapes in 'Memory as Desire' refers to a modality of time consciousness that is not silenced (or left behind). It is the world of distilled sound infiltrating walking, breathing and dreaming, reactualized every day in the metabolic spiritualization of the breath or (as happened in Lecce) an intuition of a social choreography. They talk about a 'spatial turn' in the human sciences, but when the expression 'spatial history' was coined, it referred to the temporalization of space, to the breathing paths inscribed into the maps. These paths of labour are always 'situations', and the collective translation of becoming at that place is always a more or less troubled breathing pattern.

Here, hand and mouth go together. In 'Memory as Desire', Rose's announcement of her artistic parentage morphs into a discussion of Aboriginal painting.

> Rob: In his learned work Sir George Grey reports that fifty per cent of native art consists of hands.
> Rose: Hands?
> Capt: Palm prints.
> Char: Insides made outsides.
> Rob: A signature no doubt, but art?
> Capt: A way of remembering.
> Rose: A hand.
> Char: A shell.
> Capt: A house.[49]

It turns out that the father I attributed to Rose was fictional: I had confused her husband, curtly described in the Australian Dictionary of Biography as an 'accountant', with his much better-known namesake, the Victorian/ Edwardian water colourist, Thomas Pyne (1843–1935). In a further flight of fancy – I had underestimated how many Pynes there were in the picturesque landscape – I also confused Thomas's father (James Baker Pyne, 1800–70) with William Henry Pyne (1769–1843), an illustrator and painter who specialized in picturesque settings and who wrote *Wine and Walnuts*, a collection of anecdotes about art and artists published in 1824.[50] Still, this genealogical wish-fulfilment masquerading as history had a propitious outcome, as it led me to W. H. Pyne's description of the interior of his father's weaving workshop where

> it was the custom on taking leave, when some favourite was going to a foreign clime, to chalk the inside of his hand and stamp the impression on the wall. There were few workshops without this artless memorial, under which was carved, by some shop-mate, the date of his departure, and, from time to time, the actions by sea or land in which he fought, and to some were added the melancholy gazette of death.[51]

Was this once a common custom? I have not come across other contemporary accounts. Anyway, the anecdote resurfaces in 'Memory as Desire', now told by Robinson. Charlotte wonders aloud, 'Like an outline on a map'. But no one really knows what to make of the story and it is left hanging in the air.

Hand stencils preserve indexical traces of those who make them; metonymic gestures, they assert a unique identity in the most universal of languages. They are hand writing before the hand becomes an instrument of the brain and the confrontation with them unavoidably suggests a statement that resembles perhaps a personal version of the great Delphic Know Thyself, which the urban tagger reduces to the rather more modest 'I was here'. In the case of Aboriginal rock art, they are something more: a way of sounding that is silent or whose articulation is inseparable from the hollow where they are attached. In an article called 'Red Handed: An Inquiry into the Meaning of Prehistoric Red Ochre Handprints', Kathleen Kimball asks the obvious (but usually unasked) question: 'Surely making a paint and blowing it from the mouth is more trouble than simply picking up a piece of red ochre and drawing with it, as Michelangelo did, so what is the significance of processing it, making a paint, and blowing it on the walls?' She speculates, 'Is the "breath of life" being invoked? Is there a sonorous relationship between these ideas and the early bone flutes that have been found?' Kimball speculates that the mouth-blown hand stencils were kinds of sound recording: 'If you were living 50,000 years ago, how would you capture and record a sound? How would you leave a permanent visual echo of your presence?' She cites Steven Waller's claim that rock paintings are located at points of resonance, 'painted images recalling animals are echoed in the sounds'.[52] It is easy to imagine an architecture of the pneuma, hollows that concentrate breath patterns. The outline of hands becomes the mark of an afflatus. It is a graphic shell surrounding the mortal, in whose palm the 'I' echoes. The most fleeting presence (breath) bestows on a mortal part (the hand) immortality. The hand causes the breath to spread out; the breath records the hand's fullest reach.

In terms of speech, hand stencils are purely vocalic: a steady pressure is applied to produce a consistent film of ochre around the hands. No consonantal stops are permissible, as they would interrupt the steady flow of expectorate. And this brings into focus a modern prejudice, the identification of sounding with a disembodied communication. Tom Cheetham indicates the historical impact of the invention of the alphabet on this kind of writing: 'The original Semitic alphabet shared today by both Arabic and Hebrew, had no written vowels. Only the stops. Not the breathings. The absent vowels "are nothing other than sounded breath. And the breath, for the ancient Semites, was the very mystery of life and awareness, a mystery inseparable from the invisible ... holy wind or spirit".'[53] For Cheetham, the consonantal alphabet is a halfway stage between pure orality and the modern Western alphabet (where the values of the vowels are also stipulated); it retains vocalic ambiguity, and therefore any writing in this alphabet depends on the performer for its meaning. The rise of silent writing and reading, associated with an alphabet that provided values for both vowels and consonants, paves the way for the metaphysicalization of the breath and the abstraction of sound. In the nineteenth century, Jonathan Sterne writes, 'In acoustics, physiology, and otology, sound became a waveform whose source was essentially irrelevant; hearing became a mechanical function that could be isolated and abstracted from the other senses and the human body itself.'[54] Hence, the University of Bournemouth Landscape and Perception project bases its archeo-acoustics mapping project on the proposition that 'with the ever-increasing use of mobile phones and iPods, dramatically developing digital entertainment multi-platforms, and digital navigation devices, there is a danger of the culture becoming increasingly abstracted from its primary sensory environment'.[55] The researchers identify a number of ways in which rock art/cave sites can sound: through lithophones, through echoes and wind, through water or heat expansion sounds.[56] They overlook another obvious source of sound: the production of wall paintings.

The wordlists and phraseologies transcribed in 'Memory as Desire' 'What Is Your Name', 'Cooee Song' and the like were largely collected in the early to mid-nineteenth century. In other words, the efforts of white administrators and other self-appointed interrogators to record Indigenous languages coincided with the cultural abstraction of sound from the other senses – a development critical in making the idea of mechanical sound recording meaningful. It is logical to suppose that not only their modern obsession with orality goes some way to explaining the extraordinary authority the early ethnolinguists ascribed to wordlists but also their frustration that communication with Aboriginal people was usually so enigmatic. R. M. W. Dixon cites cases of white people in Queensland in the 1970s who still had no idea that Aboriginal people spoke (i.e. had languages).[57] Presumably this was a survival of the oralist racialist fantasy: 'The value of speech was, for oralists, akin to the value of being human. To be human was to speak.' Hence, according to one oralist, 'savage races have a code of signs by which they can communicate with each other. Surely we have reached a stage in the world's history where we can lay aside the [tools] of savagery.'[58] This last remark implies a prejudice in favour of speech and against gesturalism or expressiveness of all kinds. It was a feature of sound recording technologies that they ignored these contextual aspects of communication – 'In acoustics, frequencies and waves took precedence over any particular meaning they might have in human life.'[59] Hence, the manifold signs that local people gave of belonging and of knowing who they were counted for nothing.

Advocating an 'echolocative' model of communication, anthropologist Roy Wagner writes,

> It is in sound's inability to merge with or directly encode the meanings attributed to language that it similarly becomes meaningful for human beings, allows them to listen to themselves as vectors of meaning through a medium that is not meaning. Those who wish to ground meaning in language are disposed to imagine the 'sign' through a magical precision bridging sound and sense, but such a coding, to the degree it were precise and exhaustive, would render impossible the 'play' or ambiguity, the irony of sound and meaning – would nullify sound's echolocative possibilities.[60]

The association of hand stencils with caves and other resonant overhangs suggests that the reincarnation of the great I AM in the mouth artist's breath was manifested echolocatively.

The classic site of this encounter in the early colonial literature occurs in the writings mentioned by Robinson: in Sir George Grey's account of his encounter with the Wandjina figures of the Kimberley: 'On looking over some bushes, at the sandstone rocks which were above us, I suddenly saw from one of them a most extraordinary figure peering down upon me.'[61] A peculiar feature of this and the other figures and faces subsequently located was the absence of mouths. According to Jean Gebser, who cites the mouthless Wandjina figures from the Kimberley, as well as similar mouthless works from other cultures, 'What this mouthlessness means is immediately apparent when one realizes to what extent these paintings and sculptures are expressions of the magic structure but not yet of the mythical structure. Only when myth appears does the mouth, to utter it, also appear.'[62] Communication in this pre-mythical time, Gebser speculates (following Vico), was gestural, but the more interesting speculation is about the role that sound might have played in that pre-speaking world: certain common German verbs, he notes, retain an 'acoustic-magic stress indicative of the extent to which power is expressed, not in a palpable but rather an auditory manner and appeals to the incomprehensible and pre-rational in us: to belong [*gehören*, derived from *hören* = to hear]; to obey [*gehorchen*, related to *horchen* = to hearken]; and, to submit [*hörig sein*, related to *horen* = to hear]'.⁶³ Envelopment in the auditory world is in his terms pre-temporal and spatial – 'The magic structure occurs everywhere on earth, although always at the appropriate time for any given place.'⁶⁴ Writing of this encounter in *Living in a New Country* (1992), I claimed, though, that Grey resisted that 'power', as if the Wandjinas looked at him with the same baleful glare as the classical medusa, and devised various stratagems to break the spell:

> Putting into words what he saw, Grey cast a spell of his own against the figure's numinous aura. It was not simply that the language of art enabled the figure to be assimilated to familiar categories: the image's exorcism occurred at a deeper level, in the mere act of silently vocalizing a monologue of sounds, of renaming the place in terms not its own.⁶⁵

When I came across Gebser's book later, these comments seemed to me to make fuller sense. His claim is that the magical structure possesses aura: the Wandjina figures have haloes or nimbuses: 'Man's inner potency is not externalized there via the singing voice, but by the emanation, the aura of the head and even the entire body which forms a seamless transition to the flux of things and nature with which he is merged. Where the mouth appears, the aura diminishes in strength.'⁶⁶ Gebser discerns this double effect of mythical thought and action in the etymology of *mytheomai*, which, on the one hand, means 'to discourse, talk speak' (from a root *mu-* meaning 'to sound') but, on the other hand, in the manner of primal words that are constitutionally ambiguous, is associated with 'another verb of the same root, *myein* ... meaning "to close" (with particular reference to the eyes, the mouth and wounds'.⁶⁷ So, in a paradoxical way, the opening of the mouth to speak silences the mystery. Mysticism ('speechless contemplation with closed eyes, that is, eyes turned inwards'⁶⁸) arises in resistance to the mythical ordering of society. Revisiting *Living in a New Country*, I see that Grey was a historical proxy for my own encounter with a wordless landscape of great aura: important content in that book was shared with *The Sound In-Between* (also 1992). The act of 'silent vocalization' that commences the overthrow of the spirit of place has its technological counterpart in the invention of mechanical sound recording. Paradoxically, Grey's alienation from the place stemmed from ignoring its resonance, something more than an inventory of sounds (which he also ignores). The mouthless figures were *echolocative*, articulating the resonant hollowness of the place: they did not necessarily represent anything. At 'The Cave of Hands', in the Grampians

(Victoria), I wrote in *The Sound In-Between*, 'The hand is expressed as a breath: it is the opaque soul of the speechless voice, void of volume, making its surroundings tremble. ... The hand is a roof, it holds up the roof; the cave is a mouth where the wind, blowing endlessly through the cyclone netting, casts shadows into the land that sound like sighs.'[69]

Reframing visual phenomena in auditory terms like this seems affected now – Can shadows really be compared to sighs? It was an attempt to imagine the environment in terms of a notation of resonance. The steady expectoration of liquid red ochre round the outlined hand, producing the stencilled outline on cave wall or roof, would be accompanied by a low humming – as if someone was gargling or, more controlled, maintaining a solemn drone. Humming can go either way; gathering rhythmic identity, activating other mouth movements, it can transform the mouth cavity into a resonator incubating tonally differentiated patterns (music) or vocal sequences that resemble a kind of trance talk, the outward expression of 'the constant vibration of poetry'. In resonant interiors whirring and humming may be augmented by the natural frequency of the space; infra tones may be produced, inaudible but felt as a physical presence; in special places they suggest the idea of haunting. Sonograms of these acoustic shadows would look like the sound of wind in dry grass, their flamboyant field of vertical strokes recalling a Chinese brush rendering of wind-animated reeds. Evidence of magical cave sounding is hard to come by; however, in *Songs of Central Australia*, T. G. H. Strehlow reports visiting a cave near Horseshoe Bend in 1933 where sacred *tjurungas* were housed: 'There were no songs with which these three rain *tjurunga* could be honoured while they were being refreshed with blood. The only sound permissable was a vibrating call, ho-o-o-o-o-o, ho-o-o-o-o-o, and so on, which closely resembled the normal *raiangkintja* call.'[72] The *raiangkintja* was a 'loud, vibrating call given before and after ceremonial acts'; the swinging of small bullroarers could be substituted for it.[71] The connection of this form of utterance to hand painting is suggested by the fact that when making the call 'the hand flutters before the lips'.[72]

Without invoking an esoteric system of correspondences,[73] it is obvious that my interest in hand–mouth relays embodied the feeling that full audition was kinaesthetic, and that many phenomena that were not strictly audible were vibrational in nature. The set of these vibrations would be discernible diversely, through walking, leaping and dancing as well as incantation, and such tactile activities as stroking, plucking and tracing. Gebser identifies sounding with the magical, pre-mouthing era of human existence, but something of the magical structure survives in the mythic period through poetry and song. 'Sound or tone is procreative; the ear, the projected likeness of the cavern and the labyrinth, is receptive and, consequently, is also parturient: it gives birth to the magic world.' The French '*charme*', Spanish '*encantado*' and English 'charming' basically mean 'charm' or 'spell'. He adds, 'Dance is tone become visible: the medium of conjuration and of "being heard" by the deeper reality of the world where man is united with the rhythm of the universe.'[74] Tone, like balance in dance, cannot be subtracted without the expression lapsing into representation. It is the vitality of being as a quivering state of attention or inclination: in a world of sympathetic vibrations, the one participates in the whole. The word *tonos* is

> directly related to the tuning of the Greek kithara ... tuning had to do with various degrees of the tenseness of strings, as every string player knows through his fingers. Thus the tablature, though mediated through the eye, was really a contact between the fingers and what I would like to call a musical tensor. And thus the performing subject was in direct touch with the very heart of music: with musical 'tensility', and thus with musical 'space'.[75]

In these accounts, sound descends proprioceptively and kinesthetically into the world of the senses, where it shapes the movement forms of everyday life.

Going back to childhood and Richard Jefferies's description of the sounds produced by murmurations of starlings, the 'noise made of innumerable lesser sounds, each interfering with the other', called by country folk a 'charm', I would invert this idea of descent and distribution and say that I was immersed from the beginning in a jungle of senses whose harmonization or collective signature was always impure, fractional and occasional. In this pre-musicalized sensory world, noise is the name of kinesthetically apprehended sound; and its 'language' is the memory of physical activity. For example, as a boy, I also drew with local chalk: hopscotch patterns on the pavement, faces on the door, meandering lines along the brickwork. Such lines of flight could be transposed to the landscape. The Manger below the White Horse is said to be a 'dry valley' (whatever stream created it has gone underground); the so-called Giant's Stair, the five or six steep coombes bunched in the western flank of the Manger are explained as Ice Age avalanche shutes, but I experienced these kinetically, as mad careers from summit to valley floor. In the expanding instant of headlong flight, steepness was choreographed as gravitational grace. A photograph of me competing in the school long jump exists: a local newspaper photographer, recruited for the day, had the equipment and skills we lacked to capture the instant between two strides, and the ten-year-old boy remains suspended forever in mid-flight. Clad in white vests, white shorts and white running shoes, he might be the White Horse come to life in human form. When my father died, I found this photograph pinned to a wall in the inner recesses of his garden shed; in the gloom it was easily overlooked. An 'artless memorial', established when his 'favourite' departed for 'a foreign clime', 'the suspended impulse of its lightness' functioned as a magical summons to return. Some essential movement form was discerned that impressed on the distance the rhythm of return.

This reminds me of another black-and-white photograph of my father digging the land – in those days our five-and-a-half acre plot was known simply and absolutely as 'the land'. He is paused, his right hand on the spade handle, his left hand raised to his face, wiping the sting from his eyes or the sweat from his brow. It is like Masaccio's Adam expelled from Eden, conceivably an image of lamentation. The handling of tools, my father assured me, was hereditary. His father, he said, handled tools in a way that made them fuse with his body. The plough possessed a nervature, the share directly communicating to the wrist whether the soil was of the right tilth. In soil heavy with conglomeratic grit, hoeing involves a double action; as swiftly as one cuts into the root of the weed, one must register any stony resistance that might cause the blade to skitter off and harm the crop. The good mower conserves his strength, wielding his scythe like a matador performing a sequence of passes. Such activities are auditory as well as manual: every soil slices differently – clay cuts sheer but black or warm soil tugs and hugs; the ring of loose flint advises the ploughman of anomalous conditions calling for skilful steersmanship. As the Australian poet A. D. Hope describes, mowing has its metre, 'The boy with the scythe take a stride forward, swings / From the hip, keeping place and pace, keeping time / By the sound of the scythes, by the swish and ripple.'[76] The equipment of pruning is quieter; however, I remember the saw's rasp, the click of secateurs, and the *flick* of twigs falling to form a heap at the tree's base.

Obviously, in my father's day farming was largely mechanized; as a fruit farmer, he and his father were gardeners – the use of scythe and sickle was confined to clearing weeds that could not be eradicated with the rotavator, an orange 'Charm' of the Howard series that stumbled between the gooseberry stools like a giant, drowsy beetle. A large component of soft fruit cultivation was defensive, consisting in an extraordinary dedication to the extirpation of weeds. Ours was ancestrally disturbed land; the tell-tale weeds of occupation, the phosphate-dependent stinging nettle, elderberry and goose grass, thrived in the hedgerows; ground elder (historically introduced, my father said, in a load of dung imported from Oxfordshire) furnished ideal undergrowth for war games. Our own campaign was primarily pesticidal. The forces ranged against good fruit farming read like a medieval demonology: mites, bugs, weevils, midges and spiders, caterpillars and aphids, slugs and sawflies respectively bored into heart woods, burrowed into buds, laid eggs, damaged leaves, gnawed and hollowed out the fruit. Some enemies worked 'systematically … frequently crippling the plants'.[77] Others were seasonal and opportunistic, or, attacking in a 'cottony mass', overwhelmed by sheer numbers. No wonder that in photographs from the late 1940s and early 1950s, my father appears gloved up like a nuclear leak investigator, and probing the apple foliage with his spray lance, deployed a pesticide first developed in the Second World War. This operation also had its ominous sound signature, a regular hissing that infernally mimicked watering a plant or dampening a shirt ahead of ironing. Given the inhibition of human breath associated with its application, it strangely announced a future cancer.

The point is that there were enclosures inside the Enclosures, and not all of these were undesirable. When I tried to establish a poetic history of Enclosure, I focused not only on the new assembly of songbirds but also on the altered rhythms of work. The names in the naming scene from 'What Is Your Name' were derived from colonial records – with one exception: Stephen Duck refers to the so-called thresher turned poet from Wiltshire who, in the 1740s, was associated with the hamlet of Hatford, located south-east of Faringdon in the Vale of the White Horse. I contrasted his account of working in the fields with the point of view of poetaster and poet laureate, Henry James Pye, author of the 'prospect poem' 'Farringdon Hill' (1774). Looking down into the Thames valley from his commanding height, Pye expressed imperial satisfaction,[78] eulogizing 'manly Execution' and elsewhere 'the generous swains' who 'With chearful industry / Till for their wealthy lords the peaceful plains'; he may picture master and worker sitting together after a hard day's work, drinking the cider that 'chears alike the Peasant and the Lord', but Duck presents a different view. The reapers of *The Thresher's Labour* may sit down with the master when the harvest is in and share his 'Jugs of humming ale', 'but the next Morning next reveals the Cheat, / When the same Toils we must again repeat'. Pye's middle distance 'chearful industry' feels different from the inside: 'Now in the air our knotty Weapons fly, / And now with equal Force descend from high; / Down one, one up, so well they keep the Time, / The CYCLOPS' Hammers could not truer chime ... / No Intermission in our Work we know; / The noisy Threshal must forever go.'[79]

The fact that both Duck's satire and Pye's pastorale use the same Popean couplet suggests the iron fist of ownership hidden in the velvet glove of the picturesque. As Duck observes, 'The Voice is lost, drown'd by the louder Flail', and by Pye's time, in a curious anticipation of Alfred Williams's perception of a deathly silence at the heart of unbearable noise, the countryside has been swept clean of unseemly singing, let alone the accents of suffering; property-hugging hedges have seen to that. But enclosure acts were psychological as well as political; the violence of usurpation was soon forgotten or moralized; once the new values were internalized, they were reproduced *in parvo*, not least in my grandfather's reproduction of Sutton Place in the miniature domain of the Lawn, with its croquet hoops, its arbours and herbaceous border. Inside this enclosure the rhythm of work could be set to a human pace. Listening again to the remastered tapes from 1966, I hear the distant braying of a brass band. In reality, an occasionally augmented quartet, it met most Sunday mornings, weather permitting in Elsen Foard's shed. Like much else in my father's life, his trombone playing went unremarked. His enthusiasm was unshared at home – like von Karajan's recordings of the Eighth and Ninth Symphonies purchased, as I imagine the *Encyclopedia Britannica* was, to enlarge my education, but never mentioned, which I had to discover by myself, and whose unearthly chords remained a solitary transport that was never, as I recall, communicated or even shared. In any case, in his own interior space, inside the external enclosures, as my tape recorder proves, Bill found a free voice.

'There is never a single approach to something remembered,' John Berger writes about the photograph. The same is true of sound memories: 'A radial approach has to be constructed ... so that it may be seen in terms that are simultaneously personal, political, economic, dramatic, everyday and historic.'[80] In this way, through this labour of remembering, a sound becomes personal – the 'Remember Me', say, of Purcell's *Dido and Aeneas* inside the auditorium of migration. The unheard nightingale of Knighton Crossing may have signified very differently to my father. Given my interest in tape recording, it may have been a practical gesture: another remastered tape from that period, recorded in the wood where Arturo Barea once lived, preserves through the noise the recognizable cadences of a nightingale. Unperturbed by the spectre of exile or the adolescent hunger for significance, his nightingale did not live in a golden cage in the imagination; it was perceived through the daily and seasonal rituals of work. To come back to that event, so significant in installing a sense of the unheard, is to construct a 'radial approach' that, in my father's case, was habitual. The work of preparing the soil, planting, weeding and harvesting, was repetitive and seasonal; contingent on the ever changeable weather, its details were flexible – tasks were swapped, brought forward or delayed. But there was no danger of places falling into the past or growing out of reach. In a sense, inside the stable landscape of gardening, the nightingale remained potential. Unheard here, it might be encountered somewhere else: time was ample, the environment generous and the song's coming and going independent of my anguish. In a Keatsian poem written later in Spain my alter ego dares not return until it finds 'a tune of his own pure enough to praise it'. Why so competitive? Much angst could have been saved by playing music on Sundays.

Sampled by the breeze, the hunting notes of amateur trumpet, trombone, bugle and horn were borne over the orchard. My dormer window, propped open, was a sound catcher, a kind of natural loudspeaker; out of sight, beneath its lip, the morning was a sea of birdsong. These floating airs recall another, Gluck's aria, mentioned before, inexplicably heard one morning, impressing on memory an exact visual reproduction of the street. Such auditory impressionability dictated the direction of my creative interests. Looking back on the span of notebooks filled with poetic drafts, extending from my fifteenth to my thirtieth year, I am astonished by their derivativeness. The writing of that period is more than an imitation of the styles and themes from other epochs; it is an escape from the world around. The self-appointed task of bringing back to life a kind of nature poetry in the pastoral mode produces not only critically autistic parodies of dead books but also a diaristic gap in time. Their submission to word music at the expense of intelligence represents an extraordinary running away from the dark wood of life; their interminably well-turned lines were steps on the staircase to cold mountain without the experience of passion to justify it. Psychologically, they functioned as the hours of listening to the great classics did, to pass the time. Recording words or listening to recordings, rhythm deferred the confrontation with mortality. Levinas warned against its Dionysian power: the ego succumbs to its enchantment, and in a waking dream, marching or dancing, is 'paralysed in its freedom, totally absorbed in the game'.[81] Levinas wanted to return us to the 'world of initiative and responsibility'.[82] I have repeatedly rejected this dichotomy but in those years, so far as recorded music went, it was largely true.

In my culture of classical music, little difference existed between playing a record and turning on the BBC's Third Programme. There seemed to be a natural fit between the duration of the romantic symphony and the time available for it on the radio. US commercial radio and popular music recordings had grown up together; radio ads and commentary created a seamless auditory wallpaper effect. The BBC's programming was altogether subtler, but perhaps no less subversive; live concerts and recorded music were subjected to the same durational straitjacketing, invariably lasting slightly under thirty minutes (or a multiple of this). This brick-like programming in the wall of time, taking full advantage of magnetic tape technology both in terms of audio quality and recording length, produced a new listening environment: immersive, privatized, soporific. Tuned into Delius or Sibelius, my motorways in and out of northern England's dying towns were assimilated to the filmic sublime. At home, the anthems of Handel recast late summer's thunderous clouds as proto-baroque, embryonic of angels. The great pianist Sviatoslav Richter remarked, 'I float on the waves of art and life and never really know how to distinguish what belongs to the one or the other or what is common to both. Life unfolds for me like a theatre presenting a sequence of somewhat unreal sentiments; while the things of art are real to me and go straight to my heart.'[83] What higher authority could there be to justify fleeing the dull brain's perplexities, 'as though of hemlock I had drunk'? From those years only one musical event left a lasting impression: a broadcast of Sorabji's *Opus clavicembalisticum*. This chiefly impressed me because it put creativity first. Ignoring the railway stations where other musical compositions stopped to let listening passengers alight, it continued on its own tracks for hours: magisterial, self-absorbed. I find no documentary support for this memory – which must be dated to the mid-1960s (when the composer had banned broadcasts of his work). Timeless, like the experience of flying, this spectral memory stands out when the music I can prove I heard has faded.

Regarding unheard melodies that stay in memory, Keats offered various theories for their persistence. In the 'Ode on a Grecian Urn' their continuity is temporal, deduced ecphrastically from the painting or low relief of the 'happy melodist ... for ever piping songs for ever new'. In 'The Ode to a Nightingale', the song is supposed to suffer a spatial reincarnation; the 'plaintive anthem fades' until it is 'buried deep / In the next valley-glades'. Any lasting sound requires recording technology. Asked to comment on the ideas behind 'What Is Your Name', I wanted to say that one manifestation of memory is echoic; social identity is a refraction of the voice that mirrors – but, maybe, also parodies and imprisons – one's own utterance. Instead, I came across the essay by James Clerk Maxwell, called 'Psychophysick', written the year after the invention of the Talking Phonograph. In answer to the 'celebrated question', 'What is your name?', the mathematician replies, 'My name is the Conscious Ego, one and indivisible, the Subject in relation to whom all other beings, material, human or divine, are mere Objects.' The question arises of the continuity of the Ego in time: In particular, is the persistence of consciousness tied to memory? Maxwell thinks not. Memory is simply 'a physical impression on a material system', its nature directly analogous with Edison's instrument, which 'has an ear of its own into which you may say your lesson, and a mouth of its own, which at any future time is ready to repeat that lesson'. Once consigned to 'the pages of memory', the impression, or page, no longer has anything to do with consciousness: 'Where was that page when it was out of our consciousness? Not surely in the Ego, unless there be an unconscious Ego. It must be out of the Ego, and therefore an Object.' The conclusion? Memory is not essential to individuality.[84]

Maxwell compared consciousness to the 'changes in the mode of excitation' 'essential to perfect vision'. To see, to grasp the field of vision *as it is*, involves a continuous oscillation between appearance and disappearance, attention and displacement.[85] This distinction between consciousness and memory has its analogue in poetic philosophy where a phenomenology of perception can be contrasted with an etymological approach to the meaning of experience. When Vico and his modern descendants (Heidegger and those influenced by him) derive concepts from meanings supposedly buried deep in the early usage of words, they trade in what is 'out of our consciousness'. While philosophers find these poetic sparks exciting, they are useless to the poet whose representation of memory as desire must be contained entirely inside the 'illumination' of the image itself. Yet, Maxwell's strictures may not be entirely fatal to a theory of identity produced echoically or mimetically. Once the phonograph analogy is invoked, the idea can be entertained that the brain is not only productive, to quote William James, but 'permissive or transmissive'. 'The phenomenon we think of as "mind," cognition, or mental awareness, is a consequence of the brain's behaving as a kind of receiving station to "the genuine matter of reality" transmitted by the environment.'[86] This notion of consciousness as intermediary soon finds us in the realm of contemporary neuroscientific theories of 'dynamic co-emergence', where 'living and mental processes are understood as "unities or structured wholes rather than simply as multiplicities of events external to each other, bound together by efficient causal relations"'.[87] In this updated version of double-aspect theory, 'a flock of birds, a school of fish, a moving crowd'[88] – or the murmuration of starlings – express the 'transcendental orientation' of the brain.[89]

So Maxwell was, as usual, ahead of his time. However, his initial formulation puzzled me. I wanted to ask who asked the 'celebrated question'. Maxwell seemed to say it was a non-question: 'I often catch myself, when thinking about my body or my mind, supposing that I am thinking about myself.'[90] By objectifying every perception in this way, Maxwell certainly avoided the nightmare evoked by Kierkegaard's teacher Poul Martin Møller in *The Adventures of a Danish Student* –

> [I start] to think about my own thoughts of the situation in which I find myself. I even think that I think of it, and divide myself into an infinite retrogressive sequence of 'I's who consider each other. I do not know at which 'I' to stop as the actual, and in the moment I stop at one, there is indeed again an 'I' which stops at it. I become confused and feel a dizziness as if I were looking down into a bottomless abyss.[91]

But I couldn't help thinking that, objectified as the relationship between three personae or acoustic masks (1, 2 and 3, alternatively known as Mr McGuinness, Mr Hatcher and Mac), Møller's dilemma described the situation in 'What Is Your Name'. So far as the linguistically initiated Ego goes, consciousness is invariably a listening exercise. The 'material system' corresponding to this coming to self-awareness through the other is *poetic*. It is the entire memno-technical relay of mirrored movement, reciprocated gesture and vocally mimetic behaviour through whose evolution the goal of communication is clarified.

In this evolutionary feedback loop memory and desire chase each other, but the result is not necessarily the instantiation of Lacanian Lack; the exercise of the poetic imagination makes a lasting impression. It can be compared to an exchange of personal letters. They start in the same town; they continue in foreign countries; they persist even when the correspondents have relocated and live hemispheres apart. To whom do these lyrical descriptions belong once they are posted? They belong to the receiver, not the sender. But who is the receiver? He is not the one fleeing to another world but the one who, Actaeon-like, pauses mid-flight to look back over his shoulder. On the verge of becoming a lover, he also faces self-destruction. In any case, coming half way to the other, the letter's inventory of precious details invites a comparable gesture. The writing is not the printout of the Ego but the translation of the unconscious Ego's desire. A love letter is not the transcript of a memory even when it shares a memory. It is an impersonation, a figuring forth of the person that the writer imagines will seduce the reader. Calling out to what lies beyond the Ego, its transcendental part, it locates love within the horizon of (self-) parting. Aeneas abandons Dido; Eurydice return to Orfeo; they are the mirror movements of love's primal scene. The phonograph disguises absence as presence; the letter, more honest, makes absence the occasion of remembering.

The figure of Maid Marian in 'Remember Me' speaks entirely in quotations from letters I have. My correspondence with S----- extended over thirty years. In Act One, pre-Australian quotations are used; in Act Two, I have already gone down to the Underworld. Our separation to different hemispheres is compared to the localization of speech recognition and music/noise recognition in the left and right hemispheres (respectively) of the brain. Henceforth we lived on two sides of the same magnetic tape. When in 1988 I sent the script to S-----, she replied with mixed feelings: 'I am not a disembodied voice. ... I am not always a reflection of the sky,' adding, 'If you remember me as a shadow, it is as much as I deserve.' The object was not to stir up unresolved feelings. Her letters were the incarnation in my life of that general injunction, 'Remember me! But ah! Forget my fate.' She personalized the recording, in the course of hundreds of letters tracing underneath the arc of love the secondary arc of remembering. Scattered through the dense forest of her green calligraphy was a theory of the imagination. The transformation of emotion into memory was a poetic struggle enacted in type as well as topic: the observations she caught and hauled out of time were conveyed to me in a vintage script that suggested experience doubly curated. These artefacts of our time look medieval in the era of texting, but the poetic theory they developed was inseparable from the 'material system'. And the spiritual alchemy aimed at depended on them being letters staging a drama of encounter, parting, adventure and return. They were knightly in this way, affectionately updating Arthurian tropes. Intended to bring two people who lived remote from each other close, they were to be read metaphorically as well as literally, as spells cast from the other side that swallowed up the distance. The purification intended was alchemical as it addressed the poet-alchemists themselves; the love bestowed on the details of urban and country life was to overcome 'the sleeping dragons all around' and yield the fusion in flight imagined in 'The Eve of St. Agnes'.

The theory was, indeed, decidedly Keatsian, beginning in an untrammelled appreciation of ripeness: 'Is this not how we should live, caring for the small warm details, living well into earth and sun and rain, the reflection of fire on the leather of the book and the crystal round the wine?' Extracting such details from the flow of time, she placed them in a frame of silence. The reflective zone placed around experience might be technological – 'A fox family rolling in the hedge shadow on the next slope – the vixen suddenly springing, four paws off the ground, down, to toss something small and limp – a vole or a mouse from a tussock – all in a silence; and through binoculars.' But usually it was born of her own negative capability, to be in the midst of life without the desire to possess it: 'There was a wooden knocking behind me in the apple tree. I turned my head, like a spy, and caught sight of the spotted woodpecker in the grey-green fork, splitting hazelnuts.' There are ecphrastic sound impressions out of John Constable: 'Cattle in the spinney, clumsy breaking branch and splashing water. Heard the ravens croaking, a heavier sound than the rooks.' But the value of these 'circles' cut out of the fabric of time lies in their recollection – which, she recognizes, is not a photographic reproduction but a rethreading of the image into time's texture: 'You are often in my mind, but you are always there like an underground stream. ... I saw one on Mendip today, silver and flowing steady under an icy haze of blue sky. It disappeared down into the rocks and its song was lost, but the sound of water stays with me.' Vision passes into sound, and the image, if it returns, bears the impression of its transformation and may be both familiar and strange:

> It seems strange how the people I love most will merge and blend with the patterns of leaves as I watch them, stride in the starlight, and whistle my dreaming down the wind – they haunt me like ghosts, and seeing the real people again there is an uneasy moment when the two shapes that I recognize are not one and the same person; and another moment when the ghost is real.[92]

Here is recording technology that post-produces experience, through the act of recollection altering and enriching it. The agent of this creativity is a writing that both draws near what is remembered and withdraws from it. Writing bears witness to the finitude of existence. Its looping gestures say something beyond what is said, like the spider's filaments sent out ahead of a web. Its conventions stem from its double constitution as record of loss and promise of return: hand writing, letter spacing, the tumble of lines, one breaking after another, embody memory as desire. 'I write to you on typing paper to put you on the level of my poetry – our double spacing seems self-evident. I have been living in unreality until now, and I understand that my life is to write, however insufficiently towards a clear expression of truth.' Writing separates memory from the pain that originally produced it and, again in a Keatsian way, transmutes experience into beauty: 'No resentment can shatter the world I see you in; the struggle with words to express that world is mine as well as yours. Underlying every attempt to fight, respect, or jettison both words and unworded recognitions is a central beauty which cannot be flawed.' The struggle with the recalcitrance of matter or other minds may start in physical separation, in unhappiness; but the struggle with words leads to another separation, an extraction of the memory from the noise of time. In a passage I reset in 'Remember Me', S----- wrote,

> When I am writing a letter in the sun, even as I write and hear the fountain bubbling, I also see the past, a series of episodes particularly vivid ... coloured by birds and insects. I saw kingfishers with you, flashing over an Oxfordshire river; I remember the great blue Emperor dragonfly on a New Forest afternoon; I see snow, the packed ice under my feet as I wander miserably in B------, the snow above the high plateau of L------, in Crete. I have at last come to a point where those memories which were so beautiful can be separated from the emotions which are now so painful. The shutter comes down, the picture is stilled.[93]

Perhaps the comparison of poetic transformation to the achievement of the photograph is not strictly correct for, in modern photography at least, the period of latency when the poetic image is forming is eclipsed: the stilled life of the photograph anaesthetizes time consciousness, the sensation of water bubbling underground, its unheard melody still present and giving shape to the mere passing of time. Ideally, through the struggle with words, an integration of language and rhythm is found, an image that is not Imagist but continues to flow, as a reflection furls and unfurls like a flag on the surface of the stream while keeping its place. Such a transmutation is suspended in time but retains the imprint of time passing. In the apotheosis of this process, the image no longer really matters; it is a stepping-stone to an absolute attunement where an emotional meaning beyond words is articulated. It is to turn intention into sound, to be *in tone*. It is the sigh released when the image, no longer stilled, begins to flow again.

S----- associated this second release, back into passing time, with the sound of my voice, writing in the mid-1970s:

> The marvelous sea, the foam and silver and the sense of you at my back, completeness of rhythm and suspension. It seems I do not need to draw circles around these moments; they are not capable of fragmentation ... understand in how many enchanted circles you are, and you will understand that I want your voice, not when I am frightened that they fade but when [other people] threaten to divorce me from my world.

I think this voice stayed with her even when we were not talking but no phonograph could capture it: in her reception it made sense of passing time, lifting from its tumultuous noise a murmur, a hum, a beginning song that did not have to be circled and isolated but, so long as it never went away, stayed time's dissolution.

SEVEN

Voices

One of the licences of self-writing is the defence of subjectivity. Experiences described in autobiography (or *oto*-biography) do not have to be explained; they are local knowledge whose value is personal. To come out, the personal has to assume a *persona*, a mask, a voice, a speaking position or other style; but this disguise may only be a further temptation to pass oneself off as exceptional, to claim a richness of interior experience that might, on investigation, prove to exist throughout the population. When Richard Church asserts that as a boy on the verge of puberty he used to float like a balloon under the ceilings of his house, his assertion also hovers unstably in the domain of publicly accessible knowledge.[1] When Giambattista Vico asserts that he owed his power to dive to the bottom of things and discover the foundations of human institutions to a massive *syncope*, due to a bump on the head,[2] his temporary amnesia delivering him out of this world into the next, and his return to consciousness equipping him with the irony essential to the knowledge of truth, his personal myth (or mask) has to be taken at face value. If I say that I hear sounds as voices, who is to gainsay me? When I selected certain radio scripts for publication, and recalled the ambitions I had for their production, I noticed one preoccupation held them together. Despite their thematic eclecticism, they believed that 'beyond the imperialismo of signs … there must be an art of sounds'.[3] And I also noticed that the *persona* of this art, *the personal style giving me access to this realm beyond representation*, grew out of my habitual sensation that sounds spoke to me, and, recalling one another exactly, gave me direct access to what Bruce Smith, discussing Shakespeare's Caliban, calls 'a sound world before language'.[4]

It was a sensation in two parts: of call and of recall. The birds of childhood called me to attend. All sounds had this capacity, the hush of tyres, the explosive slam of the passage door, the pitter of rain overflowing the gutter. But the percent calls of birds astonished me and demanded attention. Eventually, bending to their calls became habitual. The second part is more difficult to describe. It was the sensation of self-same recall. If I say, hearing certain bird calls, I had the sensation of another sound perfectly recalled, I give the impression of a punning aural coincidence. But when in the Himalayas I heard the first four notes of Beethoven's famous symphony in the song of the black-and-yellow grosbeak, no resemblance or ironic approximation was discerned.[5] In North Africa, when the bifasciated lark whistled the opening phrases of Dvorak's violin concerto, no transposition of musical training intervened. These were specialized acoustic experiences bound up with being an itinerant bird watcher (listener), but their extra-territoriality was not exceptional. Recognition of this kind inhabits the commonest of sounds; every day in Melbourne, I hear sounds instantly recognizable from Faringdon half a century ago. A chance involute of noise spinning through an open door is my mother's voice heard over the radio. The squeal of brakes *is* the chair leg scraping against the kitchen floor. An Australian magpie flying over degrades a theme from *Swan Lake*. It is not an involuntary Proustian memory but the thing itself. A Combray landscape *may* spread out around the sound, but is secondary, an after-effect. Local knowledge recovered through the sound association *could* have been reached otherwise (through smell or taste) and is not integral to the meaning of the sensation.

Nowadays some researchers treat this kind of acoustic identification as evidence of absolute pitch. It is recognized that defining absolute pitch musically begs the question of detecting it extra-musically. To say that someone with absolute pitch can identify the musical note corresponding to an electric shaver, a whining mosquito or a bubbling water hydrant simply widens the definition of what is musical. It does not account for the perception that different noises are identical, that resonant in one timbre may be an entire acoustic labyrinth. Perhaps the absence of a formal musical education has preserved in me this susceptibility to absolute imitation. Untrained to organize and hear sounds in musical sequences, I continued to identify their sense with their particularity.[6] The particularity of a sound object may be vast (a symphonic theme) and the number of particularities unlimited (all the birdsongs ever heard). But, whether the call is a simple pitch or a complex harmony, its recognition is not relative. Instead of conjuring up other words or tunes, it exists absolutely, as a self-same reproduction of another sound. It is not perceived as analogous. It is a noise that cannot be harmonized or generalized: the middle term of comparison – in Western music the diatonic scale, in non-tonal languages like English the phonological or articulatory-perceptual system – does not intervene. Whether or not this experience proves that my auditory cortex and dorsal frontal lobe are working closely together, who knows?[7] Whether, instead, it should be considered a benign version of Daniel Paul Schreber's impression that 'miraculously created birds' possessed of a 'natural sympathy for *similarity* of sounds', spoke to him, can again be left undecided.⁸ In an auto/oto-biography, it should be enough to describe it.

What does it mean to hear sounds as *voices*? To describe the birds as speaking suggests that they are avian ventriloquists: the exemplar of this is the talking parrot. But by the time the parrot *speaks* he is already trapped in the cage of language. He is the echo of the ego; even his song is ironic, a lament for the lost right to silence. 'I am tired, as always, of myself,' laments Rumi's Abulfulan, the Merchant: 'Day after day I face the world. It is the same, I am the same. I crave some other self. Only you understand, only you provide that other self I crave – by your voice, your beauty, your ... .' Tuti, the Parrot: 'Self I crave ....' Merchant: 'Yes, yes. You understand perfectly, as always.' Tuti: 'As always. I understand perfectly, though I only echo you.'⁹ The spiritual meaning of the allegory is that the soul seeks release from the prison of the body; literally, though, the voice seeks liberation from acting a part – from parroting. Instead of speaking through the persona of the echo, it desires to die to false selfhood and be reborn in the instant of flight. As the Tuti explains,

> I am an echo of yourself which you have caged. I have no other song to sing but songs of being caged to sing you songs of your old tired self that longs to hear some other song but can't because you have that key around your neck to keep me caged so I will sing the sweet sad song of your old tired self that can't escape yourself because you cage yourself and are afraid if you release the echo bird you'll lose yourself.¹⁰

The Merchant cannot let go ('Without you in my life, my life would be without release. We are bound together'¹¹) until the *Tuti*, hearing of his free brother's death, imitates him. 'Believing him dead, the Merchant finally unlocks the cage and casts the bird out of the window. The *Tuti* revives midair and flies away.'¹² Abulfulan had imagined 'a sweet exchange / Of our positions. I would never make you sing / But send you to your brothers .... The silence could be beautiful. ... I could rediscover what a simple thing it is to fly',¹³ but had been afraid.

While the spiritual sense of this is clear ('We are brothers, not creatures to be put in cages. / My brother taught me by his silent act / My own voice kept me a prisoner. / We are two bodies, he and I, but one soul'¹⁴), the phenomenological meaning is different. It is not the voice that imprisons the parrot but the power of speech. Naturalistic confirmation for this comes from the claim – attractive, although I have never been able to prove it – that the parrot's mimetic gifts are only displayed in captivity – yet, in the wild, parrots merrily vocalize, often in mid-flight. Such vocalizations do not represent anything but themselves; they are songs without nostalgia. Hence, in 'Memory as Desire', Robinson remarks, 'I'm not sure I don't prefer a world of sounds,' reflecting that it might be the 'cure for memory'. His wife, taking after Prospero – her Caliban, Australian Aborigines – feels it would be 'true Christian benevolence to equip them with English'. Coming back to the subject later, Charlotte Sturt says, 'They sit up all night, laughing and talking.' Robinson responds, 'It is this habit which gives them such good voices.'[15] Hinted at here are brotherly vocalizations that never fall victim to the *imperialismo of signs*. Instead of the violent subordination of one voice to another, marked by the faithful reproduction of his master's voice, a congregation of voices occupies apertures between individuals, operating as 'openings between worlds'.[16] Mimicry amidst a babble of voices is not the familiar *folie à deux* of master and slave – in reality, a loud form of silence – but the exploratory search for similarities. 'Listen to a primitive tale,' Jackson 1 urges in 'Cooee Song'. 'Of OU and EE', his double replies. 'I.e. YOU and ME', says Charlotte 2; 'Of self-sufficient sounds' (Jack 1); 'Sounds as such' (Jack 2); 'Not music, not poetry; neither thoughts nor words' (Char 1) 'Beyond these, before these'. (Jack 1).[17]

My production notes for 'Cooee Song' explain that 'the poetic principle' informing it is

> not Orphic, [not] born of nostalgia for a lost wholeness. On the contrary, it locates the emergence of identity – voices, stories, refrains – in the noise of ocean. Its poetic geography is inspired by Australia's Aegean Sea, Bass Strait, running between Tasmania and Victoria where a southern Sabir was spoken by the myriad races attracted there by the whaling and sealing. At the centre of this southern Cyclades is a kind of anti-Delos, Babel Island, named for the cacophony of sea birds, but suggesting the new confusion of the sailors' language. 'Cooee Song' is another attempt to give history's migrants a different literature of our own.[18]

Evidently, I understood speaking in voices as a different form of social communication. The primitiveness of the tale referred to the foundational nature of the discourse, a primary simplicity based on the principle of similarity. As for this 'natural sympathy for *similarity* of sounds', Schreber explained that if, 'while reeling off the automatic phrases', his birds

> perceive *either* in the vibrations of my own nerves (my thoughts) *or* in speech of people around me, words which sound the same or similar to their own phrases, they apparently express surprise and in a way fall for the similarity in sound; in other words, the surprise makes them forget the rest of their mechanical phrases and they suddenly pass over into *genuine* feeling.[19]

When they become aware of their own echoic mimicry, the birds *can get through*. Schreber thought 'that the nerves which are inside these birds are remnants (single nerves) of souls of human beings who had become blessed' – a memoirist's licence that can hardly be subjected to the laws of evidence.[20] As production notes describing the behaviour of revenants, or doubles, they would present no difficulty.

In *The Sound In-Between* I gave many examples of *social* occasions where echoic mimicry spilled over into genuine feeling. George Taplin's comment, 'The Narrinyeri [Ngarrindjeri] are skilful in the utterance of emotion by sound,' stands in for many:

> They will admire and practise the corobery (*ringbalin*) of another tribe merely for the sounds of it, although they may not understand a word of the meaning. They will learn it with great appreciation if it seems to express some feelings which theirs does not. They may not be able to define the feelings, but yet this is the case.[21]

The expressive sounds detected here are voices without language, or recognizable noises. Such communication is not primitive in the sense of conjuring up 'a sound world before language', but it does represent a fundamental difference between the ways colonizer and about-to-be colonized communicate. The British in Australia, or the Spanish in the Caribbean, will go by the book – 'language is the perfect instrument of empire,' not when it facilitates translation (or even Christian conversion) but when it subordinates the 'voice' to 'the service of the soul's thoughts.' Orthography aids orthodoxy and both, 'stabilising the modes of representing thought,' ominously thrive 'in the absence of the body.'[22] But, as we also saw, the New World strikes back, inaugurating colonial relations under the colourful aegis of the parrot. Caliban's insistence on the voices in the island places the language of merchants and colonizers between quotation marks, their summons to obey *within* the world 'full of noises' evoked, for example, in 'Columbus Echo'.

Whether or not the phenomenon of the self-same reproduction of another sound is a sign of having absolute pitch, it hypothesizes the possibility of absolute recall. In a social situation, it offers an explanation for communication improvised out of mere phonic coincidences. Two or more people who desire to communicate but have no language in common make sounds. Recognition of intention (or genuine feeling) occurs when one party hears another's utterance as their own. But underlying this theory of encounter is 'a natural sympathy for *similarity* of sounds'; otherwise, no auditory recall would be triggered when another spoke – no punning identifications would be discerned – and, the corollary of this, a dialogue predicated on echoic mimicry would not work. To an astonishing degree, my writing for radio and sound installation, and my ideas about cross-cultural communication, contemporary and historical, stem from this intuition. The notion that we can, at certain times and under certain conditions, achieve a Pythagorean 'identity of existence with essence' suggested to me that the circumstance most often associated with violence (colonial encounter) held within it an opposite promise, of liberation from the cage of analogy. Listening, like Caliban, for the voices in the noise, detected in what I called the 'sounds in-between', one retrieved an art of sounds innocent of semiotic overload. The recovery of innocence experienced in instant recall is not historical or moral: it is psychological and emotional; in the echo of ideas is grasped, Ungaretti says, our nullity – an idea faithfully reflected in 'The Letter S', my later reprise of the scripts written for Berio. 'Ou tis ou tis', a voice cries out, echoing the name Ulysses gave Polyphemus. Someone hears the cry as 'Who 'tis' or 'Who is it?' 'Nobody', comes the answer, 'nobody'.[23]

The doubles that issued from this proliferation of echoes solved for me the problem of the One and the Many. They were neither the phantasms of an imagination that confused self with cosmos, producing unsatisfactory avatars like Schreber's 'fleeting improvised men', nor the anonymous agents of Elias Canetti's crowd, first causing the self's surrender, then its subjection in the Many body, another version of Schreber's paranoid One – 'Suddenly it is as though everything were happening in one and the same body.'[24] Even the passages in the final scenes of 'What Is Your Name' lifted from Strindberg's autobiographical *Inferno*, where a stranger ('The Unknown') moves into his Paris dwelling ('He apparently occupies the rooms on both sides of me, and it is unpleasant to be beset on two sides at once'), acquired an ironic levity when disposed echoically across the trinity of 1, 2 and 3, alias Mac, Mr McGinniss and Mr Hatcher.[25] Besides, Strindberg's doubled double never speaks. In its most intimate and productive form, the doubles produced in this way become lovers. 'Cavalcanti's light signifies love's energy', NOM or Any Man says in 'On The Still Air', referring to the moment when eyes meet, in which *la donna* returns the stranger's gaze.[26] In such moments, Ungaretti writes, '*A lampi, per intuizione di una rapidità fulminea, è "rotto il gelo", e in qualche modo essa ci informe di se, ci permette di riconquistare in qualche modo innocenza.*'[27] Any Man's betrayal consists in allowing this moment to pass – 'Not to look deeply enough into your eyes. ... Not to have grasped the whole of you. ... To have trivialized the corpuscular light'. Any man has let her lapse into the realm of representation – Pasolini refers to this as 'the hand in hand of the consumerist couple.'[28] At the end of 'On The Still Air', the Trevi Fountain is turned off. In the aftermath of rivulets trickling and drying up I heard 'the echo ... the still air renouncing song' – 'The sound', Andreotti replies, 'of what might have been'.[29]

The intuition of an absolute recognition, primarily mediated through sound but synaesthetically available – one thinks how Caliban's 'twangling instruments' produce in him what might be called a tingling sensation – encouraged me to speculate about the origins of colonialism. A latter-day Vico, I tried in works like 'Mirror States' to rewrite the origins of colonial society etymologically. Instead of searching for the echoes of original ideas about the constitution of the just society, I listened *locatively*, ascribing a foundational power to the positions of the voices. In vocalizing, the place of the voice – whence it calls and whither – is unmistakably genuine, and the echo never lets you down. In that calling to come, 'existence' and 'essence' are momentarily identical. 'Mirror States' presents this poetic thesis ironically, as the echoic backdrop to the construction of a World Trade Centre in Melbourne. This initiative is characterized as an ironic *ricorso* of the original, fraudulent Batman Treaty in which local peoples allegedly ceded their land to the colonists in return for blankets, scissors and mirrors.[30] The Voices discuss the height of the new Tower of Babel, but they cannot rid themselves of historical noise, of the haunting sensation that in cutting the cloth to suit themselves, they are recapitulating the substitution of narcissism for love:

> Voice 2: And I looked up and he was measuring me for a suit. Only it was my image in the mirror he measured.
> Voice 3: Fifty scissors.
> Voice 2: It's too small, I said, but he went on cutting all the same.
> Voice 1: Let's make a name for ourselves.
> Voice 3: Sixty mirrors.
> Voice 1: We make it sixty storeys high.

Scene Eighteen is a coaching lesson for business professionals. It consists of simple and powerful body language tips. The idea is to go through the motions of communication while concealing one's true position. It is a lesson in the non-communication of genuine feeling. The recommended body language tricks ironically imitate older cross-cultural negotiations on this site (notionally on the Yarra River in Central Melbourne). Somehow, the cynicism of the tactics taught casts a backward shadow over the original negotiations between the white land-grabbers and the communities referred to at that time simply as 'the Yarra Tribe'. The conscience of this historical bad conscience is a world still full of noises – the calls of the wattlebird that continue to penetrate the anechoic chamber of neo-liberalist state capitalism:

> Voice 2: And at that time there were as many of them as there were of us.
> Voice 3: In the mornings they rose at dawn.
> Voice 1: We rose at dawn.
> Voice 4: Remember: other people will be reading you.
> Voice 2: With child-like wonder hearing the sounds of the bush.
> Voice 4: Position yourself strategically.
> Voice 1: The tinkle of the bell-birds.
> Voice 4: If you want to appear humble.
> Voice 1: The laugh of the kookaburra.

But the strain of resisting the calling to come is too great. A kind of semiotic crystallization occurs and the free flow of sounds is frozen out.

Voice 4: Seat yourself in the negative positions.
Voice 3: They called to each other across the river.
Voice 4: If your chair has arms, put your elbows on them.
Voice 3: 'He promised to buy me a bunch of blue ribbons.'
Voice 4: Do not hum.
Voice 2: 'I'll kill you, I'll kill you' (it was only a bird).
Voice 4: If your chair does not have arms ...
Voice 2: Before sounds took sides and hardened into sense.
Voice 4: Put your hands on the table.
Voice 1: The wattlebird's castanets.
Voice 4: Not in a flat clammy way, but lightly.
Voice 2: 'What is your name?'[31]

Only while writing this book did it occur to me that the immediate auditory memory that provides, as it were, the ontological ground of the writings discussed here maybe a specialized perception, one not generally diffused throughout the listening public. Besides, transposed to the realm of electro-acoustic broadcasting, it was only represented. As the passages from 'Mirror States' show, elsewhere it was simply embedded in the script like a poetic myth, and could be treated, if it was noticed at all, as a reversion fantasy, recapitulating biographically the historical notion of 'a sound world before language'. Ironically, one reason why these cultural writings did *not* fall on deaf ears in the 1980s and 1990s was the ascendancy in cultural theory of *semiotic* accounts of communication. Their appeal to the phenomenon of self-same sound was in strong contrast to the semiotic orthodoxy captured in Roland Barthes's remark – 'no sooner is a form seen than it *must* resemble something: humanity seems doomed to analogy'.[32] This is not an innocent call to exercise wit. It is an ontological claim: 'The ubiquity of tropes in visual as well as verbal forms can be seen as reflecting our fundamentally *relational* understanding of reality. Reality is framed within systems of analogy. Figures of speech enable us to see one thing in terms of another.'[33] In musical terms, this theory identifies social functionality with relative pitch acquisition – a view found formally in music education and informally in language learning, where competence in relative pitches (intonational contours, melodic phrases) transposable from instrument to instrument, speaker to speaker, pitch to pitch and scale to scale, ushers the child into the world of communication.

In fact, to tolerate a theory of the sonorous body – of the kind Harrison proposes when he speculates about 'the indicative properties of sounds [that] can open out within hearing ... quite apart from any considerations directly bearing upon meaning as such' – *dialectically* is already to bow to the semiotic supremacy, whose foundational concept is functional contrast. At the phonemic level two sound units become meaningful when juxtaposed and the sound contrast is recognized as significant within a particular language. By extrapolation, sound philosophies that question the semiotic communicational model present a logical paradox. Except as a subjective taste, or aesthetic preference, informing certain kinds of musical or poetic composition, Harrison's sonorous body, Bachelard's poetic reverie that 'revives the world of original words', or even F. Joseph Smith's 'lyrical-poetic event of *onomatopoein*, naming according to the sound of the thing we hear', strictly make no sense.[34] The closest structuralist analysis can come to the phenomenon of voices, that is, self-same noises, is to speculate that the principle of analogy or representation of one thing by another sometimes breaks down. Then, a 'positive ambiguity' corresponding to 'the "poetic function" in human communication' may occur.[35] Merleau-Ponty relates this 'good ambiguity' to 'an inclusive concern with intersubjectivity',[36] founded in the awareness of the 'body-subject' as agent of 'an ethic that is rhetorically constituted authentically'.[37] Here, the ambiguity inherent in occupying any speaking position certainly translates into an awareness that others are similarly changeable. Creating 'possibilities for things to become otherwise'[38] certainly paves the way for a different future history, one where sounds, instead of taking sides, embodied 'authentic existence',[39] but this is not the same as granting sounds their own authenticity.

To return to the phenomenon of voices that speak – albeit in the language of self-sameness: to call us or cause us to recall, they, too, must be ambiguous. A sound that calls ceases to be an unmarked molecule in the gaseous sound environment; it suggests a principle of combination, the possibility that things could be or become otherwise; when a sound is heard and immediately recognized as the self-same sound heard on another day in another place, it obviously shares with the remembered sound a quality of being sounded differently – at another time and in another place. One could say with Roy Wagner that it is echolocative. Caliban's 'thousand twangling instruments' that 'hum' about his ears sound ambiguous. They are associated with 'Sounds and set airs' and 'sometimes voices'; the voices put him to sleep and send him dreams where 'the clouds methought would open and show riches' – a fancy that recalls Sir Richard Vernon's vision of a transformed Hal in King Henry IV, who mounts his horse 'like feathered Mercury ... as if an angel dropped down from the clouds'.[40] But such auditory impressions as these are not confined to islands or prelingual monsters: they are every urban soundscape. Approaching a major building site in any city centre, pause to listen: the clangour of steel girders, the chime of aerial hammers, urban reverberations caught in stairwells, screeching brakes that decay like sirens and the serpent hiss of air suddenly released – these are our twangling instruments.

In the production of 'Mirror States' for radio, my sound recordings of the construction of Melbourne's World Trade Centre and Rialto Towers were minutely audited for traces of another, sonorous voice. As noted before, I hypothesized an auditory conscience recoverable from 'a material history of the city's soundscape.' Yet there was nothing foundational or archaeological about the urban score overheard in this way; much would depend on 'mere coincidence' embedded in the production itself: as when ambience recorded in the World Trade Centre proved on playback to preserve 'through and between the shuffle of feet and the ethereal Muzak, a conversation about a house, about a storm – themes that mimicked remarkably stories 6 and 8 of "Mirror States".' Mimicked but also reflected ironically, as 'Susannah' mimics 'Yarra', asserting an identity able 'to escape from the glassy stare of executive suites and to start to flow endlessly on the spot again.'[41]

Now I would not present this sound genealogy as proto-semiotic or pre-lingual, emphasizing instead that these different sounds belong to one timbral family. In England as in Australia, they are our everyday angels: 'Inside the metronome of passing cars … . Plucked strings of rain on the microphone … .Hiss of tyres, spool-warp, manic thrush ringing the steeple of the ear, the nightingale, the angel voice of an England England no longer hears.'[42] The recording angel in this context is the angel in the recording, that is, the voice discerned in the noise. Mediating between the divine Logos and the capacity of humans to hear the word, the angel succeeds, like Kantor's actors, in solving 'all the problems, the dilemma of autonomy and representation, with ease'.[43]

The angel identifies the echo with the consciousness of death. The angel's message is an exact repetition that mediates between the immaterial and the material. It is the repetition that *in its repetition* 'makes one think about infinity, about our life and its relationship to infinity, about this SOMETHING that approaches us, passes us by, and disappears.'[44] The angel is the ultimate passer-by. It is the figure that makes sense of Kafka's claim that 'the history of mankind is the instant between two strides taken by a traveller'.[45] Apropos the *New Angel* of Klee, Massimo Cacciari points out that the angel does not pass, transmit, intercede, but is itself the passage, an icon of the *in-stant*.[46] Cacciari suggests that the word the angel pronounces is received because we want to be called.[47] In a sense humans are constituted echoically and the quest to understand the meaning of life is, in one sense, the desire for proper pronunciation. Where better to learn this than from the primary messenger or *nuntius*, the angel? The word received in this way does not fall into historical time to be remembered at a later date: as the trace of a divine encounter, it is retained permanently, like an unheard melody.[48] When the angel joins the visible to the invisible, he does not tell of this passage but embodies it. He does not reproduce the Word, the Name, the One, but transmits it musically. It is the rhythm of the angelic souls that communicates the divine essence. The angels are types of rhythm or rhythmic types.[49] Meister Eckhart distinguishes between the word which, when pronounced, stiffens into representation and the word that endures in whomever pronounces it, as the original images of the created world remain in the Father, as Logos.[50] The angel's word must be like the grooves of the record, an impression that remains the same however many times it is played or repeated or heard. The impression is not a representation of the word, but of its track.

Descending from the realm of astral bodies,[51] and the mediation of the immediate via composite doubles both spiritual and corporeal (as the air is both breath and spirit), how is self-same sound recalled? How is the calling to come experienced as a state of being open to the future and as a definite moment of recall and renewal? Its physiological expression is well enough known. It is the tingling sensation described in the expression 'someone walked on my grave'. It is a sudden shudder, referred to in Italian as '*brivido*' or in French as '*frisson*'. In his *Etymological Dictionary*, Skeat says that Caliban's '*twang*' is a 'collateral form of tang' and he links it to *tingle*.[52] This family of words imitates a convulsive vibration associated with the involuntary recall of an emotionally charged event. It is simultaneously registered visually, aurally, indeed through all the senses, and physiologically. Should Proust's famous description of a scene from childhood brought back to him in the taste and smell of a tea-dunked biscuit be classified as a similar kind of 'involuntary memory' (the author's phrase)? Certainly it can achieve poetic expression. Ungaretti's short poem '*Su un oceano / di scampanellii / repentina / galleggia un'altra mattina*', succinctly imitates the feeling it evokes.[53] Like Caliban's 'thousand twangling instruments', his *scampanellii* (many little bells ringing), evoke voices swarming in the noise. Instantaneous recall of another scene like this involves multisensory or multimodal integration – critics do not agree about what Ungaretti's poem recalls – Is it a sound or visual image, 'the awakening of a village'[54] or 'Alexandria, grasped in the arms of the sea'?[55] Or, perhaps, the paradoxical feeling captured in *repentina* that something absent is suddenly there?[56]

Personally, I always thought of the relationship between the sound and its recall vertically; I never imagined the sound image of another morning returning to me magically from another country; even church bells unexpectedly heard in dreams seem to come up from underground, an idea the word 'suddenly' conveys with its roots in the Latin '*subire*', meaning 'to come up from underneath'. I suppose the thought is that the formerly solid ground suddenly collapses, and one is plunged down into dark depths, an experience that can go either way psychically, prefiguring self-destruction or self-renewal – or, in the case of Jung, battling to overcome the Freud–Siegfried figure in his life, both:

> In order to seize hold of these fantasies, I frequently imagined a steep descent. I even made several attempts to get to the very bottom. The first time I reached, as it were, a depth of about a thousand feet; the next time I found myself at the edge of a cosmic abyss. It was like a voyage to the

moon, or a descent into empty space. First came the image of a crater, and I had the feeling that I was in the land of the dead. The atmosphere was that of the other world.'[57]

Six days after his plummet into the darkness, he meets a 'brown-skinned man' who helps him slay Siegfried.[58] Jung interprets this silent figure as his own other, his shadow and antithesis of the Siegfried ideal; postcolonial psychoanalyst and author Octave Mannoni disagreed: 'The Jungian attitude is in effect a search for lost innocence, for a return to the happy intimacy of mother and child – and it produces an equally false picture of "savage" man as the opposite attitude. It is also equally likely to end in failure; Prospero's relationship with Ariel, for instance, was as unreal as his relationship with Caliban.'[59] A comparable attitude is evident in historical sound studies when Caliban is imagined as inhabiting 'a sound world before language'.

How different the descent into the underworld would have been if the brown-skinned man had spoken and Jung had listened. Instead of experiencing a silence violently broken by Siegfried's horn, he would have felt surrounded by the noises of the place. In the years following the burial/rebirth dreams, Jung developed his theory of active imagining whose practice James Hillman explained thus: 'If the soul wanders from the body in sleep, then our way of letting the soul return to concrete life must follow the same wandering course, an indirect meandering, a reflective puzzling, a method that never translates the madness but speaks with it in its dream language.'[60] Suppose Jung had rested on those mountain paths or in those mountain caves and fallen into a reverie, had directed his hypnagogic dreaming – a transitional state between consciousness and unconsciousness – to listen ... and then to hear, as John Ball did, 'the sound of musical instruments, pure and clear, but barely distinguishable.'[61] The Carso, where Ungaretti served as a soldier during the First World War, is also famous for its caves: perhaps his *scampanellii* heard one morning in 1917 were confused tintinnabulations issuing Python-like from a fissure in the ground. Whatever the case, Ungaretti understands the medium of recollection as liquid. Its penetration is not achieved by flinging oneself off a cliff. To enter the submarine realm of recall is to be a diver. A technique of recall is involved that can be compared to the development of active imagining or hypnagogic reverie known as lucid dreaming. In this a conscious direction is given to the dreaming. In the composition of a poem it is the exercise of the auditory imagination – as when, in Ungaretti's poem, *scampanellii* is sampled and resampled until its twanging brings into focus *galleggia*.

Olympic swimming champion Greg Louganis describes one of his virtuosic dives: 'I reached through and left the platform. I kept looking at my spot across the pool. I circled around through the air, grabbing my knees, and then picked up my spot on the surface of the water. Another rotation, spot the water, another rotation, spot the water.'[62] What appears a mere instant in the eyes of the amazed beholder is, for the diver, an evolution of poses, each flowing liquidly into the next and joined together by the thread drawn between the diving board and the 'spot'. Translated into action is, indeed, the poetry of motion in which echo and idea, recollection and renewal, fuse to produce a dive that secures the maestro's return to the surface. In the same way even an involuntary memory – the sensation of a self-same sound – may seem instantaneous, but contains within it, as the ear contains a labyrinth, a topology of simultaneity that is evolutionary. It could be compared to my childhood experience of aerialism, where alighting was always stepping down cushions of cloud and never an airless plummet into nothing.

As a matter of fact, Proust concurs with this verticalist description. His memory of 'the whole of Combray and of its surroundings', spreading out like 'the little crumbs of paper' that the Japanese dunk in porcelain bowls with water ('the moment they become wet, [they] stretch themselves and bend, take on colour and distinctive shape, become flowers or houses or people, permanent and recognisable'), is not, after all, involuntary. The initial tremor when he tastes and smells the tea-dunked biscuit may be so, but the search for its meaning involves researching a method of recall. To bring to the surface the 'dead moment which the magnetism of an identical moment has travelled so far to importune, to disturb, to raise up out of the very depths of my being', Proust improvises a mental discipline that recalls Husserl's experimental technique of *epoche*, or voluntary suspension of reliance on the given phenomena of the everyday. First, in an attempt to 'recapture once again the fleeting sensation', 'I shut out every obstacle, every extraneous idea, I stop my ears and inhibit all attention to the sounds which come from the next room.' He next lets the denied distractions back in. Then, again, he shuts them out, clearing an empty stage in front of the mind's eye. Then 'I feel something start within me, something that leaves its resting-place and attempts to rise, something that has been embedded like an anchor at a great depth; I do not know yet what it is, but I can feel it mounting slowly; I can measure the resistance, I can hear the echo of great spaces traversed.'[63]

Comparing this rendering with the original, one can see how much of *scientific* value may be lost in translation. An anchor does not rise by itself from the seabed; echoes are not among the sounds that water transmits. Nevertheless, the verticalist image is clear: whatever has been lost is imagined as surfacing out of the deep. It will not rise of its own accord.[64] He must fish for it, overcoming his aversion to hard work.

> Now that I feel nothing, it has stopped, has perhaps gone down again into its darkness, from which who can say whether it will ever rise? Ten times over I must essay the task, must lean down over the abyss. And each time the natural laziness which deters us from every difficult enterprise, every work of importance, has urged me to leave the thing alone.[65]

If we dive down to the bottom of the madeleine episode, we can recollect something else that has been forgotten, a drowned sound. It has been shown that the madeleine scene is derived from an episode in Pierre Loti's novel *My Brother Yves*, and that the coquille-shaped madeleine has its origin in the Breton *berniques* that function for Yves as memory triggers: 'The things that Yves remembers are all down at the bottom of that stairwell; some of them go even further down, like the well and the stable, which is "at the bottom" of the courtyard.'⁶⁶ The *berniques* are found in the estuary, under the tide. Proust first intended to build his famous scene around the savour of toast; it was only with the substitution of the scallop-shell-shaped madeleine that the mechanism of involuntary memory became clear, its association with de-anchoring.⁶⁷ The scallop buries its lower shell in the mud and has to be prized from its half-concealed anchorage. Brought to the surface, put to the ear, it whispers of the depths from which it has come, and (with Peter Grimes) the listener may exclaim, 'I hear those voices that will not be drowned.'

Proust makes a cameo appearance in 'Underworlds of Jean du Chas', where he is 'turned into an ancient whitethorn tree leaning on a slope'. The best lines in 'Underworlds of Jean du Chas' are riffs on, not to say rip-offs of, lines in Beckett's *tour de force* essay/lecture 'Le concentrisme' – in whose surrealistico-Ovidian turbine, '*Proust se metamorphose en aubépine à force de fumigations*'. 'Coincidence', Beckett/du Chas observes, but ironically, I felt, as a question. I remembered Verlaine's *L'échelonnement des haies*, which I always imagined as breaking waves of hawthorn blossom. Additional proof of the exultations always threatening to explode out of melancholy's flatlands was the study of Dante's *Purgatorio* that the poet undertook while he resided in the Lincolnshire Fens. For his part, Beckett had been studying *Remembrance of Things Past* and no doubt refers in the imagined metamorphosis to what might be called a proto-madeleine episode, Marcel's meditation on the odour of the hawthorns along the Méséglise way.⁶⁸ In Beckett's spoof, Proust falls victim to his own lateral olfactory stria and is turned into his own fuming cloud. It seems a poor return on genius, but consider: every time a cup of tea is poured, and a wisp of steam curls up, Marcel is there. In 'Memory as Desire', Charlotte and the Captain relate the story their Cockney nurse told them in India about her dead husband. Every Hallowe'en she poured the tea, hoping to conjure up a novel spirit companion. The outcome was always disappointing; when questioned, the spirit's ignorance of tea always gave it away. 'What is bohea? ... What is souchong?' 'Why, tea, to be sure.' 'I never tasted tea.' 'I know you then, you silly old man.'⁶⁹ Certainly, a spirit who admits ignorance of its own medium commits a serious faux-pas.

Jean du Chas is undoubtedly a spirited *plongeur* – and he plays many parts. In the First Plunge of 'Underworlds of Jean du Chas', he appears as a muscular bathing attendant therapeutically dumping girls in the surf, a gesture that some thought unmistakably erotic.[70] 'O penetration of the liquid element', Didi exclaims from his deckchair – *something* repressed is being expressed.[71] In Plunge Two, Jean joins the *Calypso* – he could be its captain, Jacques Cousteau – and takes the plunge. He likens a life to the circles radiating where the diver broke the surface – 'We are spread out voids; each plunge a cross-section of unconscious life.'[72] Deeper down, he encounters Sharon. Later she will be Charon, but down here she is the lover he missed at the station – and it happens again as trying to communicate through a full face mask and handle the buoyancy compensator so that they can stay close is fraught – 'They draw apart again as they slowly sway up from the ocean floor; a strikingly white hand flutters up to each helmet valve; they sink down again; wave in farewell.'[73] Jean's efforts to recover the meaning of a life from the seabed having failed, he takes a leaf out of Kierkegaard's book, recommending 'the rotation method' as a solution to the problem of remembering and forgetting. But, again, the results are disappointing.[74]

In the Third Plunge, the horizontal rotation curiously produces a sensation of once more diving into the dark. On this occasion, Jean experienced nitrogen narcosis and hyperesthesia, and must have recorded his experiences, as Hollow and Didi now recall it on the basis of tapes he left behind. Jean saw not only his double ('If I had a brother, another self, I think I could face it') but also 'Death's rapture'. 'I must soar out of here,' he protests, realizing that the deep sea is, indeed, his tomb and he is passing to the next world.[75] Resurrection comes to him as it did to Timarchus of Chaeroneia when, as Plutarch describes, he descended into the cave of Trophonius, and felt his soul leave him, sailing upwards to behold beneath it an ocean of coloured islands slowly revolving.[76] A vortex emerges, a chasm, vast and round as though hewn out of a sphere. 'And indistinctly, sent up from the distant depths I heard roarings innumerable and groanings of beasts, and wailings of innumerable infants, and with them mingled cries of lovers flung apart into the night.'[77] These are the sounds of his life: in the First Plunge, whirring tape cassettes in whose noise my father and son are discernible; in the Second, the submarine's constant hubble of escaping air and, in the Third, the testimony of the hydrophones trailing the hull of Charon's barge. And, ebbing and flowing throughout, the ambiguous *du-chas* of waves, of Jean's own name, the dyadic sound signature of the diver. In this dishevelled, Beckettian way, Jean du Chas manages to be the angel of his own destiny, the messenger who in speaking speaks his name.[78]

Jean du Chas is a migrant type. His 'progenitor' is 'a janitor of sorts'; a Charon figure, if you like, resembling the original Janus – who was not, as we know him, exclusively a *rector viarum*, friend of travellers, but a doubled figure associated with the management of the strong magic of running water.[79] In a life characterized as a 'nullity' – perhaps with Theodor Adorno's dictum in mind ('the past life of émigrés is, as we know, annulled') – the only disturbance to his flat existence is the erotic vortex centred on Toulouse, ancient home of the *trouvères*, the Troubadour poets who lie behind Cavalcanti and Dante; but Jean is a parrot, a literalist and translator and hears in the city's name only 'to lose', a perfect forecast of his own existence. As a diver, he performs a baroque arabesque. His is not the powerful uncoiling Michelangelo curves, but a confused divagation, noisy and dispersed, like a crowd of bubbles. He could be an illegal immigrant mingling with the crowd I recorded in Lecce – a scene later revisited in my anti-novel of remembering and forgetting, *Baroque Memories*; as Nietzsche anticipated, the anamnesic plunges of Ungaretti or Proust could equally well produce the opposite sensation. 'Cast into the abyss that which lies most heavily upon you. Let man forget … Divine is the art of forgetting. If you would raise yourself – if you yourself would dwell among the heights, cast into the sea that which lies most heavily upon you.'[80] And so on. In short, the comparison of the experience of instant recall, the sensation of hearing a self-same sound again, to diving may be peculiar to the migrant experience. There is an important distinction between the English word 'displacement' and the French word '*déplacement*': the recalled sound has not been removed from its original matrix but it *has* been heard differently – unanchored and levitated, as it were, a sensation that can be compared to the buoyancy experienced floating in the sea.

A different migrant baroque emerges from this dialectical dive: the double movement, the downward plunge and the rebound to the surface, need not produce the kind of echoic exaggeration found in R. A. Baggio's variations on the (mis-) pronunciation of his name. It might be the psychophysics of recall that is recalled. Looking at two-and-a-half-millennia-old pots in the museum at Bari, Ungaretti commented, '*il Barocco piu straordinario e piu genuino si manifesta in questi vasi rinvenuti in un ipogeo di 22 secoli fa*'[81] – acknowledging that the baroque is perhaps the signature of a creative provincialism. But a baroque style need not manifest itself in the kind of ornamental exuberance found in *Jean du Chas*; its primary characteristic is fragmentation, the necessity of parts to stand in for a lost whole, and for the artist to arrange these fragments into 'baroque' forms – which, although less in volume than the original surpass it in terms of expressive power.

In this sense a baroque poem might be identified by what it leaves out, by its pure imagism stripped of intervening linear syntax; its vocabulary might be more 'primitive', simpler, less rhetorical than that found in a classical poem. It will represent the kind of condensation associated with dreams; it will be as if the objects of memory are glimpsed at the bottom of a harbour, through veils of water; in their distortion, warped this way and that in the tide's elastic ebb and flow, they experience a creative *morcellement*, a splitting apart that is also a way of reattaching themselves, creating monstrous composites perhaps out of the strife. What is re-membered in this way will not be a lost eloquence, but the process of fragmentation and re-membering; the result may be a vision of the abyss, an intuition of endless

Empedoclean process. '*Vi arriva il poeta / e poi torna alla luce con i suoi canti / e li disperde*', wrote Ungaretti with inimitable brevity: '*Di questa poesia / mi resta / quel nulla / d'inesauribile segreto*'.[82]

Does the self-same sound have, as it were, a sound, an 'ur' sound or whispering ground out of which it grows? Does Ungaretti's *inexhaustible secret* have a secret name? It certainly has a provenance. As a schoolboy in Alexandria, Ungaretti recalls, he believed in a poetry of the inexpressible and imagined the poem as an echo of this. When his friends, Jean and Henri Thuile, told him about a pre-Alexandrian harbour buried underneath the present one, the secret acquired archaeological specificity: the place from which the poet soared back to the light was a drowned port.[83] This discovery indicated a technique. Pointing out that Ettore Serra, who organized the publication of *Il porto sepolto*, ran a diving club, Carlo Ossola thinks that Ungaretti's contemporaries imagined the poet as a diver.[84] But what was the poetic value of this identification? Was it a sound it suggested? The early poem 'In memoria' commemorates Ungaretti's friend, Mohammed Sceab, who committed suicide in Paris: stuck between cultures, he could not unloose – 'sciogliere' – '*il canto / del su abbandono*'.[85] Unreconciled to his own nullity, he could not recognize in it his subject matter, precisely 'l'inespressibile nulla',[86] for that is all the poet brings back – '*Di questa poesia / mi resta / quel nulla.*'

It is notable that, in *Il porto sepolto* the *inexpressible* undergoes a sea change, becoming the *inexhaustible*, an echoic reminiscence of the original idea whose susurrant variation recalls the hiss of the foam persisting on the beach after the wave has broken, the same oceanic recollection audible in the verb '*sciogliere*'. Eloquence here is a sibilance that alludes to but cannot represent what has been lost and found, that can only express the fact of its inexpressibility.

'The Letter S' (2006) is an outlier in the series of radiophonic works that orchestrate this book. The main sequence is spread between 1985 and 1998, starting with 'Memory as Desire', concluding with 'Underworlds of Jean du Chas'. Although they all tapped into something deeper, an auditory imagination under the spell of Mnemosyne – she who presides over the deep well of recollection, located beyond death on the other side of Lethe, the stream of forgetfulness, who symbolizes poetry's identification with the gift of recollecting the beyond – it is likely, as Virginia Madsen noted, that, but for a fortuitous cultural – technological conjunction, none of them would have been realized or even conceived. There was, perhaps, in the hands of Martin Harrison and Andrew McLennan, the producers I was mainly fortunate to work with, a Brechtian sense of radiophonic possibility. Our version of the Neue Hörspiel was quite impure. We wanted to release sound for a program of cultural renovation that extended well beyond the aesthetics of certain new effects; giving voice to 'anonymous people' was associated in our practice with a reconstruction of orphaned or anonymous sounds that carried, in our view, historical resonance. Hence our project joined up easily with other historically revisionist initiatives afoot at that time. As Madsen goes in to say,

> Musical as much as narrative, cubist and multilingual, Carter's critically significant theoretical and historical work engages silenced auditory history in Australia and repressed dialogic encounters between estranged (often indigenous) peoples and languages. Some of his work is now installed in major public spaces such as Federation Square and the Museum of Sydney.[87]

As noted earlier, works like 'Memory as Desire' were commissioned by producers familiar with the Neue Hörspiel of West German radio. At the same time, as Madsen indicates, a radiophonic culture keen to amplify Australia's historical narrative by bringing in new voices and creating new discourses of coexistence found at this time a resonance in new museum practices alive to a plurality of pasts and in the application of post-linear geometries to the design of non-hierarchically organized public space. There was, briefly, a collective excitement about the possibility of building new sense and sense of place from the ground up. However,

> the history of radio art represents a struggle to overcome the enforcement of the arbitrary boundaries drawn by the paranoid hands of the state. These boundaries stifle creativity in many ways including the political, the aesthetic, the conceptual, the sensual and the multitude of creative imaginings that shape the various modes of expression and perception in a diverse cultural terrain.[88]

Hence, the strategy Lander recommends of 'An autonomous and anarchistic cultural alternative' that refuses to collude, circumventing 'the notion that we as artists are relegated to simply "playing" with hand-me-downs from the garbage heap of military mayhem and research,'[89] a point I guess represented in 'Remember Me''s juxtaposition of NATO war games and Purcell's percussive Furies. In any case, we remained defiant. Brecht's question 'If you should think this is utopian, then I would ask you to consider why it is utopian?'[90] did not appear to us rhetorical. Why should it have been? Within four years of Brecht's call to 'change this apparatus over from distribution to communication' – he added that, in the interim, 'any attempt by the radio to give a truly public character to Public occasions is a step in the right direction'[91] – the Nazis were centralizing radio broadcasting, merging distribution with communication in a way that horrified public life. Mutatis mutandis, in the time of state-sponsored false facts eagerly amplified by the owners of the means of distribution, how much has changed?

Always fragile and vulnerable to the politics of resentment, it soon became the victim of its own pretensions. The Australian Broadcasting

Corporation withdrew its support for acoustic art, plausibly citing the migration of avant-garde composition to other digital platforms. Baroquely revisionist soundscapes drawing attention to the scaffolding of our historical representations were stripped from the museums: the replacement of the city by the building site of continuous production – whose noise, we had said, was the sound of history, once more found itself glossed over:

> Voice 1: In the building sites, I said, in the quarries.
> Voice 2: Excelling in noise.
> Voice 1: They are 'making history'.

By contrast, in 'the Museum', 'When the unexpected happens,' the heart misses a beat.

> Voice 1: How can the unexperienced ground provide a counterpoise, a ballast we can bear away?
> Voice 3: What price the ground?
> Voice 4: Or sediment of sound?
> Voice 2: You are not listening.[92]

A comparison is made here between the essential unity of a sound and its echo and the ground and its tracking; for the 'unexperienced' ground is a colonialist fantasy, like history's *terra nullius*; the foundational sensation of any migrant conscience is 'dread':

> Voice 2: The report of feet.
> Voice 1: The tread.
> Voice 2: Of borrowed ground.[93]

Anyway, it was appropriate in the sunset of this radio trajectory to look back over those works and to try once again and perhaps for the last time to articulate that inexpressible nothing which, in the context of trying to write the experience of migration into the history of becoming at that place, was everything. In Australia, poetry's unfathomable secret was also history's bad conscience. The buried life resurrected in the poem is not simply a repressed emotion but a lost collective opportunity, thrown away because we were not listening. When I listened, at least, I heard a kind of communicational cosmic microwave background irradiating our seeming illocutionary competence, or else a historico-environmental white noise maybe, a kind of low-level hiss, as of air under pressure escaping from a diver's suit. This was the unlikely subject of 'The Letter S', a partial transcription, of earlier scripts based on the colonial claim, mentioned before, that Aboriginal languages had 'no sound in their language similar' to the letter 'S'. Of course there is a theoretical explanation: every language has a different phonology and marks, or fails to mark, different sounds. But I was interested in the practical human drama that must unfold along a bilingual littoral where a sound that signifies to one side seems to the other like the sea. I put history's hard shell to my ear again and heard the sonorous body whisper, 'Beyond the imperialismo of signs … there must be an art of sounds.'

When I was imagining the afterlife of Ulysses – Berio had proposed six reincarnations across histories and cultures – I thought of Tiresias's prediction. Ulysses will not survive the wrath of Poseidon unless he makes a journey far inland and erects an altar to the Sea God there. When he meets a man on the road who mistakes the oar he carries for a winnowing shovel, he will know he has reached the appointed place. Ulysses tells Penelope about this prophecy when he gets back to Ithaca but the journey remains unfulfilled. In 'The Letter S' I overlaid onto this story the history of explorer Charles Sturt's search for an inland sea told in 'Memory as Desire'. I wondered how Ulysses, a coastal navigator whose great image of the unbounded was the sea, would have fared in the trackless wastes of a continental interior. Perhaps, like Sturt, he would have seen signs of water everywhere and fantasized navigating an archipelago from which an ancient ocean had mysteriously fled. And what of the man who had never seen an oar and could not imagine the ocean? He must have been as perplexed as Australia's inland dwellers were when they saw men laboriously hauling over the sand dunes objects that, as they later learnt, the strangers called *boats*. As Mac puts it in 'What Is Your Name', 'Before we saw them we heard them: and before we heard them, the smell of horse-dung clutched our throats. The first time I saw them, two men were hauling a canvas dinghy across the lawn.'[94] From the other side it might have been like this.

That was one beginning, to do to my own work what Berio had suggested we do to the Odyssey and recast Sturt's story in 'Memory as Desire' as a Proppian fairy tale. Here was another beginning: to imagine how 'Memory as Desire' might be heard by those who had no 'S' in their language. Introducing 'What Is Your Name' many years before, Harrison posed the question: 'Can Aboriginal Australians and white Australians hear the radio in the same way? Is it possible?' I thought it was time to listen to what he had been saying, a project I interpreted first as a new airing of the original 'Memory as Desire' broadcast he had produced. This defined the fictional *mise en scène* of 'The Letter S', the rehearsal of a new production of the old work, preceded by a playback of the 1986 production. It was a beginning conceived as a repetition of beginnings. After all, had not the beginning of radio been the transmission of the letter S across the ocean? 'It is true that the first "message" was but the constantly repeated letter "S," but to Marconi's ears it must have been soul-stirring in effect.'[95]

The same formalist rigour permitted me to overlay one *ending* onto another. The director of the new production is not the author but his son, but keeping it in the family is no guarantee of fidelity to the original. In post-Homeric lives of Ulysses and his family, roles and relationships get confused: Ulysses marries Circe but so does his son, Telemachus, while Ulysses's son by Circe, Telegonus, marries Ulysses's own wife, Penelope. In a similarly incestuous (and treacherous) fashion, Rudy intends to unthrone his father's reputation. Arguing that 'radio' is finished, his object is to bury the work. In this historical sense, 'The Letter S' is un *tombeau pour radio*, an event marked by the surcease of the father's voice (the imperialismo of signs) and the advent of an art of sound, signified here by the sea finding its voice: 'Surely you must know / who I am,' Outis cries out. But the tempest drowns his voice:

> Director: The soundwaves are his buffeteers;
>   they trompe him with their trompes.
> Rudy: Rhythm begins, you see. I hear.
>   Acatalectic tetrameter of iambs marching.
>   A fourworded wavespeech:
>   seesoo, hrss, rsseeiss, ooos.
>   Vehement breath of waters
>   amid seasnakes, rearing horses, rocks.
> Director: In cups of rocks it slops:
>   flop, slap, slop.
>   And spent its speech ceases.[96]

Since producing 'The Letter S', Christopher Williams has written about his radiophonic art practice. He takes the conventional relationship between pre-production, production and post-production and shows how a practice like his redefines it. Explaining that 'the production process in radiophonic art practice is structured around generating sonic materials, and the inscription, or encoding, of their sonic *traces* onto a flexible medium as an audio recording capable of innumerable instances of reproduction, and on which further operations may be carried out using sound studio technology', Williams defines the materials I had assembled as pre-radiophonic.[97] True, my experimental rewrites of older scripts – and beyond what I have mentioned here I also incorporated Aboriginal language transcriptions – had already created a resource that was, in his terms, located across the 'generative' and 'appropriative predeterminate'.[98] My scripts were generative and my sound recordings were appropriative. They did not exclude development but they did, to some extent, predetermine thematic and expressive parameters. But their radiophonic significance would only emerge when they had been used in 'the production of *audio* not the *re*-production of *sound*'.[99] The radiophonic artist or recording practitioner emerges at this point and is likely, in fact, to be an ensemble of listener-makers: the writer/composer who supplied the generative material, the recording artists used to make the sound, the sound recordist and the producer. Production is finished when the 'the relevant sonic materials have been assembled'.[100] Post-production refers to 'the process of editing'.[101]

It also refers to the artistic goals of the production. Williams was interested in eliciting the theatrical qualities of my material. This is clear in his general description of the purpose of the editing process, where he notes, 'Speech is usually the determining sonic material in the editing process. Other sound elements are organised around it.'[102] As he explains,

> Analysing dramatic structure will reveal a fresh impulse for a character's action within a scene, a new thought within a narrative passage; or, in a musical form, an entry of a new instrument, or the beginning of a new phrase or section. Combining selections from different takes can help develop a rich subtext in drama, and complex attitudes and unusually virtuosic musical performance. A performance may be somewhat rhythmically restructured: altering rhythms, phrasing, and pauses. Attention will be paid to time-based relations within and between sound sources, but also to the structural rhythm between phrases, movements, scenes, sequences.[103]

In finding what lay confused or concealed inside the materials he reminded me of Stanislavsky's actor, who hunts down the emotional materials hidden in the recesses of the subconscious. 'Anything that triggers the actor's imagination or entices the subconscious out of hiding can be considered a "lure"' and 'if the bird will not fly to [the hunter] by herself, then nothing will bring her from the leafy thicket. There is nothing else to do but entice the wildfowl out of the forest with special whistles, called *lures*.'[104]

A calling to come that spells death to whomever responds gives the lie to any *ethical* claim for imitation! Perhaps the fatal allure of Homer's Sirens consisted in their cultivation of exact reproduction: sailors were lured to their deaths by a sound they could not differentiate from their own desire to be called. Duck calls are effective because they accurately reproduce the mallard's *quack* (and other calls): the fact that they work shows ducks possess absolute or perfect pitch. It is intriguing to compare the radiophonic artist's hold on the generative and appropriative materials to the possession of absolute pitch: as if the actor/artist has a special gift for discerning the fundamental frequency that will cause the memories to come flooding back – in this case, the 'memories' will not, of course, be those the scriptwriter of (and sound collector for) the work intended but the hidden emotional Gestalt recessed in the sonic subconscious of the materials themselves. A kind of psychoacoustic analysis is conducted that is likely to excavate and map a labyrinth of associations unknown to the author, the actors and, indeed, unsuspected in the appropriated sound materials until they were played back together. The generalized listening involved recalls Canetti's café habitué overhearing an interplay of voices that opens up 'an overall effect unknown to the voices themselves'. Listening in noisy places myself I have often been reminded of the effect of noise vocoding on the intelligibility of human speech. To improve intelligibility, vocoders often include an 'unvoiced band or sibilance channel' for frequencies that are outside the analysis bands for typical speech but still important in speech. 'Examples are words that start with the letters 's', 'f', 'ch' or any other sibilant sound.'[105]

An 'art of sounds' emerges between speech and music. A recent study found that musicians with perfect or absolute pitch were better than the general population at segmental sound processing, and that this was seen not only in 'exceptional acuity of music processing' but in speech processing 'in the sense that AP [represents] a comprehensive analytical proficiency for acoustic decoding'.[106] Segmental speech information refers to such elements of speech as phonemes, syntactic and lexical-semantic elements. (Suprasegmental information refers to sentence level prosody.) Critically (in relation to the older literature) this study further questions a sharp distinction between left- and right-hemisphere processing functions. Surprisingly, the absolute pitch musicians did not score significantly better on suprasegmental recognition. The conclusion was that 'the auditory acuity of AP is not limited to basal auditory processing (usually conceived in terms of music processing), but extends to a more general notion of acoustic segmentation by fully integrating left-hemispheric speech-relevant networks'.[107]

Apparently, comparable conclusions can be drawn from studying birds. In a recent article, Bregman et al. challenge the view that songbirds rely on absolute pitch for the recognition of tone sequences. They tested the European starling to find out what perceptual cues were used in melody recognition. They discovered that pitch and timbre were less important than spectral shape – the overall pattern of spectral amplitudes across particular frequency bands preserved by a noise vocoder. Noting that 'in humans, speech recognition is famously robust to the pitch-degrading manipulations introduced by noise vocoders, whereas similar manipulations have severe impacts on music perception,' the authors reflect, 'Our observation that birds rely on spectral shape features to recognize sound sequences suggests a similarity to human speech recognition.'[108]

An oddity of experiments on non-human subjects is the observer's lack of self-awareness. When Michael Leaman charmingly coerced me into writing for Reaktion Books' animal series, I chose the parrot as my topic. It was the first thing that came into my head. The avian mimic par excellence, who had also inspired Sturt's fantasies of a mimic sea, it spoke to me. It solves the problems of representation and autonomy that I associate with the migrant condition. But, until I acquainted myself with the parrot literature, I had no idea that I would reach the radical conclusion that 'without parrots we cannot be human'.[109] The book included a reflection on Irene Pepperberg's forty-year-long study of her African Grey, 'Alex'. Following Rumi, I wondered whether Alex was saying to his mistress, 'I am an echo of yourself which you have caged.' Pepperberg had more recently considered the possibility that Alex was, indeed, responding critically to the experimental situation, mimicking what the scientists wanted to hear. But investigation of this kind leads into a hall of mirrors. It operates under the *imperialismo of signs*, with an abyssal conception of human (and non-human relations). Its Lockean empiricism produces logical narcissism – as Kierkegaard's moral philosophy teacher described.[110] Nothing could be further from this existential isolation than the 'inviolable presentness and simultaneity' that I have tried to evoke in this chapter, and whose cognitive mechanics Rilke suggests when he states, 'We are incessantly flowing over and over to those who preceded us, to our origins.'[111]

The fact is that avian experimenters do *not* work with 'songbirds'. They populate their laboratories with mimics – bird species that do not have a song of their own. A nightingale in a cage would be immune to their system of sonic stimuli and rewards. That birds can learn to talk intelligibly is not disputed, nor that they can repeat complex melodic phrases. But what have they learnt except how to dumb down the art of sound? As I wrote of Pepperberg's team,

> they work because of a strange untested assumption, which is that Alex will mimic his trainers knowingly, that he will internalise what they do, and identify it with his own desire, in this way learning to behave in a non-mimetic, self-motivated way. But so long as mimicry is, on the one hand, used to teach, and the other hand, discouraged in the student (the disregard of Alex's phonic sound-alikes), learning can only ever be mimicry of the other's techniques. If the parrot's own flock-oriented communicational ambitions are deliberately misunderstood, its intelligence (its gift for survival) depends on developing a certain stoical resignation, and an ironic awareness that an inverse relation exists between his success in performing them and his capacity to communicate intelligently (about the system). That is, the more he imitates their desires, the less his desire of communication can be communicated. As this desire is profoundly mimetic, originating in a doubling with interest of the other's voice, its systematic punishment represents a cruel dumbing-down. It is odd to think that the same ironies inform the public education of children throughout the Western world.[112]

In a way, the parrot is denied a voice of its own because we cannot imagine what that voice would sound like. The echoic mimicry that doubles with interest the other's voice is, in a sense, what gives voice to the voice; it perceives in the noise the other makes the desire to communicate. What comes back from this exploratory exchange is not a message, encoded, transmitted and decoded, but a sensation of existing alongside another, as Rosen suggests. An accompaniment of this kind does not go away: 'a place that we are always in'¹¹³ – 'We neither enter from some other place nor leave it for yet another place'¹¹⁴ – it redefines the present as an abiding presence 'that makes possible the articulation of past, present, and future.'¹¹⁵ Clearly, this is not a hypothesis likely to carry much weight in experimental sound studies but it makes otobiographical sense, clarifying the nature of the calling to come heard as the self-same voice in the noise. In a way Sturt was on the right track when he imagined the country beyond the horizon where the parts went. But he made a mistake when he treated them as signs of something beyond themselves (an inland sea). They were their own horizons, like angels, their own prophecies. We are told that the NeoPlatonic theurgist conducts his ritual with the aid of 'signatures' (*sunthemata*). In these the divine energy is mimetically represented or embodied. *Sunthemata* are a kind of divine alphabet. They are illuminated matter, bearing the direct 'imprint of the sun god'. This is what the parrots are: *sunthemeta*, imprints of the sun god.¹¹⁶ 'No bird so beautiful / The Northern children know. / Gently they say, he is not of the Earth, / He only falls below. / The settler's sunburnt child / In him knows all that summer ever smiled,' writes John Shaw Neilson of the 'beautiful yellow rare bird' known as the regent parrot.¹¹⁷

I mentioned the Australian poet John Shaw Neilson (1872–1942) in passing earlier on. He has been an abiding presence, but because his poetic trajectories have not crossed my radiophonic tracks, I have not said anything more. The radio work *Mac* broadcast in 2011 is a partial exception. In *Ground Truthing* (2010) I had speculated that Jowley, the historical figure on whom the 'What Is Your Name' trinity, Mac, Mr McGinniss and Mr Hatcher, had been based, and the poet might have crossed paths. I had come across circumstantial evidence – a photograph that appeared to place Jowley at Tyrrell Downs station around the time that Neilson was working there.¹¹⁸ The Voices of Scene Three announce my (by now familiar) mythopoetic project:

1. A theatre of the did-not-happen.
2. Where the fiction of possibility reigns.
1. Between the tracks cut in the Mallee the parcel of
   unsurveyed darkness.
2. The endowment of shadow.
1. Realm of strange meeting.
2. Parrot-tunneled.
1. Endowed with ghosts hanging up like fruit bats.
2. You don't see them on the stage.¹¹⁹

Christopher Williams managed to create a powerful radio drama out of my generative and appropriative materials, but I do not classify 'Mac' with the authentic series of radiophonic explorations discussed elsewhere in this book. Perhaps in response to the collapse of the listening culture that had supported the earlier works, it was too self-explanatory – the corollary is aphoristic passages that instead of suggesting overheard conversations are portentous, poetic in the bad sense. I refer to 'Mac' here solely on account of the line 'Parrot-tunneled' – because that was how I imagined a new historical horizon opening up. Parrots tunnel with their shrieks as well as their thrilling arrows of colour. Always oriented elsewhere, their calling to come cannot be anything else than prophetic. In *falling below* they bear the horizon on their shoulders.

The question of right copying had come up shortly after arriving in Australia. En route to Wilpena Pound in the Flinders Ranges, we had stopped off to look at the Yourambulla Caves Aboriginal rock paintings, executed presumably by ancestors of the present-day Adyamathanha people at an unspecified time in the past. The relationship between the mainly black designs – polydactylic 'plants', split arrows, radiant 'sun' patterns and pennants – and the surfaces where they occur struck me. Cracks in the rock ceiling, seepage patches, as well as the cloudlike bulge and hollow of the walls inspired a distribution of patterns and variety of pattern treatment that suggested something like an art of morpho-chromatic mimicry. If natural weathering processes were the other hand (and mouth) in this art, it was reasonable to ask what was being represented. Beyond the self-doubling through the gesture, what did the marks mean? Were they indexically related to human gestures or conventionalized symbols? Could they be read or sounded? Was their social expression music, dance or song? Or something else – as I speculated – a description of the view, of the flat country at the base of the cliff that Sturt and his boat had traversed 140 years before? If so, they were *sunthemata* too, graphic diagrams of energy forms occasionally incarnated at that place – like the sun rising or the arms of a constellation leaning out of the dark sky. As sacred graffiti, like the inscription of the Delphic Oracle, their message was in the summons to attend. Whatever the case, right copying would have depended on understanding the spirit in the gesture; painting over and renewing an earlier mark could either sap the energy latent there or recharge it. There was a self-same reproduction of the gesture which, if it failed, was only a mechanically executed copy.

Following my thoughts about the Yourambulla rock paintings is a handwritten notebook entry made a couple of days later at the summit of St Mary Peak in Wilpena Pound. In my personal literary geography, the hilltop meditation on a possible, non-European poetics of place marks the source of the headwaters from which *The Road to Botany Bay* eventually flowed. However, off to one side, as I see now in revisiting that notebook, another creek bed could be seen wriggling away. The passage in question reads, 'Curious how many bird call notes in Australia seem to mimic human sounds, syllables, cries, whistles and tones from Europe.' Here is the beginning of the *second* side of the sound history whose first side was the auditory impressions laid down in childhood and strangely revisited and deepened in Spain and Italy. That notebook is a Janus-faced document and herein lies its authenticity. In it side-by-side lie thoughts in passing on the character of Scarlatti sonatas ('express exactly that life between events, the promenade rather than the pictures'), odd lexical items (*sbianchire* – 'to be revealed in your real person. Used by actors when another actor reveals their true identity'), and, of course, possible projects, amongst which, 'the supreme challenge' is already identified. It is 'to characterise the sea'.[120]

When, a decade later, I started 'Enclosure Acts', an autobiographical essay about growing up in Faringdon, I took Herbert Read to task for his common sense assertion that 'all life is an echo of our first sensations, and we build up our consciousness, our whole mental life, by variations and combinations of these elementary sensations',[121] I was in a sense right. My first sensations in the new country were not an echo of what had gone before. The reverse, rather: initiated into these new, elementary associations through the 'second birth' of migration I could hear primary echoes that communicated an inviolable presentness and simultaneity. When from the underworld I conducted an acoustic archaeology on the upper world, I did so like Charon sifting the sounds for what could be carried over. As Mac remarks, 'One thing I'll say for Heaven: they fix your pronunciation. Down there, the confusion of sounds was unimaginable.' 'The tongue of our natives was soft, euphonious' could be heard as 'De-dunk offer neighed eaves oft you phoney as', while to a people who had inherited the belief that 'Ithaca itself was scarcely more longed for by Ulysses, than Botany Bay by the adventurers who had traversed so many thousands of miles to take possession of it', the description of Aboriginal speech, 'The syllables blurred together', could easily be mistaken for 'Ilium blue with heather'. In any case, the confusion was never going to be resolved: ruled over by the parrot-god Guacamayo, it was always a struggle between the littoral and the literal, between the wash of noise and astonished mishearing – as they reflect in 'The Letter S'.

> Author: After you find home–
> and I don't say you will,
> home being a shadow
> of what might have been –
> you will go away,
> on business, pleasure,
> and meet, far from the sea,
> a man, who is a liar like me.
> Actor 5: *Anku-ntja*.
> Actor 4: Go.
> Actor 5: *Wiya*.
> Actor 4: Don't.
> Actor 5: *Nyuntu*.
> Actor 6: You, why don't you go? Are you going?[122]

EIGHT

Callings

Although shaped by a radio trajectory, it is obviously inaccurate to link a personal sound history exclusively to a technological moment. Radio scripts may be aligned to radiophonic composition and even articulate the expressive values of an Ars Acustica in the phonetic and phonemic woof and warp of the language, but, as noted before, they are, from a compositional point of view, inessential until translated into a recording. Even in realization, passages from these scripts test actorly ability and (in post-production) beg the question of expressiveness. Edgard Varèse notoriously disregarded the idiomatic: determined to use 'electrical means' to project his ideas, he is said not to have cared that much 'about certain practicalities': his demand for an oboe played louder than a trumpet springs to mind.[1] Of course, no competition of talent is implied by this comparison. I just wanted to draw attention to the sense in which these writings for radio voice regularly projected beyond the recordable to take back something that, at least in media theory, is regarded as dead on arrival when recorded. Gregory Whitehead has written alarmingly of radio as thanatology. In its broadcasting of 'disembodies', 'the original speech act begins to disintegrate as soon as it comes to grip with its schizophonic double.'[2]

So voicings were certainly part of the motivation, and while I fully understood that once vocalizations were recorded and repeated, they inevitably forfeited their bodily impersonation and became signifiers of something, I was interested in what happened off microphone: even if the recorded performance eliminated the rush and brush of breath, certain expulsive punctuation marks, as David Appelbaum puts it, could set the tone for the communicational ecology nurtured in post-production.[3] This was evident in 'The Calling to Come', where the drama corresponded to close-miking, incorporating the inarticulate sounds of intimacy into the symbolic fabric of discourse. Again, a media orthodoxy was perhaps being questioned. The glossolalic notion of voice as 'an infantile disorder of language … a confused and shapeless substance located in the body, the mere expression of emotion, a pure excess that may suddenly spout out of the speaking subject'[4] – this characterization of free vocal practice with idiocy failed to resonate with me. Any voice, it seemed to me, solicited a call back; it surely hearkened to a call and it intended to communicate, even if it said nothing. I used to cite the disappearing habit of whistling, a rather pastoral or meditative form of musical talking aloud: suspended between introspection and birdsong, it shared its expressive genealogy with the siren.

'The Calling to Come' was certainly a work of solicitation: it represented an intimacy that could not be put into words. The chief historical source for it, Williams Dawes's manuscript 'Vocabulary and Grammar' of the so-called Sydney Language, was exceptional for its acoustic sensibility.[5] Appelbaum writes melodramatically of the threat to communication posed by the cough: 'Smooth, continuous, and unbroken, the voice seems to vocalise the sounds of angelic space. … Throw in the lurching, gyrating garble of the cough, and the presumption is hard to maintain.'[6] But Appelbaum is talking about the erasure of bodily sensation found in public speaking and broadcasting. Outside this forum, the cough signifies an interruption creating a threshold to listen – it is a call that cannot be answered except perhaps through an access of self-awareness (stopping one's own chatter) or (as was the case with my mother) hastening to give succour. In fact, Appelbaum's proclamation can be reversed. If the angelic address is distinguished by communicating on its own terms – in Heidegger's terms, it calls beings to Being – then integral to it is a *lack* of eloquence. 'The call does not seek to be listened to except as such: no transcendent solicitation is seeking to open a path by means of it to the core of conscience. The call is not a calling of anything, and nothing is calling. Or rather something is!'[7]

Portentous, no doubt, but not unnecessarily enigmatic, the anti-transcendental call issues from the body of the stranger. It is the call to care, reflected in Dawes's notes in his growing closeness to Patyegarang. A distinguishing feature of Dawes's auditory environment is, as I say, its intimacy, as if he worked close-up to the lips of those who speak to him, mimics the movement of lips and tongues, anxious to capture the body of their speech. And the first result is to relocate communication in vocalization. Dawes reports at least two words signifying 'to cough' and 'to talk'; he records terms signifying 'to breathe', 'to swallow', 'to yawn', 'to sneeze', 'to blow the nose'.[8] Contrast this sensitivity to the body of speech with the lexicon of the exactly contemporary *Anonymous Vocabulary*, where only three somatic interruptions are recorded – 'he snores', 'to whistle' and 'to laugh'.[9] A word signifying 'silence or hush' and 'uttered in a whisper' is reported, but one wonders if the writer has simply extrapolated from the mode of delivery. This by now familiar cultural phonophobia extends to the non-human environment, where only two sounds are recorded, for 'thunder' and 'a bird with a shrill note'.[10]

The problem with the Heideggerian formulation is its fairy-tale appeal to authenticity, when the essence of the anti-angelic vocalization is, in fact, its historical irony. In this regard, Dawes is far more intelligent. Glossing the phrase 'I speak falsely in jest or to make believe,' Dawes carefully corrected himself, 'I only make believe; I did not tell a lie.'[11] This attitude uncannily recalls the advice that Scarlatti offered the performer of his *Essercizi*: authenticity consists in playing a part, in using irony to unmask authority. Remembering that George Worgan, the author of those impulsive epistolary exercises to his brother, was Dawes's First Fleet contemporary, one suspects the coexistence alongside the official history of colonization of another, unofficial record, exilic, echoic and, by its wit, offering the most serious, if subtle, resistance to the violence of arbitrarily assumed power. 'L'ironie est encore plus sérieuse que le sérieux.'[12] Ernest Goffmann reports, as an example of ritual insubordination relying on irony, prisoners naming the punishment block the 'tea garden'.[13] It would certainly be ironic if Dawes's and Patyegarang's mutual preoccupation with 'slurred speech' were mistaken for political disengagement; as Dawes demonstrated by his later refusal to join Governor Phillip's man hunts, he fully foresaw the prison house about to enclose Patyegarang and Yalgear as they sipped their tea.[14]

Another objection to framing a personal sound history in terms of the rise and fall of radiophonic art is that it is unduly pessimistic. It is true, as Anna Friz writes in the recent compilation *Re-Inventing Radio*, that 'twenty-first-century radio artists' still struggle with 'issues of state and corporate control, generic and syndicated media formats, public radio sectors that are steadily eliminating the few programs with an interest in experimental work, and the public expectation that radio will consist of music, news, weather and advertisements'.[15] Friz, though, claims that in a post-radiophonic era radiophonic art migrates to what she calls 'community radio'. She describes an interactive practice where, to some extent, Brecht's call for listeners to become producers is worked out: audience members can come to the radio station and become programmers – 'a kind of circle of transception between the listening community and the radio station' is initiated.[16] Friz backs a new 'radiophonic subjectivity'. 'We are not confronted with the voices of the dead, but the traces of the living,' their physicality resonating in 'the timbre, tone, and intonation of the voice'.[17] But this optimism cannot be theoretically valid: at best the new radio represents the voice, admits it into the listening room. In the same way, the admission of 'the lip-smacking, sniffing, coughing, breathing, pauses, filler words; the uneven mic technique, the limits of cheap equipment, and the sound of the environmental context – be that a living room, the street corner, or a forest'[18] may represent a satisfying attack on the centralized media and a victory for what Whitehead memorably calls 'the wireless imagination', but it does not solve the problem of the call.

The truly emancipated radio imagination would, presumably, embrace the enduring mystery of the electromagnetic continuum, its potential for an endless messaging without audience. In this truly angelic art practice, any response to the calling to come would be entirely a matter of chance, comparable to finding a message in a bottle on the beach. Even if the origin of the message were declared in the message, its meaning would reside mainly in the encounter there and then; and the response to its solicitation would not be a rescue mission but rather a new self-consciousness, one aptly associated with the edge of the sea, whose endlessly slurred syllables unmistakably pronounce the sound of finitude. And, as is indicated elsewhere in *Re-Inventing Radio*, this revived 'fascination of Hertzian space – information floating invisibly through space' does not really depend on 'traditional radio technology' at all.[19] As a progressive project, I think this transcendental radio aesthetic is equivalent to the political (and poetic) project of materializing public space, a dimension of shared human experience that will remain phantasmatic so long as access to the airwaves (*qua* media of encounter) is policed and punishable.

Going back to the question of voice, if any lesson is to be derived from my experience of making radio works, it is that easy assumptions about the authenticity of the voice are questionable. Vocalizations of most kind are self-consciously produced and their origin is never singular. Firstly, listen to yourself talking: Where does the voice come from? The larynx, the mouth, the ear, the chest or all of these in a resonating feedback loop. Then, listen to yourself listening (as happens in talking to yourself, speaking under your breath or in the soundless speech where the lips still move), and, inevitably, you find yourself possessed by the voice of the stranger, desirable, more outspoken and eloquent. I am reminded of the schizo man, Louis Wolfson, discussed by Chantal Thomas, who, to block out his mother's voice, carried a radio around with him, and playing music at full blast imposed it on passers-by under the impression 'it were perhaps he, and not the radio, who made the music'.[20]

Vocalizing purely for the purpose of hearing oneself speak may not only drown out other voices but also open the way to auditory glossolalia, as what are meaningful sounds in one language are mere clicks and hisses in another: from a writerly point of view, passages in 'Columbus Echo' exploited the phenomenon known as 'Raudive voices': I listened to the white noise of the babble – as most of the languages I was transcribing were unknown to me, I was free to improvise their phonologies, and had the sensation that murmurations of other voices lay secreted within the vocalic crackle and pop. These voices didn't say much but seemed to supply the *beat* of the sound, its rhythmic ostinato, something like periodic radio interference or the rhythmic flutter and wobble of the revolving vinyl disc.

In short, the voice is always a calling: just as I originally experienced the birds as an orientation, so with the voice – coercive or convivial, it takes me out of myself. Involved in conscience from the beginning, it therefore dictates a way or, better, advertises a crossroads and a choice. One way implies that the first 'other' is 'a primordial relation between me and my speech. ... Through this relation, the other myself can become other.' In this understanding, writes Maurice Merleau-Ponty, 'the common language which we speak is something like the anonymous corporeality which we share with other organisms'.[21] The other way suppresses, enslaves and ultimately silences this relation; its amplifications aim at self-possession, and the mythic authority that authenticity bestows. Its ideal communication is the perfect repetition of his master's voice: the technologies of voice reproduction fulfilled this solipsistic fantasy, but they did not invent it.

My favourite example of the latter impulse comes from the controversial nineteenth-century German archaeologist Heinrich Schliemann. In his autobiographical memoir, Schliemann recalls how, as a shipping-clerk in Amsterdam, he taught himself foreign languages. His method consisted 'in reading a great deal aloud, without making a translation [and] taking a lesson every day'. Supplied by a teacher with the rules of pronunciation, he would recite them aloud, committing whole novels to memory; when, as a result, he suffered from insomnia, he employed the time 'in going over in my mind what I had read on the preceding evening', commenting, 'The memory being always much more concentrated at night than in the daytime, I found these repetitions at night of paramount use.' No teacher was available to help him with Russian, so he read out aloud his chosen text (a Russian translation of Fénelon's *Les Aventures de Télémaque*) relying on pronunciation tips derived from a grammar. He appears not to have seen any flaw in this method, instead informing his public that thinking 'I should make more progress if I had someone to whom I could relate the adventures of Telemachus, ... I hired a poor Jew for four francs a week, who had to come every evening for two hours to listen to my Russian recitations, of which he did not understand a syllable.'[22]

Schliemann was a classic narcissist, habitually confusing his own will with the desire of the world. Pretty much all his biographical, historical and archaeological claims have to be taken with a grain of salt, if not dismissed as lies. But if his prodigious memory feats are even half-true they are remarkable. He must have possessed an extraordinary affection for the sound of his own voice, which, one surmises, he raised in order to drown out other voices that might threaten his ego integrity. 'My recitations', he wrote, 'delivered in a loud voice, annoyed the other tenants, who complained to the landlord, and twice while studying the Russian language I was forced to change my lodgings.' But, he added, 'these inconveniences did not diminish my zeal'.[23] Without psycho-pathologizing this behaviour, its social impact must have been obvious to everyone: even when addressing another, Schliemann was talking to himself. He spoke by rote with the object of bending his interlocutor to his will. The power of his eloquence was proportional to the silencing of objection, interruption and even of the voice (as it might express itself through accent, intonation or tone).

But why moralize? Until recently, the Schliemann language-learning method was conventional school pedagogy: witness the evidence of the Philips EL 3586 recordings, which (ironically, as it now seems) I excavated and transcribed in the hope of retrieving some trace of my former self, mysteriously audible in the grain of the voice. I certainly lacked Schliemann's genius for picking up languages; going over ancient Greek verbs again and again was imagined as deepening a memory groove in the brain; instead, it had the opposite effect – like a record played too many times, my memory was worn smooth and retained nothing. Multiple ironies hovered around those recordings, their rediscovery and their incorporation into a radio work. They included the strange moment in the French dialogue where I ask, 'Quel est vôtre nom? ... Quel est vôtre prénom ... Avez-vous un surnom?' and decline to reply – like the recalcitrant native who will not become a colonial informant. My pidgin French is odd too, implying the very societal immersion I singularly lacked, and was curiously echoed in translations of the 1986 script, 'What Is Your Name', which, in (Mauritian-inflected) French became 'Quel nom toi' and in German, 'Wie ist dein Name' I have taken one of the parts in a live reading of 'What Is Your Name', and performed in the first radio production of 'The Native Informant' – a guest appearance which wove into the auditory memory another ironic inflection: Why, reciting the dialectally sumptuous names of the villages around where I was brought up, did I imitate our vinyl recording of Dylan Thomas reading *Under Milk Wood*?[24]

Pairing 'The Calling to Come' and 'What Is Your Name' suggests a persistent, perhaps emotionally distinctive, motivation of these radio works (which at one time I wanted to collect under the rubric 'works for voice'). Out of the same archive of colonial materials could come a call that was communal or a call that was coercive. At the heart of this distinction was the role allotted to irony – manifested in performance as echoic mimicry. While the elaboration of sound-alike phrases in 'The Calling to Come' recalls those sugar–barley columns characteristic of southern baroque architecture, one snake of desire winding round the other to extend a twining that is clearly erotic, the same echoic principle of sense formation deployed in 'What Is Your Name' facilitates an interrogation in which the informant's first imprisonment is the call as a demand to speak, that is, to render themselves in language. Odysseus might outwit Polyphemus by announcing himself as Outis (Nobody) but, in 'What Is Your Name', Cretan paradoxes of this kind cut no ice.

One manifestation of this is the colonization of the voice. Hence, the syllables that in 'The 7448' playfully riff on the auditory picturesque assume in 'What Is Your Name' a decidedly sombre aspect. 'La, la, la', the anonymous Chorus of 'The 7448' sings, imitating the rhythmic work song of rowers but, transposed to the interrogation room, we find:

2. The affixes which signify nothing are *la*.
1. La.
2. Wider please.
1. Leh.
2. Leah.
3. Lear?
2. Leah: And Leah said, 'A troop cometh, and she called his name Gad': you know.
1. *Ma*.
2. Wider, please.
1. Me.
2. Meah.
3. Promiscuous syllables.[25]

'Promiscuous phrases' were a commonplace of colonial word books and signified Aboriginal expressions taken out of context and used purely to illustrate a grammatical principle. They were verbal structures that could be applied anywhere. When combined with the language theory of Hermann Ritz, however, they turn out to govern vocalization as such: the very act of opening one's mouth to make a sound that is not meaningful is interpreted as sluttish, as if phonology were a fatal Siren only admissible when subordinated to the legitimation of the lexicon.

In performance a script is never simply the reproduction of an idea. Temporarily, at least, it sketches a new society. The effect of the performance is a combination of directorial/authorial intent and performer interpretation. As an aesthetic object, the production is a translation; as a dramaturgical object, however, its actors represent nothing but themselves. Even where the theme is the silencing of the voice, the experience of realizing this intention serves to enable the actors to find their own voices. Among the many conundrums of translation encountered in developing a multilingual production script for 'Wie is dein Name' was the rendering of 'You play ball with us ... we'll play ball with us'. In one Russian version, it was rendered (at least when translated back into English) as 'If you sing with us, we will sing with you'. However, this rendering of the figurative sense was undercut when the same translator rendered the line 'We will play at ball' as 'We will play at *a* ball'. Additional semantic interference occurred when the cricket equipment 'stumps' produced, possibly via the German '*stumpf*', a word that in Russian suggested the idea that the debutante should beware of 'dumb blokes at the ball'. Through a concomitant mode of production, these pre-production workshops authentically reproduced an original scene of impossible translation: 'Discussing this spreading web of associations, the Russian speakers in our group could not decide whether it was a misunderstanding of the English, an attempt to preserve its punning logic or a clever and ironic joke. We concluded that the case was undecidable.'[26]

The losses in translation that occur in pre- and post-production are frequently where the real work of representation occurs: the faithful production is one that is true to the promiscuity of meanings that, in obscuring the original sense, bring it out.[27] These mechanics internal to the presentation also have an outside. In the suspension of a definite meaning, the participants in the project are brought together. The undecidability of the translation creates a situation where everyone has a voice. The wall between the inside of the theatre and, say, the outside of the urban neighbourhood dissolves, and drama (from direction to set design) yields to dramaturgy, the constant movement back and forth between inside and outside. When this happens an odd consequence follows: a reversal of the migration from radio to theatral space occurs and the work (now a complex meeting place of creative interactions) operates in a post-radiophonic environment – somewhat in the manner of Friz's 'community radio'. I had an inkling of this in 2004 when I found outside our rehearsal venue in Koppenplatz, Mitte, Karl Biedermann's Kristallnacht memorial *Der verlassene Raum*, whose ensemble of a table and two chairs (one pushed over and lying on its back) exactly replicated (or anticipated) the essential stage set I had proposed for 'Wie ist dein Name'.

2. Look, if this was a prison-camp, we wouldn't detain you.
3. Or a hospital.
2. If this was a mission station in 1838.
3. You would be free to go.
2. But from this place.
3. There's no escape.
2. Not for me.
3. Not for me.
2. Not for you.
3. It is, as you might say, a universal condition.[28]

If 'What Is Your Name' 'Wie ist dein Name' name represented a 'universal condition', it was because it was localizable everywhere. Its nowhere would never evaporate into the Cartesian orthogonals of general planning but would always be an exceptional circumstance of responsibility, where a calling, good or bad, had been heeded. Perhaps this implication had particular appeal for a migrant who, as *Amplifications* repeatedly demonstrates, has predicated finding his own voice on not giving up on the translation between upper and under worlds (psychic as well as geographic). More territorialized sensibilities have perhaps found this self-spatialization or – distribution difficult to place: when in *Ground Truthing* I asserted a dramaturgical identity between the choreography of a photograph by John Hunter Kerr of a Djadja Wurrung ceremony from mid-Victoria (dated late 1850s) and the 'Black Bacchants', an iconic graffito at Tacheles on Oranienburger Strasse, Mitte (since destroyed), the territorial ideology of the local (and unique) was offended.[29] Substituted here was an archipelagic historical sensibility alert to the rise of self-forming communities wherever the calling to subordination or liberation is heard and, one way or the other, must be heeded. In the post-Berlin Wall context, these were images of ecstatic solidarity that gave the wall a good name, for they contested the idea that the stage was an empty place, materializing instead the walls that everywhere, visibly or invisibly, divide or confine us.

Friendships formed during that production have endured. One of the student actors in the 2004 production is now a successful film director. His current film overlays the stories of the actors onto the history of Berlin. The figure of arch treasure hunter Heinrich Schliemann makes an important appearance. In the Battle of Berlin what remained of his Trojan relics (ceramics mainly, as the gold had already been removed) was reduced to shards even smaller than the original fragments – and, we have surmised, an archaeological dig conducted in one of the post-war rubbish hills might, in theory, recover it all again. Shot across eighty eight venues, the film merges psychological drama and urban dramaturgy. For my part I have begun to improvise what I call sound graffiti: during my recent residencies in the city, I have taken to improvising in the streets, talking aloud to myself and recording the results. I have also been recording street musicians at selected venues, and overlaying my own voice onto the recordings. I am listening for secret structural resemblances, as if in the interference pattern created certain acoustic place signatures can be detected, anonymous vocalizations, overheard rhythmical and intonational patternings, which belong there only in the sense of having been heard, heeded and amplified.

In the context of the endless production of impressions of Berlin by visiting novelists, artists and musicians, the vocal mediation of my experience is, I believe, novel. The experience itself, though, is not novel: it is the haunting presence of the past, palpably present in, for example, Gunter Demnig's *Stolpersteine*, stumbled across along the most innocuous of suburban streets. How are absent voices represented without falling into the pornography of representation? It was precisely the question raised by 'What Is Your Name'. A clue exists in contemporary graffiti and post-graffiti practices, which habitually recall mandated public spaces to their role in civic conscience as sites where conflicts have taken place and where things continue to happen. In summer 2016 a new tag appeared on a rubbish bin next to Biedermann's sculpture: 'It's after the end of the world.' A year later, this pronouncement is almost invisible under two new spray canned calligrams: the announcement of the post-histoire epoch has become historical. On the other side of the rubbish bin the head of an ancient Greek warrior has been stencilled. Street furniture for the collection of rubbish and the image of a Hollywood Achilles are conjoined. Of course, the conjunction does not imply a conscious intention to *memorialize Schliemann*: in Berlin, the over-determination of motifs is habitual. However, such mere or promiscuous coincidences suggest the environmental channelling of a historical unconscious. Something is in the air – as I felt in 2004.

Even then the invitation of the sculpture was obvious: our stage set was already provided; why seek to imitate it? If Biedermann's memorial so clearly fulfilled our ambition not to reduce history to theatrical representation but instead (like radio) to redeem it through the radicalization of public space, why not perform 'What Is Your Name' at that site? In this regard, the wall-less interior evoked by *Der verlassene Raum* had another attraction; it turned the rectangular platz, bordered on four side by four or five storey residential and institutional buildings, into a theatre – a transformation facilitated by the landscape design itself which, dating from 1927, proposed a miniature version of Bernini's colonnade and rotunda scheme for the forecourt to St Peter's in Rome. Over the years this design has decayed (if it was ever fully implemented). However, it continues to determine the planting scheme, the ranks of benches and the military order of the linden trees, as well as the enclosure set aside as a children's playground at the southern end.

Biedermann's work plays to this theatrical crowd but also locates it ironically, not appearing, as Klaus Neumann writes, to be 'erected to be seen' – at least not as a memorial. '*Der Verlassene Raum*, after all, looks from afar like nothing more than a table and two chairs. From a distance, there are no signifiers that point to its function as a memorial.'[30] Located at the very oculus of the park's axialization (thinking of Bernini's scheme), the work tries not to be a focal point. Its disconcertingly oversized furniture feels awkward, as if embarrassed to be looked at – and, typically, even though it stands centre stage, visitors to the platz overlook it while residents ignore it.

In this theatre of the overlooked the traditional room of the theatre has also been abandoned; a theatre without walls where the exits and entrances trail away north-south from Tor Strasse down Grosse Hamburger Strasse, its empty stage is an indistinctly edged region, a vaguely territorialized neighbourhood in which the sculpture is one potential event among many. The informal comings and goings of passers-by have a diurnal rhythm; and the park certainly fosters normative behaviours – the park benches periodically attract people on lunch breaks, book readers and oddly nervous smokers. But non-normative events occur: a wedding party arrives like a flock of the doves, and the bride insists on lying down in the fallen chair; the table's impassive bronze surface accommodates diaper-changing, impromptu picnics and lovers clinging to each other as if facing a precipice. My observation over time is that *Der verlassene Raum* operates as a strange attractor, as a calling to come uttered in a voice too low to hear. It is like Barea's public radio installed in the local bar. In *The Telephone Book*, Avital Ronell memorably correlated Heidegger's call to conscience with the invention of the telephone. Henceforth we could never switch off; even when the telephone was not ringing, we were waiting for it to ring; henceforth we were always *on*, ready to receive. But somehow *The Abandoned Room* is a radio set turned off.

Responding to the euphemistically framed East German government competition to create a memorial 'for the contributions of Jews in Berlin' on the fiftieth anniversary of Kristallnacht, Biedermann explained, 'The pieces of old-fashioned furniture ... are to point to *irretrievable* losses that occurred because a large group of people, their way of life and their culture, are missing.'[31] What *can* be retrieved, then, he seems to suggest, may be the experience of the irretrievable. Without trivializing the seriousness of the intent, one of the irretrievable losses is the wireless – which, in 1930s promotional literature, is invariably depicted theatrically as a device that draws people to it. Often, it sits like an enigmatic doll's house on the kitchen table, the family old and young craning their necks to decipher the human voice amidst its crackle. I have watched kids with their soundboxes perch on the table checking their iPhones. In one take on Biedermann's sculptural ensemble, the source of the shock registered by the fallen chair might have been a public radio announcement.

From this point of view, Friz's community radio may be far too prescriptive: 'Community radio must not be merely representational, but enact community,'[32] she writes, but a formulation like this begs the question of interference, risking an entrainment that leaves little room for transitory and reversible syncopation. Communication is improved, even solidarity, but so perhaps is the likelihood of collective madness, as Hermann Gutzmann discovered. A feature of Biedermann's ensemble is its parquet floor, around whose edges are cast lines from Nelly Sachs's poem 'O die Schornsteine' (from her no less pertinently titled 1946 book *In the Habitations of Death*). Defamiliarized in this park setting, it occurs to me that this flooring evokes the older association of parquet with that anomalous zone in baroque theatre between the amphitheatre and the stage where the distinction between the actor and the onlooker was confused and all might participate in the realization (or overthrow) of the playwright's intentions. I like the way that any visitor curious about the work's significance has to walk round the room to read the inscription, like cemetery visitors who avoid treading on the grass. To get inside this work you stay outside it, filing along the edge or standing at the border.

Independent radio makers resisting the elimination of experimental work from the public sector often compare themselves to graffiti artists. Broadcasting from its antenna in Nancy, Radio Graffiti, for example, '*établit sa programmation essentiellement autour des artistes indépendants en mettant en avant les groupes de la region,*' asserting '*Vous aimez écouter radio Graffiti parce qu'elle est libre et indépendante. Sa tonalité est unique ainsi que sa programmation.*'[33] The idea is to diffuse subversive soundwaves, to infiltrate the urban auditory unconscious and leave a unique tonal tag. Radio is re-territorialized – a region of shared cultural production values is proposed – but its collective consciousness is post-theatral, free of explicit program or identifiable voice. Yet post-radio radiophonic art might as easily produce *dramaturgy graffiti*, decentralized movements of people in time and space, regionalized by the antenna of a public artwork. If the democratization of the media means downscaling the potentially cosmic outreach of electromagnetic communication in the interests of realizing 'the potentiality of personalized expression',[34] then one quickly finds oneself in the realm of 'tactical media' whose low wattage transmissions may penetrate no more than a few city blocks.[35] Here, certainly, the 'singular glance of authority has been fragmented, democratised by the multitudes that now participate in its formation'.[36]

And here, certainly, as radio elides into the production of public space it becomes functionally indistinguishable from a new form of urban dramaturgy. Obviously, pop-up pavement-side household goods, recycled furniture and 'outdoors' interior décor anticipate this subtle disruption: they are everyday rubbish hills and can easily be augmented with other objects (miniature table and chairs, etc.) that tread the boundary between chance and design. Likewise, the affective qualities of place, evoked by the term ambience, may be intrinsically performative: 'The word *amble* implies a departure from the straight and narrow, a certain locomotory self-awareness; one might sway from side to side; one who ambles takes notice of things about them; their progress is measured, animated and responsive to the affordances the environment offers.'[37] The dramaturgy of turbulence can be compared to the white noise of the radio oscillating between stations, out of which periodically, through the mere coincidence of entrainment, a group recognizably motivated by the same interests emerges. The challenge of producing a radio script in this post-radiophonic environment might be to produce this effect.

Whether or not this outcome is plausible, or even desirable, remains to be seen, but I wanted to conclude this book with a scenario that loosens the bonds of cultural–technological determinism that risked framing the narrative. If the tradition of scripting precedes radio, it also postdates it: and radio, too, its potential to produce new idealizations of public space, rooted in self-realizing communities that are archipelagically related, was, perhaps, always inherent in the dynamics of the meeting place and remains potentially fertile as a metaphor of new performatively defined political communities.

'What Is Your Name' is not formulaic any more than *Der Verlassene Raum* is generic. They are instances that find their place, I am suggesting, by considering them as part of the larger, invisible and turbulent public region. Nevertheless, this idea only has any value if it is catalytic, leading to a new understanding of what public space might communicate. All the radio scripts discussed in this book are variations, in one way or another, on the theme announced in the title of the earliest of them. They all explore the relationship between memory and desire, particularly from the migrant perspective where, in the absence of any filial connection to the place, the historical identification must be affiliative. In Berlin, how does the outsider remember what has been lost? How does this desire – which public memorials to the Holocaust presumably aim to stimulate (witness the almost immediately adjacent Memorial outside the Jewish cemetery further down Grosse Hamburger Strasse) – affiliate to a past without turning into either historical voyeurism or unconscionable self-justification?

One answer has to be to draw attention to the estrangement of the tourist: if the chief audience of these public artworks appears to be visitors from out of town, whether from regional Germany or internationally, one of their purposes is to connect what happened here to a universal condition. This does not lie in the past but surrounds the policed territories of free passage in the form of detention camps where illegal immigrants are held. It is chilling to see footage of seaside tourists sunbaking adjacent to human beings detained largely because they have experienced the horrors of genocide. It is curious that those most qualified to identify with the experiences evoked in these places are not able to frequent them. It is not a question of colonizing their sorrow, inviting them, once again, to answer the colonial demand for translations that can represent the unspeakable eloquently; simply of acknowledging that, in relation to these continuing dictatorships of the border, it is the outsiders who are insiders when it comes to interpreting what is really going on.

And the interpretation is partly performed along another border – between language and speech, and in the deformation (or reformation) of both. 'What is your name?' the phrase occurs in 'The 7448': these scripts for radio exist intertextually, an archipelago of radiating and intersecting dramatic regions; their borders are permeable, and new arrangements of their collective material are perfectly possible. Somewhere, in this intertextual region, the anti-baroque minimalism of 'What Is Your Name' produces an expressive frustration that cannot any longer be contained. In reaction, there is a euphuistic out-spilling, evident not only in 'The 7448' but also in 'Underworlds of Jean du Chas', one of whose functions is to override correct pronunciation and, indeed, the controls that language attempts to place on expression. Here the pidginization of languages is pre-emptive rather than reactive in its resistance to translation.

And these are not simply aesthetic glosses on mutual incomprehension in the context of terror: they report real situations. 'A case in point is represented by the musical band of African immigrants who, in Lampedusa, advertised the local policy of Responsible Tourism.' In a reggae song, 'a Caribbean, foreign musical genre', they borrowed the conventions of classical 'epic' literature not belonging to the immigrants' cultural schemata, to represent their own journey. 'Row, row, to Lampedusa we go, / Go, go, for a better life we row, yeah, / O dolce Musa, portami a Lampedusa.'[38] Or, as we might say:

> Voice 10: Archipelagos of consonants.
> Voice 3: And the wind, the ocean currents.
> Voice 11: Variable vowels: O and E.
> Voice 4: Again and again, breaking.[39]

Oddly, language used in refugee camps may resemble the 'ad hoc language arrangements of groups in contact' in 'superdiverse megalopolises'.[40] Therefore, bizarre as it may seem, these American sounds may be perfectly at home in Koppenplatz – as I discovered only the other day, picking up on my iPhone Palestinian-American rap star D. J. Khalid's *Suffering from Success*, emanating from a soundbox hung under a basketball hoop, creating a kind of rhythm into which the couple idly bouncing the ball fitted themselves. Imagined in this noisy world of multiple callings the name of the work – its precarious linguistic identity – exceeds anything that the place can contain or muffle. Embryonic themes of voicing and languaging explored through echoic mimicry, questions of translation and translatability and the posing of a counter-discourse of sensitivity (play that brings about its own rules of respect and civility) – these are all concerns of asylum seekers and those who seek to help them. They are usually sidelined or unacknowledged. Communication that is open to the liberating propensities of language is never in these circumstances escapist: it plots a fundamentally different political language.

When I first experimented with speaking aloud at Koppenplatz, I was concerned that my voice did not carry. I wondered whether I was self-censoring. I was nervous that my experiments in languaging, letting sense repeatedly fall through the oubliette of sound into a glossolalic underworld, would be found offensive. I reckoned that the disturbance would be tolerated if, like the pub singer, I recited in a way that no one could hear. Now, the fact that it dies away appeals to me as a measure of the physical limits to what can be said. Equally, when I improvise centre stage, riffing on passages from 'What Is Your Name', 'Wie ist dein Name' and 'Quel nom toi', while sitting at Biedermann's table, I experience a kind of invisibility, as if my textual interpellation erects protective walls. Outside its reach, the distraught voices of the local environment seem to become audible. To detect their natural frequencies, and to perceive in their rhythmic ruins an alternative community not walking in step, it is, I am finding, necessary to stay untuned. When inner voice and outer voice elide, no direction is given; although the ghosts are present, often disguised as passers-by, they make no demands; beyond the occasional enquiry, pointing to where you have come from, or the polite request for a photograph, they continue to talk among themselves. But, provisionally at least, that is something.

Collect

Sound works referred to in *Amplifications*

1. 'Memory as Desire' (1986)

Recorded in the studios of the Australian Broadcasting Corporation (Sydney) with actors Ron Haddrick (Charles Sturt), June Salter (Charlotte Sturt), Don Pascoe (George Augustus Robinson) and Jane Harders (Rose Robinson). Production for ABC-FM's program *Images*: Martin Harrison; audio recording and post-production: Phillip Ulman, Ann Winter. First broadcast: 9 June 1986; second broadcast (revised version): 21 May 1987. Duration: 55 minutes.

See also Paul Carter, 'Desire of Dialogue, Radio Writing and Environmental Sound' in *Uncertain Ground*, edited by Martin Thomas (Sydney: Art Gallery of New South Wales Publications, 1999), 143–60.

Paul Carter, 'Birdsong from *Memory as Desire*', *Scripsi*, vol. 5, no. 3, pp. 235–44.

2. 'What Is Your Name' (1986)

Recorded in the studios of the Australian Broadcasting Corporation (Sydney) with actors Terry Reid, John Dicks and Robert Menzies. Production for ABC-FM's program *Surface Tension*: Martin Harrison; audio recording and post-production: Craig Preston and Andy Henley. First broadcast: ABC Radio National, 'Surface Tension,' 13 September 1986. Duration: 28 minutes.

'What Is Your Name' was an ABC entry for the Prix Futura (French translation, 'Quel nom toi' by Philippe Tanguy). In 1991 Westdeutscher Rundfunk, Köln made a German production of the work 'Wie Ist Ihr Name'. First broadcast: 10 December 1991. In 2004 Prompt! Berlin created a multilingual theatre production, 'Wie ist dein Name', presented at Theaterdiscounter, Mitte, Berlin, 1–3 June.

See also Paul Carter, *Ground Truthing, Explorations in a Creative Region* (Perth: University of Western Australia Publishing, 2010), 246–64.

'The Empty Space is a Wall, The Role of Theatrical Translation in the Public Reinscription of the Other', *Performance Research*, 2005, Routledge (Taylor & Francis), 'In Form' issue.

3. 'Scarlatti' (1986)
Recorded in the studios of the Australian Broadcasting Corporation (Sydney) with actors Andrew Tighe, José Farinas. Production for ABC-FM's program *Surface Tension*: Martin Harrison. First broadcast: ABC Radio National, 'Surface Tension,' 13 September 1986; second broadcast: 18 August 1987; third broadcast: ABC-FM, 'The Listening Room', 18 January 1988. Duration: 28 minutes 50 seconds.

See also Paul Carter, 'Territorialising Atmospherics: The Radiophonics of Public Space,' *Architecture and Culture*, vol. 3, no. 2, 2015, pp. 245–62.

4. 'Remember Me' (1988)
'Remember Me' was commissioned by Martin Harrison for the fourth in the series Poets Reading, which he, Ruark Lewis and Ann Berriman curated for the 1988 Sydney Biennale. It was first performed by the author on 25 June 1988, at the Art Gallery of New South Wales, using a 1952 recording of *Dido and Aeneas* by Henry Purcell and a digitally remastered version of the same recording (1988).

The radio production was recorded in the studios of the Australian Broadcasting Corporation (Sydney) with actors Matthew O'Sullivan (Sound Engineer), Judy Nunn (Mrs Ear), Edgar Metcalfe (Mr Ear), David Nettheim (Bernard Miles and Walter Legge), Robyn Nevin (Kirsten Flagstad), Ivar Kants (Robin Hood), Susan Lyons (Maid Marion), Edgar Metcalfe (Henry Johansen) and David Downer (Henry Purcell). Production for ABC-FM's program *The Listening Room*: Andrew McLennan. First broadcast: ABC-FM's 'The Listening Room'. 14 August 1988. Duration: 25 minutes.

'Remember Me' won a 1989 HiFi Drama Award.

See also Paul Carter, 'Remember Me', script and notes, *Meanjin*, vol. 48, no. 1, 1989, pp. 116–34.

Paul Carter, 'Remember Me/Mimicry: a short performance history', *Art & Text*, no. 31, December 1988–February 1989, pp. 43–9.

5. 'Mirror States' (1989)
Although unrealized at its intended site, 'Mirror States' has been produced for radio in a joint production of ABC-FM's 'The Listening Room' (production: Tony MacGregor) and Westdeutscher Rundfunk Köln (production: Klaus Schöning) with actors Peter Murphy, Eugenia Fragos, Michael Fry, Kylie Belling and Kim Trengrove and post-production by Les Gilbert. The first

broadcast (9 October) coincided with the installation of the work at the ARX-89 Festival, Perth Institute of Contemporary Art, Perth, Western Australia. Duration: 31 minutes 28 seconds.

See also Paul Carter, 'Mirror States, A Site-Specific Sound Installation,' *The Sound In-Between, Voice, Space, Performance* (University of New South Wales Press/ New Endeavour Press, 1992), 94–114.

6. 'On The Still Air' (1990)

Recorded in the studios of the Australian Broadcasting Corporation (Sydney) with actors Antonio Palumbo (Andreotti), Domenico Gentile (Pasolini), Simon Chilvers (Ezra Pound), David Downer (T. S. Eliot), Paolo Totaro (Dante), Pino Bosi (Cavalcant.i), Livia Bosi (Questa), Paul Carter (Nom). Production for ABC-FM's program *The Listening Room*: Andrew McLennan. Australian Broadcasting Corporation. First broadcast: The Listening Room, ABC Fine Music, 3 September 1990. Duration: 35 minutes.

7. 'Cooee Song' (1990)

At the time of writing (September 2017), an interpretation of this work is in development as part of an ambient sound and light composition for Cato Square, Prahran (Melbourne). An interesting feature of this interpretation is its return to a MIDI keyboard controller technology roughly contemporary with the original conception of 'Cooee Song' to process recordings and generate outputs that can be performed and manipulated in real time.

See Paul Carter, 'Cooee Song, A Performance Work for Two Actors and Their Voices' in *The Sound In-Between*, 52–69.

8. 'Named in the Margin' (1991)

Electro-acoustic installation, script commissioned by Sound Design Studio, Melbourne, for Hyde Park Barracks, Sydney (a property of the Historic Houses Trust of New South Wales).

'Named in the Margin' was a Sound Design Studio production, recorded at Studio 52, Richmond, with actors Bruce Kerr, Roy Baldwin, Stanley McGeogh, Jeff Hodgson and Helen Noonan. Museum installation for Sound Design Studio: Les Gilbert. Linear duration: 1 hour 5 minutes.

See also Paul Carter, 'An Innovative Approach to History', in *Thumbnail Dipped in Tar*, publication of the National Book Council, vol. 18, no. 5, 1991.

Paul Carter, 'Performing History: The Hyde Park Barracks Voice Collages', *Transition*, nos. 36/37, 1992, pp. 5–11.

9. 'Columbus Echo' (1992)

Electro-acoustic installation, script commissioned by Sound Design Studio (Melbourne) in association with Cambridge Seven Architects, as part of the design for the Aquarium of Genoa.

'Columbus Echo' was a joint production of Sound Design Studio and ABC-FM's 'The Listening Room'. It was recorded in the Melbourne studios of the ABC on 18–20 March 1992 with actors Poni Poselli (Genoese, Italian), Carlos Sanchez (Colombian Spanish and Guarani), Joseph Ghiocas (Italian, Greek), Gabriella Caroscalao (Portuguese, Indonesian, Timorese languages), Alberto Vila (Catalan, Spanish), Paul Karo (English, Japanese, and a very sprightly Nahuatl!), Grant Smith (English, French, German, Yiddish), Helen Noonan (English, Italian), Manuella Carluzzi (Italian), Easter Wu (Mandarin), Lili Chen (Cantonese), Carlos Espinosa (Spanish), Roberto Bertini (Italian). Linear duration: 1 hour 30 minutes.

See also Paul Carter, 'Emergency Languages: Echoes of Columbus in Discourses of Precarity,' in *Migration and the Contemporary Mediterranean*, edited by C. Gualtieri (Bern: Peter Lang, 2018).

10. 'The 7448' (1992)

Commissioned by Westdeutscher Rundfunk Köln and ABC-FM's 'The Listening Room' and recorded in the studios of the Australian Broadcasting Corporation (Melbourne) with actors Frank Witting, Ursula Grawe, Paul Karo, Manuela Carluzzi, Grant Smith, Lisa Rodgers, Hartley Newnham, Roy Baldwin, Gabriella Carascalao, Alberto Vila, Carlos Sanchez, Glen Riddle and Natalie Lin. The English segments of the script were translated into German by Frank Witting and Ursula Grawe. The work was broadcast in Germany as '7448, Eine Kolumbische Phantasie'. Production for the ABC-FM program *The Listening Room*: Andrew McLennan; audio-recording and post-production: Philip Ulman. First Australian broadcast: 'The Listening Room, ABC Fine Music, September 1992. Duration: 48 minutes.

The script for *The 7448* was a winner of the Ars Acustica Prize 1991 jointly sponsored by ABC-FM's 'The Listening Room' and WDR Köln. *The 7448* was featured at the 4th Exposition Sonore Internationale, Arles, France, July 1995. A selection from the work is released on CD Vergo under the auspices of WDR's Akustische Kunst Studio.

See also Paul Carter, 'Emergency Languages: Echoes of Columbus in Discourses of Precarity,' in *Migration and the Contemporary Mediterranean*, edited by C. Gualtieri (Bern: Peter Lang, 2018).

11. 'The Native Informant' (1993)

Commissioned by the Australian Broadcasting Corporation. Recorded in the studios of the ABC (Sydney) with actors Paul Carter, Andrew

Ford, David Wicks and Helen Noonan. Production for the ABC-FM program The Listening Room: Andrew McLennan; audio recording and post-production: Phillip Ulman. First broadcast: 6 December. Duration: 28 minutes.

See also Paul Carter, 'Ambiguous Traces, Mishearing, and Auditory Space,' in *Hearing Cultures, Essays on Sound, Listening and Modernity*, edited by V. Erlmann (Oxford: Berg, 2004), 43–64.

12. 'Tuned Noises'/'Abgestimmte Gerausche' (1994)
Commissioned by Studio Akustische Kunst, Westdeutscher Rundfunk Köln. Regie: Hein Bruehl. First broadcast, 24 May 1994. Duration: 29 minutes 30 seconds. Regie: Hein Bruehl.
  In 1998 'Tuned Noises' was commissioned by the Australian Broadcasting Corporation. Production for the ABC-FM's program The Listening Room: Andrew McLennan. Narration: Paul Carter. First English broadcast, May 1998. Duration: 28 minutes 30 seconds.

13. 'The Calling to Come' (1995)
Electro-acoustic installation, script commissioned by the Museum of Sydney.
  'The Calling to Come' was a co-production of the Museum of Sydney and ABC-FM's 'The Listening Room'. It was recorded in the studios of the ABC on 22 February 1995 with actors Arthur Dignam and Kristina Nehm. Production for the ABC: Andrew McLennan; sound recording: Russell Thomson; audio post-production: Phillip Ulman. Museum installation for Magian Multimedia (Melbourne): Nigel Frayne; audio post-production: John Campbell. Sound design consultant for the Museum installation: Andrew McLennan.
  'The Calling to Come' was broadcast on ABC Classic FM's 'The Listening Room' on 22 May 1995. Duration: 20 minutes.

See also Paul Carter, *The Calling to Come* (Sydney: Historic Houses Trust of New South Wales, 1996).
  Paul Carter, 'Speaking Pantomimes: Notes on *The Calling to Come*', *Leonardo Music Journal*, vol. 6, Summer 1996–1997, 95–8.
  Paul Carter, *Material Thinking, the Theory and Practice of Creative Research* (Carlton, VC: Melbourne University Publishing, 2004), 153–76.

14. 'Lost Subjects' (1995)
Electro-acoustic installation, script commissioned by the Museum of Sydney.
  'Lost Subjects' was a co-production of the Museum of Sydney and ABC-FM's 'The Listening Room'. It was recorded in the Melbourne (Southbank) studios of the ABC on 1–3 October 1994 with actors Beverley Dunn, Diane Flatley, Geoffrey Hodgson, Paul Karo, Bruce Kerr, Stanley McGeagh, Rachel Morris and Grant Smith. Production for the ABC: Andrew McLennan; sound

recording: Russell Thomson; audio post-production: Phillip Ulman. Museum installation for Magian Multimedia (Melbourne): Nigel Frayne; audio post-production: John Campbell. Sound design consultant for the museum installation: Andrew McLennan. Linear duration: 1 hour 35 minutes.

See also Paul Carter, *Lost Subjects* (Sydney: Historic Houses Trust of New South Wales, 1999).

15. 'Underworlds of Jean du Chas' (1998)

Recorded in the studios of the Australian Broadcasting Corporation (Sydney) with actors Barry Otto, Gillian Jones, Tim Elliott, Jeremy Sims with the assistance of the Newtown High School for the Performing Arts. Production for ABC-FM's program *The Listening Room*: Andrew McLennan; sound engineering: Phillip Ulman. First broadcast: 18 May 1998. Second broadcast: 3 May 1999. Duration: 52 minutes.

16. 'Light' (1996)

Sound installation, sculptural setting, performance commissioned by Adelaide Festival of the Arts, with Hossein Valamanesh, Chandrabhanu and Andrew McLennan.

The sound for the sound installation of 'Light' was recorded in the studios of the Australian Broadcasting Corporation (Sydney) with actors Barry Otto (Light), Patrick Dickson (Double, Another), Peter Carroll (Another, Double), Helen Noonan (Mary), Gillian Jones ('E'), Linda Cropper (Maria) and singers Angus Stewart, Michael Carmody and Stephan Newton. Sound production, sound installation and site design: Andrew McLennan; audio-engineering and post-production: Phillip Ulman. First radio broadcast on the ABC-FM program *The Listening Room* coincided with opening of installation: 4 March. Duration (in radio version): 52 minutes.

See also Paul Carter (with Chandrabhanu), 'Light', *Visual Arts Program catalogue*, Telstra Adelaide Festival 96, pp. 12–13.

17. 'Old Wives' Tales' (1997)

Script commissioned by Chandrabhanu and the Bharatam Dance Company. First performance: C.U.B. Malthouse, Melbourne, 25 September–4 October. Duration (performance): c. 1 hour 20 minutes.

18. *Out of Their Feeling* (1999)

Electro-acoustic sound sculpture, Great Irish Famine Commemoration (An Gorta Mor) Sculpture, Hyde Park Barracks, Sydney, commissioned by Hossein & Angela Valamanesh. Duration: 25 minutes.

Recorded with actors Judy Pindar, Mary O'Connell and musicians: Mico Russell, Sean Ac Donnca. Sound recordist: Brendan Frost.

See also 'Unwonted Silence: An Gorta Mor and the Presence of Sound' in *Hearing Places: Sound, Place, Time and Culture*, edited by R. Bandt, M. Duffy and D. MacKinnon (Cambridge Scholars Publishing, Newcastle, UK, 2007), 202–13.

19. 'The Letter S' (2005)
Commissioned by the Australian Broadcasting Corporation with actors Paul Blackwell, Lizzie Falkland, Caroline Mignone, Nathan O'Keefe, Rory Walker. Sound Engineer: Simon Rose. Recorded 19–21 July 2005. Production for the ABC program *Airplay*: Christopher Williams. First broadcast 15 and 16 September. Duration: 29 minutes.

20. 'Mac' (2012)
Commissioned by the Australian Broadcasting Corporation with actors Jermain Hempton and Matt Hein (Jowley), Lisa Flanagan (Jowley's Birth Mother), Eileen Darley (Jowley's Adoptive Mother), Rory Walker (Neilson), Nathan O'Keefe (Doctor), Kate Wadey (Nurse). Production for the ABC program *Airplay*: Christopher Williams; recording engineer: Andrea Hensing. Recorded at ABC Collinswood, SA. December 2010. First broadcast 11 September 2011 on ABC AM *Airplay*. Duration: 42 minutes.

The major unpublished scripts discussed in *Amplifications* are collected in Paul Carter, *Absolute Rhythm: works for minor radio* as follows: 'Memory as Desire', 14–48; 'What Is Your Name', 49–69; 'Cooee Song', 69–93; 'Scarlatti', 94–114; 'Remember Me', 115–37; 'On The Still Air', 138–64; 'The Native Informant', 165–87; 'The Calling to Come', 188–204; 'The 7448', 205–62; 'Underworlds of Jean du Chas', 263–301.

NOTES

Prolude

1. Paul Carter, 'Mirror States', in *The Sound In-Between, Voice, Space, Performance* (Sydney: New Endeavour Press/ University of New South Wales Press, 1992), 93–114, 102.
2. Watkin Tench, 'A Narrative of the Expedition to Botany Bay', in *Sydney's First Four Years*, ed. L. F. Fitzhardinge (Sydney: Library of Australian History, 1979), 31.
3. David Collins, *An Account of the English Colony of New South Wales* (London: T. Cadell & W. Davies, 1802), 2 vols, vol. 1, 192.
4. Arthur Schopenhauer, *Parerga and Paralipomena*, trans. E. F. J. Payne (Oxford: Clarendon Press, 1974), 2 vols, vol. 1, 77.
5. Schopenhauer, *Parerga and Paralipomena*, vol. 1, 180–1.
6. Tench, 'A Narrative of the Expedition to Botany Bay', 79.
7. Emily Dickinson, *The Poems of Emily Dickinson*, ed. T. H. Johnson (Cambridge, MA: The Belknap Press, 1955), 2 vols, vol. 2, 647–8, no. 870.
8. Paul Carter, 'Memory as Desire', 18.
9. By now I was performing, directing and involved in post-production.
10. Paul Carter, 'Remember Me', 116, 118.
11. Paul Carter, 'The Native Informant', 166.
12. Francis Bacon, *Philosophical Works*, ed. J. Spedding, R. L. Ellis and D. D. Heath, vol. 2 (Cambridge: Cambridge University Press, 1857/2011), 427.
13. Bacon, *Philosophical Works*, vol. 2, 427.
14. Dr John Worgan, George's uncle, was a Scarlatti devotee. He copied a group of sonatas belonging to an organist of the Chapel Royal, Madrid. Charles Burney writes of him in this connection in his *A General History of Music*, remarking of the sonatas, 'Few have now perseverance sufficient to vanquish their peculiar difficulties of execution.' (Keith Anderson, cover notes, *Scarlatti, D.: Keyboard Sonatas* (complete), vol. 9, 2008 (https://www.naxos.com/ma insite/blurbs_reviews.asp?item_code=8.570368&catNum=570368&filetype =About%20this%20Recording&language=English)), viewed 8 June 2013. George brought a piano with him to the colony but his repertoire is unknown.
15. George Worgan, *Journal of a First Fleet Surgeon* (Sydney: Library Council of New South Wales in association with the Library of Australian History, 1978), 26–7.

16 In a way Worgan's peroration proposed founding the colony in echoic mimicry: *Suscitate*, to set in rapid motion, to call into being, to give life to, to animate. The colony calls the capital to hearken, to lend an ear to 'a string of little Transactions, Occurrences, Excursions & Adventures', as Worgan says.

17 Discussed in Paul Carter, 'Emergency Languages: Echoes of Columbus in Discourses of Precarity', in *Migration and the Contemporary Mediterranean: Shifting Cultures in 21st-Century Europe*, ed. C. Gualtieri (Bern: Peter Lang Publishers, 2019), 285–304.

18 Paul Carter, *Lost Subjects* (Sydney: Historic Houses Trust of New South Wales, 1998), 93 with Kurt Schwitters particularly in mind.

19 Virginia Madsen, 'Written in Air: Experiments in Radio', in *Experimental Music: Audio Explorations in Australia*, ed. Gail Priest (Sydney: UNSW Press, 2009), 154–74, 169.

20 Tench, 'A Narrative of the Expedition to Botany Bay,' 293.

21 Ibid., 189. See also Jakelin Troy, *The Sydney Language* (Canberra: Panther Publishing, 1998, 1998), 20.

22 Carter, *Lost Subjects*, 178. Quoting Francis Grose's 1811 Dictionary of the Vulgar Tongue, the commentary in *Lost Subjects* adds, 'To Top. To cheat, or trick: also to insult: he thought to have topped upon me. Top; the signal among taylors for snuffing the candles: he who last pronounces that word, is obliged to get up and perform the operation. – to be topped, to be hanged' (245).

23 In introducing Vico's autobiography in the first issue of *Raccolta d'opusculi scientifici e filologici*, Calogera 'attributes the idea of such a text on one's own life to Lodoli, who coined from the Greek the term *periautografia*, using *peri-* (around, what surrounds or encloses) rather than *bios* as the term to connect to *autos* and *graphe*. *Peri* is a life-term, often used as a prefix to anatomical terms to characterise what surrounds a designated organ.' Donald Verene, *The New Art of Autobiography, An Essay on the Life of Giambattista Vico Written by Himself* (Oxford: Clarendon Press, 1991), 66.

Chapter 1: Charms

1 Richard Jefferies, *Wildlife in a Southern Country* (London: Thomas Nelson & Sons, 1937), 236.

2 Gaston Bachelard, *The Poetics of Reverie*, trans. D. Russell (Boston: Beacon Press, 1969), 188.

3 *Letters: Summer 1926, Boris Pasternak, Marina Tsvetayeva, Rainer Maria Rilke*, trans. M. Wettlin, W. Arndt and J. Gambrell (New York: New York Review of Books, 2001), 117.

4 Roberta Reeder, *Anna Akhmatova, Poet & Prophet* (London: Allen & Busby, 1995), 489.

5   Reeder, *Anna Akhmatova, Poet & Prophet*, 139.
6   Quoted by R. Murray Schafer, *The Tuning of the World* (New York: Knopf, 1977), 62.
7   Scott G. Bruce, *Silence and Sign Language in Medieval Monasticism* (Cambridge: Cambridge University Press, 2007), 13.
8   From the Latin 'occidere', 'to kill'; or rather the old French 'occire', 'occis', denoting the doom which the nightingale imprecates or supplicates on all who do offence to Love. Hence the nightingale's call can be declined like a verb (see *Rossignol, an edition and translation*, ed. J. L. Baird and J. R. Kane (Kent: Ohio, Kent State University Press, 1978), 78).
9   Claire Nahmad, *Angel Messages: The Oracle of the Birds* (London: Watkins Publishing, 2010), unnumbered.
10 Richard Rolle, *The Fire of Love or Melody of Love and The Mending of Love*, translated into modern English by Evelyn Underhill (London: Methuen, 1920), 61.
11 Rolle, *The Fire of Love*, 140.
12 Ibid., 141.
13 Sukanta Chaudhuri, 'The Renaissance God as Man of Letters,' in *Renaissance Themes: Essays Presented to Arun Kumar Das Gupta*, ed. S. Chaudhuri (New Delhi: Anthem Press, 2009), 10.
14 There must have been exceptions. In *The Village Book* (London: Jonathan Cape, 1930), based on his residence in Georgeham, North Devon, Henry Williamson describes Grannie Parsons's appreciation of 'the little grass-bird – the chiff-chaff, the celandine among birds, whose plain-song is so precious as it comes hopefully over the border of winter' (199). However, Richard Jefferies strongly reinforces my impression: remarking that 'it is curious that, though singled out as a first sign of spring, the chiffchaff has never entered into the home life of the people. ... I never once heard a countryman, labourer or farmer, or anyone who was always out of doors, so much as allude to it' (Richard Jefferies, *Nature Near London* (London: John Clare Books, 1980), 146–7).
15 After all, a hatred of nightingales was not solely tied to the cultivation of prayer. W. H. Hudson reports an incident around 1850 where a 'gentleman' who occupied a house in Ringmer (famous for its associations with Gilbert White) 'had all the nightingales frequenting the grounds destroyed. Their late singing disturbed his rest' (*Nature in Downland* (London: Madonald Futura, 1981; orig. pub. 1923), 225.)
16 William Marshall, *The Rural Economy of Norfolk* (2 vols, London, 1787), 303. Quoted by Penny Williams, *The Later Tudors, 1547–1603* (Oxford: Oxford University Press, 1995), 9.
17 Writing in the 1920s Anthony Collett attributed the multiplication of small song birds to the 'widespread extermination of all beasts and birds reputed foes to grouse, partridge or pheasant'. Now, as a result, he writes, 'In a pheasant-cover of not more than two acres, one might hear at least four or five

nightingales, and blackcaps and whitethroats and garden warblers, too many to distinguish' (*The Changing Face of England* (London: Nisbet & Co., 1926), 269–70).

18 *Cobbett's Weekly Register* (London: John M.Cobbett, 1821), Vol. XL (21 July–29 December 1821), 1208.

19 Jean-Pierre Vernant, *Myth and Thought among the Greeks* (London: Routledge & Kegan Paul, 1983), 53.

20 Richard Church, *Over the Bridge, an Essay in Autobiography* (London: The Reprint Society, 1956), 191–2, 222.

21 Aldous Huxley, *Eyeless in Gaza* (London: Chatto & Windus, 1938), 607.

22 Huxley, *Eyeless in Gaza*, 619.

23 Alfred Williams, *Folk Songs of the Upper Thames* (London: Duckworth, 1923), 24.

24 Gilbert Keith Chesterton, *A Short History of England* (London: Chatto & Windus, 1917), 166.

25 Alfred Williams, *Songs in Wiltshire* (London: Erskine MacDonald, 1909) http://www.alfredwilliams.org.uk/thepoems.html, viewed 14 April 2013.

26 Williams, *Folk Songs of the Upper Thames*, referring to the 'Introduction', especially pages 23–9.

27 Martin Harrison, *Ancient Noise*, unpublished ms. available by arrangement through The Martin Harrison Archives, Institute of Postcolonial Studies, Melbourne.

28 Maurice Merleau-Ponty, *The Visible and the Invisible*, trans. A. Lingis (Evanston: Northwestern University Press, 1968), 159.

29 https://www.bu.edu/cas/magazine/fall09/wagenknecht, viewed 8 August 2016.

30 Paul Carter, *Parrot* (London: Reaktion Books, 2006), 73–4.

31 'Soviet atomic bomb project', https://ipfs.io/ipfs/QmXoypizjW3WknFiJnK LwHCnL72vedxjQkDDP1mXWo6uco/wiki/Soviet_atomic_bomb_project. html, viewed 4 March 2016.

32 John DeBusk, 'Part Three – My Experience at the Nevada Test Site-(Area 7)', https://www.linkedin.com/pulse/my-experience-nevada-test-site-area-7-john-d ebusk-part-fortin-2, viewed 4 March 2016.

33 Alfred Williams, *Life in a Railway Factory* (Gloucester: Alan Sutton, 1984), 75.

34 Barry Truax, 'Electroacoustic Music and the Soundscape: The Inner and Outer World', in *Companion to Contemporary Musical Thought*, ed. J. Paynter, R. Orton, P. Seymour and T. Howell (London: Routledge, 1992), 374–98.

35 http://acousticstoday.org/wp-content/uploads/2015/06/Concorde-Booms-an d-the-Mysterious-East-Coast-Noises-Peter-H-Rogers-and-Domenic-J.-Maglieri. pdf, viewed 4 March 2016. To 'ensonify' means to irradiate a medium (usually water but also earth) with sound in order to locate or image objects.

36 https://www.gov.uk/government/uploads/system/uploads/attachment_data/f ile/278275/NASM_Brize_Norton_OEM_04_14.pdf, viewed 4 March 2016.

37 One performance was broadcast and has been issued on CD by Walhall (WLCD 0186): see http://www.classicalsource.com/db_control/db_cd_review.php?id=5003. In the Walhall catalogue, however, the date of the performance is given as 1 October 1951. http://www.musicweb-international.com/classRev/2008/July08/Purcell_Dido_5096902.htm#ixzz4EJODrSKx

38 Todd Decker, 'A Waltz with and for the Greatest Generation: Music in *Band of Brothers* (2001)', in *American Militarism on the Small Screen*, ed. Anna Froula and Stacy Takacs (New York: Routledge, 2016), 93–110, 106.

39 Decker, 'A Waltz with and for the Greatest Generation: Music in *Band of Brothers*', 107. A further irony, or coincidence, is that *Band of Brothers* is based on the wartime experience of the Easy Company, whose name is curiously picked up in 'Easy,' the name of the 5 November 1951 bomb.

40 Kai Vogeley and Christian Kupke, 'Disturbances of Time Consciousness from a Phenomenological and Neuroscientific Perspective', *Schizophrenia Bulletin* 33, no. 1 (2007), 157–65, 159.

41 Harrison, *Ancient Noise*.

42 George Petrie, *The Ancient Music of Ireland*, http://www.toad.net/~sticker/nosurrender/primary.html, viewed 18 August 1998. This reflection became the poetic core of 'Out of Their Feeling,' a sound installation forming part of Hossein and Angela Valamanesh's Memorial to the Great Irish Famine at Hyde Park Barracks, Sydney (see Paul Carter, 'Unwonted Silence: An Gorta Mór and the Presence of Sound', in *Hearing Places, Sound, Place, Time, Culture*, ed. R. Bandt, M. Duffy and D. MacKinnon (Newcastle, UK: Cambridge Scholars Publishing, 2007), 202–13). Track 19 of the CD released with this volume features an extract from the composition.

43 Truax, 'Electroacoustic Music and the Soundscape: The Inner and Outer World,' 376.

44 Ibid., 377.

45 Ibid., 375.

46 As reformulated by Gareth Stedman Jones, 'Introduction', in Karl Marx and Friedrich Engels, *The Communist Manifesto* (London: Penguin, 2010), 1–185, 130ff.

47 George Fox, http://www.lancaster.ac.uk/quakers/spence/Sp_39a_el.html, viewed 14 May 2016.

48 George Fox, 'Introduction', *The Journal of George Fox*, ed. N. Penney (Cambridge: Cambridge University Press, 2 vols, 1911), xxvi.

49 Fox, *The Journal of George Fox*.

50 Ibid.

51 Maurice Merleau-Ponty, *The Visible and the Invisible*, 159.

52 Ibid., 155.

53 V. Khlebnikov and A. Kruchonykh, 'The Letter as Such' (1913) in *Collected Works of Velimir Khlebnikov*, vol. 1, Letters and Theoretical Writings, ed. Charlotte Douglas, trans. P. Schmidt (Cambridge, MA: Harvard University

Press, 1987), 257–8, 258. See https://monoskop.org/images/6/6a/Khlebnikov_V elimir_Collected_Works_1_Letters_and_Theoretical_Writings.pdf
54 2 Cor. 12.2–4.
55 George Sturt, *A Small Boy in the Sixties* (Cambridge: Cambridge University Press, 1927), 137.
56 Richard Jefferies, 'Bevis', in *The Jefferies Companion*, ed. S. J. Looker (London: Phoenix House, 1948), 265–6.
57 Harrison, *Ancient Noise*.
58 R. J. E. Riley, 'Insect Traffic: Protest, Activism and the Swarm', *Performance Research* 16, no. 5 (2014), 41–8, 43.
59 Sturt, *A Small Boy in the Sixties*, 161.
60 Joseph Arch, *The Story of His Life Told by Himself* (London: Hutchinson, 1898), 134–5.
61 Arch, *The Story of His Life Told by Himself*, 136–8.
62 Ibid., 241, quoting Thorold Rogers, *Six Centuries of Work and Wages*, itself quoted in the *Labourers Union Chronicle*, 19 September 1874.
63 E. J. Hobsbawm, *The Age of Revolution: 1789–1848* (New York: Mentor, 1962), 49.
64 Arch, *The Story of His Life Told by Himself*, 63.
65 Ibid., 213.
66 Ibid., 214.
67 Winifred Holtby, *South Riding, an English Landscape* (London: the Reprint Society, 1949).
68 Arch, *The Story of His Life Told by Himself*, 65.
69 Ibid., 73.
70 Hudson, *Nature in Downland*, 22.
71 Thomas Hughes, *The Scouring of the White Horse* (Cambridge: Macmillan, 1859), 92.
72 Hughes, *The Scouring of the White Horse*, 142.
73 Ibid., 215.
74 Ibid., 301.

Chapter 2: Returns

1 R. Murray Schafer, *The Soundscape: Our Sonic Environment and the Tuning of the World* (Rochester, VT: Destiny Books, 1994), 271.
2 Albert Bregman, *Auditory Scene Analysis: The Perceptual Organization of Sound* (Cambridge MA: The MIT Press, 1990), p. 73, quoting Roger Shepard.
3 Roman Jakobson, *The Sound Shape of Language*, vol. 8 of *Selected Writings* (New York: Mouton de Gruyer, 1988), 34.

4   Cited by Jakobson, *The Sound Shape of Language*, 34.
5   Stephen Handel, *Listening, an Introduction to the Perception of Auditory Events* (Cambridge, MA: The MIT Press, 1989), 463. For a research agenda arising from these pioneering studies, see Robin Barger et al., *Sonification Report: Status of the Field and Research Agenda*, Prepared for the National Science Foundation by members of the International Community for Auditory Display, 1997. http://www.icad.org/websiteV2.0/References/nsf.html, viewed 4 March 2010.
6   Handel makes this suggestion (Ibid., 463), but see also, more recently, Michela C. Tacca, 'Commonalities between perception and cognition', *Frontiers in Psychology* 2, (November 2011), Article 358. https://www.frontiersin.org/research-topics/265/pdf, viewed 8 September 2016.
7   Again, Handel advances this idea (Ibid., 541), but see also, for an interesting suggestion that excessive lateralisation may be associated with schizophrenia, 'a left hemisphere disorder,' see Rachel L. C. Mitchell, Tim J. Crow, 'Right hemisphere language function and schizophrenia: the forgotten hemisphere?', *Brain* 128, no. 5 (1 May 2005), 963–78.
8   Julian Jaynes, *The Origin of Consciousness in the Breakdown of the Bicameral* (London: Penguin, 1990), 325.
9   Helga Kraft, 'Staging a Critique of Modernism: Elias Canetti's Plays', in *A Companion to the Works of Elias Canetti*, ed. Dagmar C. G. Loretz (Rochester, NY: Camden House, 2004), 137–56, 144.
10 Elias Canetti, *The Play of the Eyes* (London: Granta Books, 2011), 262. Someone's acoustic mask, according to Canetti, is the entirety of their vocal behaviour; as such it takes account of 'a language that betrays itself not by what but by how something is said or covered up. Everything becomes a matter of style and the lack of it betrays itself precisely in its various linguistic disguises.' (Gitta Honegger, 'Translator's Note', in Elias Canetti, *The Wedding* (New York: PAJ Publications, 1986), 5–6).
11 Antigone Kefala, *Alexia, a Tale of Two Cultures* (Sydney: John Ferguson, 1984), unnumbered.
12 Kefala, *Alexia, a Tale of Two Cultures*, unnumbered.
13 Hermann B. Ritz, 'The Speech of the Tasmanian Aborigines', *Proceedings of the Royal Historical Society of Tasmania* for 1909, pp. 45–81, 53–4.
14 Max Müller, *Lectures on the Science of Language* (New York: Scribner's, 1862), 385.
15 James Dawson, *Australian Aborigines* (Canberra: Australian Institute of Aboriginal Studies, 1981, orig. pub, pub. 1881), lvi.
16 Dawson, *Australian Aborigines*, lvi.
17 Ibid.
18 Ibid., li.
19 Glenda Sluga, 'Bonegilla Reception and Training Centre,' MA Thesis, University of Melbourne, 1985, 253.

20 J. L. Hammond and Barbara Hammond, *The Village Labourer, 1770–1832* (London: Longmans, Green & Co, 1911), 211.
21 John Shaw Neilson, *The Collected Verse, a Variorum Edition*, ed. M. Roberts (Canberra: Australian Scholarly Editions Centre, 2003), 252.
22 Alexander Herzen, *Childhood, Youth and Exile*, trans. J. D. Duff (Oxford: Oxford University Press, 1980), 58.
23 'Notes on the Natural History of South Australia', No.2, *South Australian Magazine*, vol. 1, no. 6, 1842.
24 T. Horton Jones, *Six Months in South Australia, with Some Account of Port Philip (sic) and Portland Bay* (London: J. Cross, 1838), 161.
25 G. H. Haydon, *The Australian Emigrant* (London: Arthur Hall, Virtue, and Co, 1854), 77.
26 J. J. Pascoe, *History of Adelaide and Its Vicinity* (Adelaide: Hussey and Gillingham, 1901), 13.
27 C. J. Ellis, A. M. Ellis, M. Tur and A. McCardell, 'Classification of Sounds in Pitjantjatjara-Speaking Areas', in *Australian Aboriginal Concepts*, ed. L. R. Hiatt (Canberra: Australian Institute of Aboriginal Studies, 1978), 67–80, 69.
28 Ellis, Ellis, Tur and McCardell, 'Classification of Sounds in Pitjantjatjara-Speaking Areas', 70.
29 Ibid.
30 Ritz, 'The Speech of the Tasmanian Aborigines', 53–4.
31 Thomas De Quincey, *The Collected Writings of Thomas De Quincey*, ed. D. Masson (London: A. & C. Black, 1896–7), 14 vols, vol. 1, 287.
32 De Quincey, *The Collected Writings of Thomas De Quincey*, vol. 1, 287.
33 Gabriele D'Annunzio, *Il Fuoco* (Milano: Fratelli Treves, 1900), 4–5. (The noise of cheering rose from the San Gregorio ferry, and echoed along the Grand Canal bouncing [re-echoing] off the precious discs of porphyry and serpentine bejewelling the Ca' Dario.)
34 D'Annunzio, *Il Fuoco*, 7. (The black and dense crowd swaying in the interval, the rooms [*vani*] of the Ducal loggias were filled with a confused roar [*romorio*] like the rushing sound [*rombo*] that animates the volutes of seashells [or when a seashell is put to the ear].)
35 Paul Carter, *The Lie of the Land* (London: Faber & Faber, 1996), 343–63.
36 Thomas de Quincey, *Confessions of an English Opium Eater and Other Writings*, ed. A. Ward (New York and Scarborough, Ontario: New American Library, Signet Classics, 1966), 129–30. No single edition of De Quincey is authoritative or easily available. I have consulted many over the years.
37 Edgar Allan Poe, 'Introduction', *The Conchologist's First Book* (Philadelphia: Haswell, Barrington, and Haswell, 1839), 16.
38 Poe, 'Introduction', 16.
39 Ibid., 17.
40 Ibid.

41 Sigmund Freud, *The Complete Letters of Sigmund Freud to Wilhelm Fliess, 1887–1904*, trans. J. Moussaieff Masson (Cambridge, MA: Harvard University Press, 1985), 207.
42 A random walk is defined as a process where the current value of a variable is composed of the past value plus an error term defined as a white noise (a normal variable with zero mean and variance one). http://cmapskm.ihmc.us/rid=1052458884462_996058812_7176/randomwalk.pdf, viewed 10 October 2016.
43 Roger Ascham, 'Toxophilus', in *English Works*, ed. W. A. Wright (Cambridge: Cambridge University Press, 1970), 112.
44 An unpublished manuscript called 'The Lance and the Shield'. In Pound's rendering, these passages are respectively 'And airs grown calm when white the dawn appeareth/ And white snow falling where no wind is bent' (*Sonnets and Ballate of Guido Cavalcanti*, introduced and translated by Ezra Pound (London: Stephen Swift and Co), 1912, Sonnet XVIII, 51) and 'Who is she coming, drawing all men's gaze,/ Who makes the air one trembling clarity' (*Sonnets and Ballate of Guido Cavalcanti*, Sonnet VII, 29).
45 Paul Carter, 'On the Still Air', 138–64.
46 Carter, *The Lie of the Land*, 327.
47 Joscelyn Godwin, *Harmonies of Heaven and Earth, Mysticism in Music from Antiquity to the Avant-Garde* (London Rochester, Vermont: Inner Traditions International, 1987), Kindle Edition, Chapter 4: Music and the currents of time. Location: 2090 of 4472.
48 Jean-Luc Marion, *Being Given: Toward a Phenomenology of Givenness*, trans. J. L. Kosky (Stanford, CA: Stanford University Press, 2002), 213.
49 Robert Brough Smyth, *The Aborigines of Victoria* (Melbourne: The Government Printer, 1876), 2 vols, vol. 1, 428.
50 See Paul Carter, 'Mirror States' and 'This Other Eden, Performing Space', in *The Sound In-Between Voice, Space, Performance* (Sydney: New Endeavour Press/University of New South Wales Press, 1992), 71–92.
51 Paul Carter, 'Underworlds of Jean du Chas', 263–301.
52 Samuel Beckett, 'Le Concentrisme', *Disiecta* (London: John Calder, 1983), 35–42.
53 Fernando Pessoa, *The Book of Disquiet*, trans. A. Mac Adam (New York: Pantheon Books, 1991), 19.
54 See Brill's, *Companion to Greek and Latin Epyllion and Its Reception*, ed. M. Baumbach and S. Bär (Leiden: Brill, 2012), xiii.
55 Luciano Berio, pers. comm.
56 Abbie Conant and William Osborne, 'Grock and the Berio Sequenza V' (http://www.osborne-conant.org/Grock.htm) provide invaluable information about the broader context of Grock's question, viewed 4 April 2016. I wish I had known all of this then.
57 Francis M. Cornford, *Principium Sapientiae: A Study in the Origins of Greek Thought* (Cambridge: Cambridge University Press, 1952), 102.

58 H. Munro Chadwick and N. Kershaw Chadwick, *Growth of Literature*, vol. III, Part 1, 'The Oral Literature of the Tatars' (Cambridge: Cambridge University Press, 1940), 181–2.
59 Chadwick and Chadwick, *Growth of Literature*, vol. III, Part 1, 182.
60 A metaphor Berio used in our correspondence.
61 Paul Carter, 'Winged Words', letter to Andrew McLennan, 18 February 2005, 1–3, 1.
62 Antonin Artaud, *Le théâtre et son double* (Paris: Gallimard, 1993), 88.
63 Paul Carter, 'Duende Imagines the Piazza', in *Baroque Memories* (Manchester: Carcanet, 1994), 157–76, 158–9.
64 See https://www.jstor.org/stable/40169795?seq=1#page_scan_tab_contents, viewed 7 June 2016.
65 Quoted by Vicente L. Rafael in 'Gods and Grammar: The Politics of Translation in the Spanish Colonisation of the Tagalogs of the Philippines', in *Notebooks in Cultural Analysis*, ed. N. F. Cantor and N. King (Durham, NC: Duke University, 1986), vol. 3, a special issue on 'Voice', 97–133, 98.
66 See Carter, 'Emergency Languages: Echoes of Columbus in Discourses of Precarity'.
67 Paul Carter, 'The 7448', 203–62. This idea is anticipated in *The Lie of the Land*, in the discussion of the consonant–vowel combination 'ca' (187–91).
68 Stephen Greenblatt, *Marvelous Possessions* (Chicago: University of Chicago Press, 1991), 98. These paragraphs pass through Paul Carter, 'Ambiguous Traces, Mishearing, and Auditory Space', in *Hearing Cultures, Essays on Sound, Listening and Modernity*, ed. V. Erlmann (Oxford and New York: Berg, 2004), 43–64.
69 Greenblatt, *Marvelous Possessions*, 99.
70 O. Mannoni, *Prospero and Caliban, The Psychology of Colonisation*, trans. P. Powesland (London: Methuen, 1956), 100.
71 Emphasis mine. Sylvia Cauby Novaes, *The Play of Mirrors, The Representation of the Self as Mirrored in the Other*, trans. I. M. Burbridge (Austin: University of Texas Press, 1997), 143.
72 Julio F. Guillén Tato, *La parla marinera en el Diario del primer viaje de Cristóbal Colón* (Madrid: Instituto Historico de Marina, Consejo Superior de Investigacines Cientificas, 1951), 130.
73 Guillén Tato, *La parla marinera en el Diario del primer viaje de Cristóbal Colón*, 131.
74 Correctly observing that the Taino 'language … is identical in all of these islands of India, and they all understand one another and go to all the islands in their canoe', Columbus contrasts this with the situation in Guinea where 'there are a thousand kinds of language and one does not understand'. (Christopher Columbus, *The Diario of Christopher Columbus's First Voyage to America, 1492–1493*, trans. O. Dunn and J. E. Kelley, Jr. (Norman: University of Oklahoma Press, 1989), 147.)

75 Columbus, *The Diario of Christopher Columbus's First Voyage to America, 1492–1493*, 63.
76 Miguel A. Asturias, 'América, la engañadora', in *América, fábula de fábulas y otros ensayos*, ed. R. Calan (Caracas: Monte Avila Editores, 1969), 343–7, 343.
77 Francisco Maldonado de Guevara, *El Primer Contacto de Blancos y Gentes de Color en America* (Valladolid: Cuesta, 1924), 54. A follower of Lévy-Bruhl, de Guevara exaggerated the antithesis between 'primitive' and non-primitive mentality.
78 See also Carter, *Parrot*, 93.
79 Columbus, *The Diario of Christopher Columbus's First Voyage to America, 1492–1493*, 143.
80 Gerald Sider, 'When Parrots Learn to Talk, and Why They Can't: Domination, Deception, and Self-deception in Indian-White Relations', *Comparative Studies in Society and History*, 29 (1987): 3–23, 15.
81 Columbus, *The Diario of Christopher Columbus's First Voyage to America, 1492–1493*, 213.
82 Ibid., 221.
83 Frederick G. Cassidy, *Jamaica Talk* (London: Macmillan, 1961), 304.
84 Columbus, *The Diario of Christopher Columbus's First Voyage to America, 1492–1493*, 45.
85 Ibid., 55.
86 John Milton, *Paradise Lost*, Bk IV, 602–3.
87 Quoting Charles Sturt, *Narrative of an expedition into Central Australia during the years 1855, 5, and 6* (London: T. & W. Boone, 1849), 2 vols, vol. 1, 63.
88 See https://blogs.crikey.com.au/northern/2009/05/02/bird-of-the-week-panpa npalala-crested-bellbird-oreoica-gutturalis/, viewed 18 April 2016.
89 *The Works of Dugald Stewart* (Cambridge: Hilliard & Brown, 1839), 7 vols, vol. 3, 'Elements of the Philosophy of the Human Mind', 404.
90 Mannoni, *Prospero and Caliban*, 200; see also Paul Carter, *Material Thinking, the Theory and Practice of Creative Research* (Melbourne: Melbourne University Press, 2004), 42.
91 Roy Wagner, *An Anthropology of the Subject, Holographic Worldview in New Guinea and Its Meaning and Significance for the World of Anthropology* (Berkeley: University of California Press, 2001), 137.
92 Gerhard Rühm, quoted by Andrew McLennan, 'Formes Circulaires: A Journey Around the Radio Works of Paul Carter', *Southerly* 66, no. 2 (2006), 81–102, 85.
93 Klaus Schöning and M. E. Cory, 'The Contours of Acoustic Art', *Theatre Journal* 43, no. 3: Radio Drama (1991), 307–24, 322. I am indebted to Christopher Williams for this quotation.
94 Andreas Hagelüken, 'Acoustic (Media) Art: Ars Acustica and the Idea of a Unique Art Form for Radio – An Examination of the Historical Conditions in Germany', *World New Music Magazine* (Saarbrücken) 90, no. 102 (July 2006), 5.

95 Artaud, *Le théâtre et son double*, 89.
96 See Carter, *Material Thinking*, 73.
97 See Carter, *Lost Subjects*, 167. As is noted in the *Lost Subjects* commentary, one of the functions of colonial science was to evacuate atmosphere from discourse: the natural alliance or shared breathing pattern of trees and clouds was interrupted; the tangled bush was reduced to a lifeless stump, and laid up as timber supporting technological enclosure (240).
98 Tadeusz Kantor, *A Journey through Other Spaces, Essays and Manifestos, 1944–1990*, ed. and trans. M. Kobialka (Berkeley: University of California Press, 1993), 99.
99 Kantor, *A Journey through Other Spaces, Essays and Manifestos, 1944–1990*, 101.
100 Ibid., 100.
101 Ibid., 101.
102 R. A. Baggio, *The Shoe in My Cheese* (Melbourne: R. A. Baggio, 1989), 15.
103 E. J. Drechsel, 'Metacommunicative Functions of Mobilian Jargon, an American Pidgin of the Lower Mississippi River Region', in *Pidgin and Creole Languages*, ed. G. C. Gilbert (Honolulu: University of Hawaii Press, 1987), 433–44, 434.
104 D. H. Whalen, 'The Native Speaker and Indeterminacy', in *A Festschrift for Native Speaker*, ed. F. Coulmas (The Hague: Mouton, 1981), 263–78, 274.
105 Whalen, 'The Native Speaker and Indeterminacy', 272.
106 Ibid., 266.
107 Paul Carter, *Material Thinking*, 2004 and *The Calling to Come* (Sydney: Historic Houses Trust of New South Wales, 1996).
108 W. von Raffler-Engel, 'The Native Speaker in His New Found Body', in *A Festschrift for Native Speaker*, ed. F. Coulmas (The Hague: Mouton, 1981), 263–78, 299.
109 For a brief survey of these and other radiophonic works, see Andrew McLennan 'Formes Circulaires: A Journey around the Radio Works of Paul Carter'.
110 http://en.wikipedia.org/wiki/Whistler_(radio), viewed 14 June 2016.
111 Artaud, *Le théâtre et son double*, 96. My translations.
112 Virginia Madsen, 'The Call of the Wild', in *Uncertain Ground, Essays between Art and Nature*, ed. M. Thomas (Sydney: Art Gallery of New South Wales, 1999), 29–49, 32.
113 David Tomas, *Transcultural Space and Transcultural Beings* (Boulder, CO: Westview Press, 1996), 120.
114 Tomas, *Transcultural Space and Transcultural Beings*, 121.
115 Brandon LaBelle, 'Misplace – Dropping Eaves on Ethics', in *Hearing Places: Sound, Place, Time, and Culture*, ed. R. Bandt, M. Duffy and D. MacKinnon (Newcastle: Cambridge Scholars Publishing, 2007), 8–17, 14.
116 Luciano Berio, *Luciano Berio, Two Interviews* (New York: Marion Boyars Publishers, 1985), 112–13.

117 Theodor W. Adorno, 'Vierhändig, noch einmal', in *Musikalische Schriften IV, Gesammelte Schriften Band 17*, ed. R. Tiedemann (Frankfurt am Main: Suhrkamp, 1982), 303–6. These quotations from: William Lockhart, 'Listening to the Domestic Music Machine: Keyboard Arrangement in the Nineteenth Century,' PhD Dissertation, Humboldt-Universität, 2012. http://edoc.hu-berlin.de/dissertationen/lockhart-william-2012-02-14/PDF/lockhart.pdf

118 http://mcpress.media-commons.org/piracycrusade/chap2/music-in-the-air-radio-and-the-record-industry/, viewed 18 May 2017.

119 Bertolt Brecht, 'The Radio as an Apparatus of Communication'. [Originally 'Der Rundfunk als Kommunikationsapparat' in *Blätter des Hessischen Landestheaters*, Darmstadt, No. 16, July 1932]. Accessed at http://telematic.walkerart.org/telereal/bit_brecht.html, viewed 18 May 2017.

120 Brecht, 'The Radio as an Apparatus of Communication'.

121 http://www.classical.net/music/recs/reviews/c/cap60012a.php, viewed 23 May 2017.

122 Andreas Hagelüken, 'Acoustic (Media) Art: *Ars Acustica* and the Idea of a Unique Art form for Radio – An Examination of the Historical Conditions in Germany,' 9–11.

123 See discussion in Irna Priore, 'Vestiges of Twelve-Tone Practice as Compositional Process in Berio's *Sequenza I* for Solo Flute', in *Berio's Sequenzas, Essays in Performance, Composition and Analysis*, ed. J. K. Halfyard (London: Routledge, 2016), 191–208. References are to Umberto Eco, *The Open Work*, trans. A. Cancogni (Cambridge, MA: Harvard University Press, 1989).

124 Umberto Eco, *The Limits of Interpretation* (Bloomington: India University Press, 1990), 141.

125 Nick Nesbitt and Brian Hulse, eds, *Sounding the Virtual: Gilles Deleuze and the Theory and Philosophy of Music* (Farnham and Surrey: Ashgate, 2010).

126 Eco, *The Limits of Interpretation*, 142. See also Venn, 'Proliferations and Limitations: Berio's Reworking of the Sequenzas', 172–90, especially his discussion (173ff) of Jean-Jacques Nattiez's critique of the theory of the 'open work' in *Music and Discourse, Toward a Semiology of Music*, trans. C. Abbate (Princeton NJ: Princeton University Press, 1990), 82–7.

127 Friedrich Kittler, *Gramophone, Film, Typewriter*, trans. G. Winthrop-Young and M. Wutz (Stanford, CA: Stanford University Press, 1999), 87–8.

128 Kittler, *Gramophone, Film, Typewriter*, 88.

129 Beckett, *Disiecta*, 49.

130 James Martin Harding, *Adorno and 'A Writing of the Ruins': Essays on Modern Aesthetics and Anglo-American Literature and Culture* (New York: SUNY Press, 1997), 56.

131 Dan Lander, 'Radiocastings: Musings on Radio and Art', in *Radio Rethink: Art, Sound and Transmission*, ed. D. Augaitis and D. Lander (Banff and Alberta: Banff Centre for the Arts, 1994). Accessed at http://cec.sonus.ca/econtact/Radiophonic/Radiocasting.htm, viewed 8 May 2017.

132 Ralph Kirkpatrick, *Domenico Scarlatti* (Princeton NJ: Princeton University Press, 1953), 167.

133 Paul Carter, 'Scarlatti', 94–114.

134 Ezra Pound, *Antheil and the Treatise on Harmony* (Chicago: P. Covici, Inc, 1927), 129.

135 See Kirkpatrick, *Domenico Scarlatti*, 252–79.

136 Ibid., 275.

137 Fernando Pessoa, *Fernando Pessoa & Co, Selected Poems*, ed. and trans. R. Zenith (New York: Grove Press, 1998), 58.

138 Kirkpatrick, *Domenico Scarlatti*, 42.

139 Giuseppe Mazzotta, *The New Map of the World, The Poetic Philosophy of Giambattista Vico* (Princeton, NJ: Princeton University Press, 1999), 206.

140 James Olney, *Memory and Narrative: The Weave of Life-Writing* (Chicago: University of Chicago Press, 1998), xv.

141 James Goetsch, *Vico's Axioms, The Geometry of the Human World* (New Haven, CT: Yale University Press, 1995), 42–3. *Ingegno*, Goetsch sums up, 'environs the whole world'. For a chorographical and geographical application of this idea, see Paul Carter, *Dark Writing: Geography, Performance, Design* (Honolulu: University of Hawai'i Press, 2008), 43–5.

Chapter 3: Rattles

1 Apart from the scarcely accessible performance note (Paul Carter with Chandrabhanu, 'Light', *Visual Arts Program Catalogue*, Telstra Adelaide Festival 96, 1996, 12–13) nothing has been written about this production. Important historical and cultural context occurs in Carter, *The Lie of the Land*, 203–90 and the slightly earlier article 'Ageless Light: A Tubercular History', *Journal of Australian Cultural Studies* (1995), 186–207.

2 As I thought then; in fact, Laennec's *Treatise* was first published in English translation in 1829.

3 René Théophile H. Laennec, *A Treatise on the Diseases of the Chest*, trans. J. Forbes (London: Thomas & George Underwood, 1829), 56. See also Hugh M. Kinghorn, 'The Classification of Râles', *The Canadian Medical Association Journal* (April 1932), 438–45.

4 Laennec, *A Treatise on the Diseases of the Chest*, 56.

5 Ibid., 57.

6 Carter, 'Underworlds of Jean du Chas', 282.

7 Gary Shapiro, *Alcyone, Nietzsche on Gifts, Noise, and Women* (Albany, NY: SUNY Press, 1991), 71–2.

8 The diary has been published. *William Light's Brief Journal and Australian Diaries*, ed. D. Elder (Adelaide: Wakefield Press, 1984).

9 Laennec, *A Treatise on Diseases of the Chest*, 53.
10 Ibid.
11 Ibid., 57.
12 Ludwig Friedrich Kraemtz, *A Complete Course of Meteorology*, trans. C. V. Walker (London: H. Baillière, 1845), 492.
13 Kinghorn, 'The Classification of Râles,' 442.
14 John Ball, 'Zermatt in 1845', in *Peaks, Passes and Glaciers, a Series of Excursions by Members of the Alpine Club*, ed. J. Ball (London: Longman, 1860), 108–34, 121–2.
15 *Novalis: Philosophical Writings*, trans. and ed. M. M. Stoljar (New York: State University of New York Press, 1997), 49.
16 Laennec, *A Treatise on Diseases of the Chest*, 56–7.
17 Lord Faringdon [Gavin Henderson], 'Fairy Story of Faringdon's Little Theatre', *The Wiltshire Herald and Advertiser,* 14 December 1953.
18 Constantin Stanislavsky, *An Actor Prepares*, trans. E. R. Hapgood (London: Geoffrey Bles, 1937), 37.
19 Huxley, *Eyeless in Gaza*, 607.
20 Stanislavsky, *An Actor Prepares*, 29.
21 Gilbert White, Letter XVII, *The Natural History of Selborne* (London: Oxford University Press, 1951), 167.
22 Robert Graves, *The White Goddess* (London: Faber & Faber, 1956), 191.
23 Graves, *The White Goddess*, 384.
24 Ibid., 315.
25 Ibid., 371.
26 Ibid.
27 Ibid., 489.
28 Ibid., 179.
29 Francesco Colonna, *Hypnerotomachia Poliphili, The Strife of Love in a Dream*, trans. J. Godwin (London: Thames & Hudson, 1999), 36.
30 Caitlin Thomas, *Leftover Life to Kill* (London: Putnam, 1957).
31 Thomas Hardy, *A Pair of Blue Eyes* (London: Tinsley Brothers, 1873), 93.
32 J. F. Kirkaldy, *Minerals and Rocks in Colour* (London: Blandford Press, 1963), 84.
33 Emanuel Swedenborg, *Arcana Coelestia*, trans. J. F. Potts (West Chester, PA: Swedenborg Foundation, 2009), volume 1, 518.
34 See respectively *The Methodist Hymn Book with Tunes* (London: Methodist Conference Office, 1933), 393 and John Dryden, *Poetry, Prose, and Plays*, ed. D. Grant (London: Hart-Davis, 1952), 48–50.
35 Enid Blyton, *The Valley of Adventure* (London: Macmillan, 1947).
36 *Novalis: Philosophical Writings*, 46.

Chapter 4: Sirens

1. John Lort Stokes, *Discoveries in Australia* (London: T. and W. Boone, 1846), vol. 2, 316–17.
2. Paul Davies, *The Mind of God* (London: Penguin, 1993), 20.
3. W. H. Auden, 'Dingley Dell and the Fleet', in *The Dyer's Hand* (New York: Random House, 1948), 409.
4. See Thomas K. Hubbard, 'Utopianism and the Sophistic City', in *The City as Comedy*, ed. G. W. Dobson (Chapel Hill: The University of North Carolina Press, 1997), 23–50, 41, note 3.
5. See Maurizio Bettini, Luigi Spina, *Il mito delle Sirene. Immagini e racconti dalla Grecia a oggi* (Torino: Einaudi, 2007).
6. George Eliot, *Silas Marner*, ed. Q. Leavis (London: Penguin, 1967), 51.
7. Arturo Barea, *Lorca, the Poet and His People* (London: Faber & Faber, 1944).
8. Arch, *The Story of His Life Told by Himself*, 392.
9. García Lorca, Federico, *Prosa*, ed. M. García-Posada, Obras completes, vol. 3 (Barcelona: Galaxia Guternberg/Círculo de Lectores, 1997), 256.
10. Walter Rose, *Good Neighbours* (Cambridge: Green Books, 1988), 5.
11. Brecht, 'The Radio as an Apparatus of Communication.'
12. Kirkpatrick, *Domenico Scarlatti*, 167.
13. Ibid.
14. Ibid.
15. Antonio Bordonau, 'In the Aids Age.' This testament was posted in 2001 and is no longer current. The text (in English) is preserved in 'End: Turbulent Dramaturgy,' a workshop about the 'end' of radio, Theater der Welt, Mannheim World Theatre Festival, 2014, 1–14. The workshop was located in the Nationaltheater Mannheim Werkhaus, Mozartstr. 9 and occurred on 6 (5.00–7.00 pm), 7 (2.00–5.00 pm) and 8 (4.00–7.00 pm) June 2014. Participants came from French, Spanish, German and Indigenous language backgrounds.
16. Kirkpatrick, *Domenico Scarlatti*, 202–3.
17. Most probably, our passage refers to a period of writing in Rome in the summer of 1948 (Kirkpatrick, *Domenico Scarlatti*, vi).
18. Barea, *Lorca, the Poet and His People*.
19. Ibid., 33.
20. Ibid.
21. Carlos Saura, 'Flamenco' (1955) caption, Museo Nacional Centro de Arte Reina Sofia, viewed 21 September 2014.
22. Considered as an exceptionally creative discourse, the openness to interpretation which the gaps between concepts in the *coplas* made possible – gaps whose sole logic was the rhythm – corresponds to the transitional discourse of innovation described by Derek Pigrum, 'The "potential space" of transitional creative

notation', accessed at www.textjournal.com.au/speciss/issue13/Pigrum.pdf, viewed 15 June 2015.
23. Georges Didi-Huberman, *Le Danseur des Solitudes* (Paris: Éditions de Minuit, 2006), 160.
24. Jacqueline Ogell, 'The Portuguese Scarlatti', CD notes, Australian Broadcasting Corporation, 2007.
25. George Steiner, *Lessons of the Masters* (Cambridge, MA: Harvard University Press, 2003), 131.
26. David Conte, 'The Teaching Methods of Nadia Boulanger' at www.davidconte.net/Boulanger.pdf, viewed 19 July 2017.
27. Although not in Boulanger's case, as her call to music was triggered by a siren: 'One day I heard a fire bell. Instead of crying out and hiding, I rushed to the piano and tried to reproduce the sounds' (Léonie Rosenstiel, *Nadia Boulanger: A Life in Music* (New York: W.W. Norton & Co, 1982), 26). In 1908, her cantata, *La Sirène*, was placed second in France's Prix de Rome competition.
28. Kimberley DaCosta Holton, 'Fado Historiography: Old Myths and New Frontiers,' *Portuguese Cultural Studies* (Winter 2006), 1–17, 2.
29. According to the founder of the *saudisismo* movement, poet Teixera de Pascoas. (Kimberley DaCosta Holton, 'Fado Historiography: Old Myths and New Frontiers', 5.)
30. 'Pothos, as the wider factor in eros, drives the sailor-wanderer to quest for what cannot be fulfilled …' (James Hillman, 'Pothos: The Nostalgia of the Puer Eternus', *Loose Ends*, Primary Papers in Archetypal Psychology (Washington DC: Spring Publications, 1975), 54.)
31. Hillman, 'Pothos: The Nostalgia of the Puer Eternus', 58.
32. Ibid., 59.
33. Holton, 'Fado Historiography: Old Myths and New Frontiers'.
34. Ibid., 10.
35. Joaquim Pais de Brito, 'O Fado: Etnografia na Cidade', in *Antropologia Urbana e Sociedade no Brasil e em Portugal* (Rio de Janeiro: Jorge Zahar Editor, 1999), 24–42, 29–30. Quoted and translated by DaCosta Holton.
36. Kirkpatrick, *Domenico Scarlatti*, 303.
37. Sacheverell Sitwell, *Southern Baroque Art* (London: Duckworth, 1931), 190–1.
38. Holton, 'Fado Historiography: Old Myths and New Frontiers', 5.
39. Vladimir Jankélévitch, *Une Vie en toutes lettres: Correspondance*, ed. F. Schwab (Paris: Liana Levi, 1998), 170.
40. Jankélévitch, *Une Vie en toutes lettres*, 109.
41. Vladimir Jankélévitch, *L'Ironie* (Paris: Flammarion, 1964), 35.
42. There is a learned debate about Scarlatti's attitude to the pianoforte but in 'Scarlatti' the Radio Voices commenting on the sound action remark: 'At the depression of a pedal to prolong the pain. ... This decadence destroys feeling.' (Carter, 'Scarlatti', 110.)

43 Jankélévitch, *L'Ironie*, 36 and also footnote.
44 Ravel/Scarbo/Scarlatti from http://archive.org/stream/ravel00jank/ravel00 jank_djvu.txt, viewed 21 September 2013.
45 Ibid.
46 Ibid.
47 Ana Padilla Mangas, 'Introducción', Antonio y Manuel Machado, *La Lola va a los Puertos* (Madrid: Biblioteca Nueva, 2010), 9–29, 27.
48 Vicente Llatas, *El Hablar del Villar del Arzobispo y su comarca* (Valencia: Institución Alfonso el Magnanimo, 1959, 2 vols). Barea likewise experienced this: 'He witnessed the burning of a church and of the Jesuit Escuela Pía which he had attended. He later wrote: "I went home in profound distress. It was impossible to applaud the violence. I was convinced that the Church of Spain was an evil which had to be eradicated. But I revolted against this stupid destruction"'. (William Chislett, 'Arturo Barea: From Civil War Madrid to Exile in Oxfordshire, FT Weekend Oxford Literary Festival', at http://www.williamchislett.com/2015/03/arturo-barea-from-civil-war-madrid-to-exile-in-oxfordshire-ft-weekend-oxford-literary-festival, viewed 18 January 2016.)
49 Ausiàs March, 'Io só aquest que em dic Ausias' (Ausiàs March, *Poesia*, a cura de J. Ferrante., Barcelona Editions 62, 1999, 114, l.88, 190). The translation is by Arthur Terry, whose *Ausias March, Selected Poems* (Edinburgh: Edinburgh University Press, 1977) was published opportunely in the year I was putting March's poems into rhymed stanzas – see 132 of that edition. A suitably speculative but interesting riff on March's line is Josep Piera, *Jo sóc aquest que em dic Ausiàs March* (Barcelona: Barcelona Edicions, 2002), 62.
50 Antonio Machado, 'Retrato', in *Campos de Castilla* (Madrid: Rinacimiento, 1912), XCVII.
51 Brenda Cappuccio, 'Gloria et al.', in *In Her Words: Critical Studies on Gloria Fuertes*, ed. Margaret Persin (Lewisburg: Bucknell University Press, 2011), 40–51, 47.
52 John Shaw Neilson, 'To an Elusive Maiden', in John Shaw Neilson, *The Collected Verse, a variorum edition*, 929. The similarity has been noticed by others (see Paul Carter, *Ground Truthing, Explorations in a Creative Region* (University of Western Australia Publishing, 2010), 301, note 19).
53 See the commentary on Robert Browning's Fra Lippo Lippi at http://www.victorianweb.org/authors/rb/lippi/text.html, viewed 12 March 2016.
54 Arthur Terry, *Ausias March, Selected Poems*, XCII, ll.131–2, 93.
55 Ibid., CI, 17–24, 113.
56 Ibid., XCII, 55–8, 89.
57 Dante Alighieri, *Purgatorio*, 19.58–59. Tomás Antonio Valle, 'A New Perspective on Dante's Dream of the Siren', *The Oswald Review* 15, no. 1 (2013), 1–16, 1.
58 A linguistic role reversal as I was in the time referred to translating 'la Mort falaguera … Braços oberts és eixada a carrera …' in Terry's prose translation,

'Death has come out into the street with open arms, her eyes weeping from excess of joy …' (*Ausias March, Selected Poems*, XI, 8–10, 40–1).

59 Bordonau, 'In the AIDs age.' Formerly available on-line; now partially preserved in the workshop briefing papers.

60 Paul Carter, 'End: Turbulent dramaturgy,' a workshop about the 'end' of radio. http://www.geisteswissenschaften.fu-berlin.de/en/v/interweaving-performance-cultures/events/workshops/workshop_paul-carter_theater-der-welt.html.

61 T. S. Eliot, 'East Coker', Four Quartets, *The Complete Poems and Plays* (London: Faber & Faber, 1969), 182.

62 www.online-literature.com/elbert-hubbard/journeys-vol-eight/5, viewed 8 August 2016.

63 Paul Carter, *Meeting Place, the Human Encounter and the Challenge of Coexistence* (Minneapolis: University of Minnesota Press, 2013), 101. See also related discussion in Paul Carter, *Repressed Spaces, the Poetics of Agoraphobia* (London: Reaktion Books, 2002), 56–9.

64 Richard Oulahan, 'The Prince Who Will Be King,' *People*, 12 August 1974, https://people.com/archive/the-prince-who-will-be-king-vol-2-no-7/, viewed 14 February 2013.

65 Kirkpatrick, *Domenico Scarlatti*, 319.

66 Under the Nazis: '"Doing things" meant reporting on others, for instance: So-and-so was listening to a foreign radio station. Ideally, Quangel would have packed up all the radios in Otto's room and stashed them in the basement. You couldn't be careful enough in times like these' (Hans Fallada, *Alone in Berlin*, trans. M. Hofmann (London: Penguin, 2009), 10).

67 Xavier Valiño, 'La censura radiofónica en la producción discográfica durante el franquismo', *Historia Actual Online* 42, no. 1 (2017), 87–97, 87.

68 See Geoffrey B. Pingree, 'Franco and the Filmmakers: Critical Myths, Transparent Realities.' www.publicacions.ub.edu/bibliotecadigital/cinema/filmhistoria/Art.%20pingree.pdf, viewed 18 April 2016.

69 See Pingree, 'Franco and the Filmmakers: Critical Myths, Transparent Realities'.

70 *The Selected Prose of Fernando Pessoa*, ed. and trans. R. Zenith (New York: Grove Press, 2001), xxi.

71 *The Selected Prose of Fernando Pessoa*, 290–1.

72 Pingree, 'Franco and the Filmmakers: Critical Myths, Transparent Realities'.

73 Lander, 'Radiocastings: Musings on Radio and Art', 19.

74 Ibid.

75 See Paul Carter, 'Exploding Scarlatti', at http://www.geisteswissenschaften.fu-berlin.de/en/v/interweaving-performance-cultures/events/workshops/workshop_paul-carter_theater-der-welt.html

76 Radio 'should step out of the supply business and organize its listener as suppliers. … Any attempt by the radio to give a truly public character to public occasions is a step in the right direction' (Brecht, 'The Radio as an Apparatus of Communication').

77 Kirkpatrick, *Domenico Scarlatti*.

78 Ibid., 102–3.

79 Giambattista Vico, *On the Most Ancient Wisdom of the Italians*, trans. L. M. Palmer (Ithaca, NY: Cornell University Press, 1988), 96–7. See also Carter, *Dark Writing*, 22–5.

80 Luisa Morales, 'Domenico Scarlatti in Spain: An Introduction', in Domenico Scarlatti en España/ Domenico Scarlatti in Spain, Proceedings of FIMTE Symposia, 2006–7, ed. L. Morales (Almería: Asociación Cultural LEAL, 2009), 15–27, 16. Importantly, Morales considers that three clavichords recorded in the Will were harpsichords *not* forte pianos.

81 Morales, 'Domenico Scarlatti in Spain: An Introduction', 18.

82 Jean-Pierre Vernant, *Myth and Thought among the Greeks* (London: Routledge & Kegan Paul, 1983), 80.

83 Lope de Vega, *Fuenteovejuna, A Dual-Language Book*, trans. S. Appelbaum (New York: Dover Publications, 2002), no page numbers.

84 http://plopes.org/project/future-places, viewed 17 July 2017.

85 Elizabeth Shillito Walser, Eurof Walters and Peter Hague, 'Vocal Communication between Ewes and Their Own and Alien Lambs', *Behaviour* 81, no. 2/4 (1982), 140–51, 140.

86 Judith Peraino, *Listening to the Sirens: Musical Technologies of Queer Identity from Homer to Hedwig* (Berkeley, CA: University of California Press, 2006), 16.

87 Lope de Vega, *Fuenteovejuna, A Dual-Language Book*.

88 Edward A. Roberts, *A Comprehensive Etymological Dictionary of the Spanish Language* (New York: XLibris, 2014), 162.

89 Horace, *Satires. Epistles. The Art of Poetry*, trans. H. Rushton Fairclough, (Cambridge, MA: Harvard University Press, 1926) Epistle 1, ll. 44–45, 385.

90 Christoph Cox, 'Beyond Representation and Signification: Toward a Sonic Materialism', *Journal of Visual Culture* 10, no. 2 (2011), 145–62, 147.

91 Cox, 'Beyond Representation and Signification: Toward a Sonic Materialism', 155.

92 Ibid., 156.

93 Ibid., 157.

94 Ibid., 158.

95 Jacques Attali, *Noise, The Political Economy of Music*, trans. B. Massumi (Minneapolis: University of Minnesota Press, 1985), 6.

96 David Garcia and Geert Lovink, The ABC of Tactical Media, 1997, www.t acticalmediafiles.net/article.jsp?objectnumber=37996, viewed 8 April 2008.

97 Eric Kuitenberg, *Legacies of Tactical Media* (Amsterdam: Institute of Network Cultures, Amsterdam, 2011), 19.

98 Again, this notion is well-known within the utopian discourse of neue Hörspiel. Heisenbüttel makes the point that the Neue Hörspiel represented 'an "open" dramaturgy,"' a critique of traditional 'narrative Hörspiel dramaturgy'. There emerges in relation to this 'a kind of "Hörspielmacher", who often carried over the creative process into the production studio as director. This new kind of writer–composer–director expanded the inventory of artistic tool available and put them to new tests' (320) (Klaus Schöning and Mark E. Cory, 'The Contours of Acoustic Art', *Theatre Journal* 43, no. 3 (October 1991), 307–24, 320).

99 Randall Collins, *Interaction Ritual Chains* (Princeton: Princeton University Press, 2004), 77.

100 Collins, *Interaction Ritual Chains*, 76.

101 Ibid., 44.

102 Kai van Eikels, 'From "Archein" to "Prattein" – Suggestions for an Un-creative Collectivity' 5–22, in *Rehearsing Collectivity: Choreography Beyond Dance*, ed. E. Basteri, E. Guidi and E Ricci (Berlin: Argobooks, 2012), 5–19, 11.

103 Martin Clayton, Rebecca Sager and Udo Will, 'In Time with the Music: The Concept of Entrainment and its Significance for Ethnomusicology', *European Meetings in Ethnomusicology* 11 (2005), 3–142, 20. http://oro.open.ac.uk/2661/1/InTimeWithTheMusic.pdf, viewed 20 June 2016.

104 William Hellerman and Don Goddard, *Catalogue for 'Sound/Art' at The Sculpture Center* (New York City, 1–30 May 1983 and BACA/DCC Gallery 1–30 June 1983).

105 Kevin Kelly, 'Out of Control', quoted by Franco 'Bifo' Berardi Bifo, 'Swarm Rhythm Refrain Singularity', 92–7, 93, in *Kukkia*, ed. K. Kucia and T. Nauha, c.2008. http://www.academia.edu/7817648/Kukkia_book, viewed 21 June 2016.

106 'Bifo' Berardi Bifo, 'Swarm Rhythm Refrain Singularity', 92–7, 93.

107 Eliot, *Silas Marner*, 22.

108 Didi-Huberman, *Le Danseur des Solitudes*, 30.

109 García Lorca, 'Theory and Play of The *Duende*', trans. A. S. Kline, 2007. http://www.poetryintranslation.com/PITBR/Spanish/LorcaDuende.htm, viewed 8 September 2016.

110 See Thomas Beard, 'Jose Val del Omar: Across Three Avant Gardes', www.museoreinasofia.es/images/descargas/pdf/2010/10TB_en.pdf.

111 See www.valdelomar.com/cine3.php?lang=en&menu_act=5&cine1_cod=&cine2_cod=11&cine3_codi=66.

112 Pass, like passage, is a 'middle voice' concept. You pass someone but you also 'pass the ball'. You make a pass, a flirtatious move, like the toreador shaking the red blanket perhaps. Or you 'pass', staying out of the game. If you have a pass, you have either free entry or permission not to take part. Similarly, a passage both implicates and liberates – a 'dead end' is a passage that has lost its potential to discover a new beginning in the ending.

113 Didi-Huberman, *Le Danseur des Solitudes*, 123.

114 Ibid., 124.

115 'Tirar tu reloj al agua: Variaciones sobre una cinegrafía intuida de José Val del Omar', www.valdelomar.com/prensa/prensa_14.pdf, viewed 18 June 2026.

116 Ibid.

117 Yusimi Rodriguez, 'A Unique Musician on the Streets of Havana', *The Havana Times*, 26 December 2012. https://www.havanatimes.org/?p=84166, viewed 4 June 2016.

118 'Hang (instrument)', Wikipedia entry, https://en.wikipedia.org/wiki/Hang_(instrument), viewed 4 June 2016.

119 Eugeni Bonet: compartir la duda, Una entrevista por Antonio Orihuela at http://www.zemos98.org/festivales/zemos988/pack/eugenibonetcompartirla duda_latelevisionnolofilma.pdf, viewed 5 June 2016.

120 Eventually, nine of the ten poems forming the 'Scarlatti in Lisbon' sequence were set in this way (Paul Carter, *Ecstacies and Elegies* (Perth: UWAP, 2013), 59–72). They were uploaded to Soundcloud on 23 May 2016. https://soundcl oud.com/thestrangemechanism/scarlatti-in-lisbon/s-JDr2M

121 Nicole Loraux, *The Experiences of Tiresias: The Feminine and the Greek Man*, trans. P. Wissing (Princeton: Princeton University Press, 1995), 220.

122 Luis Cernuda, 'Rio Vespertino', *La Realidad y el Deseo (1924–1962)* (Madrid: Alianza Literaria, 2012), 234–5.

123 Luis Cernuda, 'Historial de un Libro', *La Realidad y el Deseo (1924–1962)* (Madrid: Alianza Literaria, 2012), 381–420, 409.

124 See Aileen Anne Logan, 'Memory and Exile in the Poetry of Luis Cernuda', PhD Dissertation, University of St Andrews, 2007, 142; Luis Cernuda, *Poesía y Literatura* (Barcelona: Editorial Seix Barral, 1960), 252.

125 William Whewell, *The Philosophy of the Inductive Sciences, Founded Upon Their History* (London: John W. Parker, 1840), 2 vols, vol. 2, 65.

126 Pablo Uribe, 'The War in Spain – A Basque boy tells his story', Basque Children of '37 Association UK, *Newsletter*, No. 16, December 2011, 6. Originally published in *The New Leader*, 20 August 1937.

127 Cernuda, 'Historial de un Libro', 401.

128 Ibid., 408.

129 William Sansom, 'The Naked Flame', *Fire and Water: The London Firefighters Blitz 1940–42 Remembered*, quoting Rilke (*The Notebook of Malte Laurids Brigge*, trans. J. Linton (Oxford: Oxford University Press, 1985, [orig. pub. 1910], 5).

130 Bill Law, 'Fire Fighting in Faringdon, A brief history of Faringdon Fire Station', 1–17, 7. http://www.faringdon.org/uploads/1/4/7/6/14765418/farin gdon_fire_station_history_2015.pdf. Accessed 18 June 2016. Previously, during the Civil War, Henderson had worked as a volunteer ambulance driver in Teruel. (See Martin Murphy in *Blanco White, El Rebelde Ilustrado*, ed. A.

Cascales Ramos (Sevilla: Fundación Publica Andaluza Centro de Estudios Andaluces, 2009), 105.)

131 'Lord Faringdon, Buscot and the Spanish Civil War', http://www.basquechildren.org/ Newsletter no. 2, September 2004. More seriously, lodged on the outskirts of the Buscot Estate, Garfías interpreted *'la verde lengua de parques y jardines'* – which made no sense to him and which in his poem 'Primavera en Eaton Hastings' are transformed into *'mi blanca Andalucia'* – as *'la conquista del paisaje natural como analoga al imperialismo britanico'* (Murphy, 'Blanco White y otros anglófilos españoles,' 97–108, 106.

132 Martin Heidegger, 'The Turning', in *The Question Concerning Technology and Other Essays*, trans. W. Lovitt (New York: Garland Publishing, 1977), 36–49, 45–6.

133 'Unamuno remarked that by studying his own language, a man would become aware of "*lo inconsciente en nosotros*", that is, of the common, unconscious psychic bond he shared with all other human beings' (Gayana Jurkevich, 'Unamuno's Intrahistoria and Jung's Collective Unconscious: Parallels, Convergences, and Common Sources,' *Comparative Literature* 43, no. 1, 43–59, 48).

134 Jurkevich, 'Unamuno's Intrahistoria and Jung's Collective Unconscious: Parallels, Convergences, and Common Sources', 49.

135 Paul R. Olson, *The Great Chiasmus: Word and Flesh in the Novels of Unamuno* (West Lafayette, IN: Purdue University Press, 2003), 2.

Chapter 5: Echoes

1 Paul Carter and Martin Harrison, 'Application to Develop Spring Song', 1990, 1–12, 5. In author's possession. As a linguistic descriptor, 'Tasmanian' is a misnomer: at least four different languages appear to have been spoken across the island.

2 Ibid., 6.

3 Paul Carter, 'Cooee Song', in *The Sound In-Between, Voice, Space, Performance* (New Endeavour Press/University of New South Wales Press, 1992), 52–69.

4 Martin Harrison, *The Distribution of Voice* (St Lucia, QLD: University of Queensland Press, 1993).

5 Umberto Eco, *The Open Work*, trans. A. Cancogni (Cambridge, MA: Harvard University Press, 1989), 1.

6 Carter, 'Ambiguous Traces, Mishearing, and Auditory Space', 43–64.

7 See Kantor, *A Journey through Other Spaces, Essays and Manifestos, 1944–1990*, 101 and Carter, 'Ambiguous Traces, Mishearing, and Auditory Space', 48.

8 Tomas *Transcultural Space and Transcultural Beings*, 1.

9 Ibid.

10 Quotations from Carter, *Material Thinking*, 2004, chapter 6.
11 Julio Marzán, *The Numinous Site: The Poetry of Luis Palés Matos* (Cranbury, NJ: Associated University Presses, 1995), 65.
12 Massimo Cacciari, *L'angelo necessario* (Milano: Adelphi, 1986), 23–4.
13 'Relay' is discussed in 'Trace: A Running Commentary on *Relay*', chapter 7 of Carter, *Dark Writing*, 203–27. The monograph *Mythform: The Making of Nearamnew at Federation Square, Melbourne* was published in 2005. 'Relay for Radio', subtitled 'a hocket for public radio', was first broadcast in September 2000, 'Nearamnewspeak' was first broadcast in December 2002. Both were produced for the Australian Broadcasting Corporation by Andrew McLennan.
14 Carter, *Dark Writing*, 272.
15 LaBelle, '*Misplace* – Dropping Eaves on Ethics', 12.
16 Horace Smith clarifies, 'We had a real excellent time here. You will see the photograph taken in the very act of singing'. (Murray J. Longman, 'Songs of the Tasmanian Aborigines as recorded by Mrs. Fanny Cochrane Smith', *Papers and Proceedings of the Royal Society of Tasmania* 94 (1960): 79–86, 84.)
17 Hermann B. Ritz, 'An Introduction to the Study of the Aboriginal Speech of Tasmania', *Papers and Proceedings of the Royal Society of Tasmania*, 1908, 73–83, 83.
18 Ferdinand Kriwet, quoted in K. Schöning and M. E. Cory, 'The Contours of Acoustic Art', *Theatre Journal* 43, no. 3 (Radio Drama 1991), 307–24, 321.
19 Alice M. Moyle, 'Tasmanian Music, an Impasse?', *Records of the Queen Victoria Museum* 26 (1968), 1–21, 1.
20 Longman, 'Songs of the Tasmanian Aborigines as recorded by Mrs. Fanny Cochrane Smith', 85.
21 Harrison, *Ancient Noise*, 139.
22 Ibid.
23 Ibid., 152.
24 Ibid., 151.
25 Ibid., 139.
26 Paul Carter, 'Travelling Blind: Notes for a Sound Geography', in *The Sound In-Between: Voice, Space, Performance* (Sydney: University of New South Wales Press/ New Endeavour Press, 1992), 117–38, 133.
27 Ibid., 124.
28 Paul Carter, 'Towards a Sound Photography: A Lake Eyre Notebook', in *Living in a New Country* (London: Faber & Faber, 1992), 78–97, 91.
29 Harrison, *Ancient Noise*, 148.
30 Ibid., 140.
31 Ibid., 151.

32 Ibid., 172.
33 Martin Harrison, 'Red Marine', in *Wild Bees, Selected Poems* (Perth: UWAP, 2008), 16–17.
34 Gaston Bachelard, *The Poetics of Reverie*, trans. D. Russell (Boston: Beacon Press, 1969), 189, where, of course, *La gorge* in French means both 'gorge' and 'throat'.
35 Mary LeCron Foster, 'The Symbolic Structure of Primordial Language', in *Human Evolution: Biosocial Perspectives*, ed. S. L. Washburn and E. R. McCown (Menlo Park, CA: Benjamin/Cummings Publishing Company, 1978), 117. Note: these sentences are reproduced from Carter, *Meeting Place*, 43. They are crossroads where two different stories *intersect*.
36 Bachelard, *The Poetics of Reverie*, 189.
37 René Descartes, *Philosophical Letters*, ed. and trans. A. Kenny (Oxford, 1970), 23–4.
38 Gottfried Wilhelm Leibniz, *Leibniz: Philosophical Writings*, trans. M. Morris (London: J. W. Dent, 1961), 236.
39 Carter, *The Lie of the Land*, 291.
40 Carter, *The Road to Botany Bay*, chapter 6.
41 Paul Verlaine, *Oeuvres Poétiques* (Paris: Bordas, 1967), 140. See also http://www.lieder.net/lieder/get_text.html?TextId=16258 for translation by Peter Low viewed 8 December 2012.
42 Harrison, *Ancient Noise*, 141.
43 Reported, not without irony, in *The Sound In-Between*: 'You ask for practical advice. A friend of mine once advertised a practical examination for poets. One of the questions was: "Read Lamartine on the Murray. The rustle of the reeds might be European, but *Le reel est étroit, le possible est immense.* ... "Comment"' (185).
44 Neilson, *The Collected Verse, A Variorum Edition*, 30.
45 See Carter, *Ground Truthing*, 218–19. 'I usually found a stanza suitable at once. It is rarely that I have had to alter a verse form' (Neilson, *The Collected Verse, A Variorum Edition*, 31).
46 Allan Marett, *Songs, Dreamings & Ghosts: The Wangga of North Australia* (Middletown, CT: Wesleyan University Press, 2005), 27–8.
47 *The Communism of John Ruskin*, ed. W. D. P. Bliss (New York: The Humboldt Publishing Co., 1891), 164 footnote.
48 *The Poems of Sappho, with Historical and Critical Notes*, ed. and trans. E. M. Cox (London: Williams & Norgate, 1924), 51.
49 Charles H. Cox and Jean W. Cox, *Wittgenstein's Vision* (San Diego: Libra Pub, 1984), 45.
50 Cox and Cox, *Wittgenstein's Vision*, 52.
51 Stanley Rosen, *Metaphysics in Ordinary Language* (New Haven: Yale University Press, 1999), 31.

52 William Desmond, *Desire, Dialectic, and Otherness, An Essay on Origins* (New Haven: Yale University Press), 1987.
53 Paul Carter, 'The Migrant's Vision', in *Mythform, the making of* Nearamnew *at Federation Square, Melbourne* (Carlton, VIC: Miegunyah Press, 2005), 78.
54 Carter, 'The Maker's Vision', *Mythform*, 47. I had invoked the same second creation story in 'Mirror States', published ten years earlier in *The Sound In-Between*, see especially scenes 24–32 (110–14).
55 See the discussions in Carter, *Meeting Place*, 6–8.
56 Carter, *Mythform*, 78.
57 Frances, Language, Identity and Land in the East Kimberley of Western Australia', in *Maintaining the Links: Language, Identity and the Land*, ed. J. Blythe and R. McKenna Brown (Bath, UK: The Foundation for Endangered Languages, 2003), 41–7, 43.
58 *The Poems of Sappho, with Historical and Critical Notes*, 20.
59 Ibid., 42.
60 Ibid., 15.
61 Ibid., 21.
62 Carter, *Mythform*, 86, footnote 1.
63 *The Poems of Sappho, with Historical and Critical Notes*, 108.
64 Cox and Cox, *Wittgenstein's Vision*, 49.
65 Ibid., 43.
66 Alex Kozulin, 'Vygotsky in Context', in *Thought and Language*, ed. Lev Vygotsky and trans. A. Kozulin (Cambridge, MA: The MIT Press, 1986), xi–lxi, xxvii.
67 Kozulin, 'Vygotsky in Context', xxxvii.
68 Ibid.
69 See Carter, 'Mirror States', *The Sound In-Between*, 100. The many meanings of Yarra are discussed in Carter, *Living in a New Country*. For a more up to date summary, see Ian D. Clark and Laura M. Kostanski, 'An Indigenous History of Stonington, A Report to the City of Stonnington', School of Business, University of Ballarat, 30 June 2006, 1–214, 24.
70 Bachelard, *The Poetics of Reverie*, 189.
71 Christopher Smith, 'Nietzsche's Recovery of Primary Acoustical Experience: *The Birth of Tragedy out of the Spirit of Music* as "Fundamental Ontology"'. *Essays in Sound* 3 (1996), 81–91, 89.
72 Quoted by Marysia Lewandowska, 'Speaking, the Holding of Breath', in *Sound by Artists*, ed. D. Lander and M. Lexier (Toronto: Art Metropole, 1990), 55–62, 55.
73 Smith, 'Nietzsche's Recovery of Primary Acoustical Experience', 88.
74 Ibid.
75 Carter, 'Underworlds of Jean du Chas', 266.

76 See Walt Whitman, 'Sea-Shore Fancies', in *Complete Poetry and Selected Prose and Letters*, ed. E. Holloway (London: The Nonesuch Press, 1967) and, in the same volume, 'A Backward Glance O'er Travel'd Roads', 874.
77 Alain Corbin, *The Lure of the Sea: The Discovery of the Seaside in the Western World, 1750–1840*, trans. Jocelyn Phelps (Berkeley: University of California Press, Cambridge, 1994), 74.
78 Carter, 'Underworlds of Jean du Chas', 273.
79 Ibid., 266.
80 Carter, *The Sound In-Between*, 95–6.
81 Quoted by Max Bruinsma, '*Notes of a Listener*', in *Sound by Artists*, ed. D. Lander and M. Lexier (Toronto: Art Metropole and Walter Phillips Gallery, 1990), 90.
82 Carter, *The Sound In-Between*, 89.
83 Vygotsky, *Thought and Language*, 182.
84 Ibid.
85 Edmund Spenser, *The Poetical Works* (Oxford: Oxford University Press, 1965), 600–2.
86 'The Ferryman's Vision', in Carter, *Mythform*, 102.
87 Spenser, 'A View of the Present State of Ireland'.
88 Ibid.
89 Ibid.
90 Andrew Hadfield, *Edmund Spenser's Irish Experience: Wilde Fruit and Savage Soyl* (Oxford: Oxford University Press, 1997), 18.
91 'Mirror States', in Carter, *The Sound In-Between*, 96.
92 Adrian Burrow et al., 'The Architecture of Contact: Unearthing a Cultural Landmark of Early Melbourne'. At https://www.australianarchaeologicalassociation.com.au/wp-content/uploads/2014/11/Griffiths-Billy.pdf. In *A Bend in the Yarra: a history of the Merri Creek Protectorate Station* (Canberra: Aboriginal Studies Press, 2004), Ian Clark and Tony Hebdon give her name as 'Susan' (18).
93 Mark Amsler, *Etymology and Grammatical Discourse in Late Antiquity and the Early Middle Ages* (Amsterdam: J. Benjamins Pub. Co., 1988), 238.
94 Hélène Cixous, 'Foreword', in Clarice Lispector, *The Stream of Life*, trans. E. Lowe and E. Fitz (Minneapolis: University of Minnesota Press, 1989), xx.
95 Carter, 'Mirror States', *The Sound In-Between*, 110.
96 Ibid., 114.
97 The story of the production is told in Carter, *Material Thinking*, chapter 4.
98 'Old Wives' Tales', Chandrabhanu and the Bharatam Dance Company, Old Wives' Tales, C.U.B. Malthouse, Melbourne, 25 September–4 October 1997.
99 Recalled and discussed in Carter, *Dark Writing*, 214–15.

100 Jan Critchett, 'Kaawirn Kuunawarn (1820–1889)', Australian Dictionary of Biography, National Centre of Biography, Australian National University, http://adb.anu.edu.au/biography/kaawirn-kuunawarn-13018/text23537, published first in hardcopy 2005, viewed 4 April 2012.

101 Jan Critchett, *Untold Stories: Memories and Lives of Victorian Kooris* (Carlton, VIC: Melbourne University Press, 1998), 124–5. This was admittedly unusual – as was the fact that Patrick Mitchell acknowledged fathering 'an Aboriginal child'.

102 Dawson, *Australian Aborigines*, iii.

103 Barry J. Blake, 'Dialects of Western Kulin, Western Victoria Yartwatjali, Tjapwurrung, Djadjawurrung', La Trobe University, 2011, 12. www.vcaa.vic.edu.au/alcv/DialectsofWesternKulin-WesternVictoria.pdf, viewed 14 March 2013, 12.

104 Spellings continue to be subject to change: Dawson's 'Chaap wuurong' corresponds to 'Tjapwurrung' and the currently preferred 'Djabwurrung'. Clark recommends that 'Gundidjmara' be replaced by 'Dhauwurd Wurrung' (see *Aboriginal Languages and Clans: An Historical Atlas of Western and Central Victoria, 1800–1900*, Monash Publications in Geography, Monash University, Melbourne, Victoria, 3168, 1990, 23). However, most writers continue to use 'Gundidjmara' or 'Gunditjmara'.

105 Dawson, *Australian Aborigines*, v.

106 Vygotsky, *Thought and Language*, xxxvi.

107 Blake translates it as 'mouth' (*Dialects of Western Kulin, Western Victoria, Yartwatjali, Tjapwurrung, Djadjawurrung*, 11), a rendering that is not matched in Dawson. It is also to be noted that in Djabwurrung (the language spoken around Camperdown), the same word was used, but in a dialectally different form: *Wuuro*.

108 I have published the materials relating to this name choice and, as yet, no qualified ethnolinguist has contested my reading; in any Australian sound history it remains, therefore, exceptional. See 'Lips in language and space: imaginary places in James Dawson's *Australian Aborigines* (1881)', *Spatiality and Symbolic Expression: On the Links between Place and Culture*, ed. B. Richardson (New York: Palgrave Macmillan, 2015), 105–29; 'The Enigma of Access: James Dawson and the Question of Ownership in Translation', 'Law and Its Accidents', a special symposium issue of the *Griffith Law Review* 22, no. 1 (2013), 8–27.

109 Dawson, *Australian Aborigines*, lxxx, lxxxi.

110 Ibid.

111 Ibid., lxxix.

112 Franca Tamasari and James Wallace, 'Towards an Experiential Archaeology of Place: From Location to Situation through the Body', in *The Social Archaeology of Australian Indigenous Societies*, ed. B. David, B. Barker and I. J. McNiven (Canberra: Aboriginal Studies Press, 2006), 209. See also the discussion in Paul Carter, *Decolonising Governance, Archipelagic Thinking* (London: Routledge, 2018), 156–7.

113 Dawson, *Australian Aborigines*, 54.
114 Ibid., 99.
115 Ibid., viii.
116 Ibid., xxxv.
117 J. Hillis Miller, *Ariadne's Thread* (New Haven, CT: Yale University Press, 1992), 155.
118 Miller, *Ariadne's Thread*, 155.
119 Ibid., 55. See also Carter, *Meeting Place* for this passage in context, 108–14.
120 Smyth, *The Aborigines of Victoria*, vol. 1, 444.
121 Ibid., 435.
122 Ibid., 439.
123 Ibid., 444.
124 Cornford, *Principium Sapientiae: The Origins of Greek Philosophical Thought*, 191.
125 Smyth, *The Aborigines of Victoria*, vol. 1, 435.
126 Bachelard, *The Poetics of Reverie*, 176.
127 T. C. Lethbridge, 'Witches' (*RLE Witchcraft: Investigating an Ancient Religion*, Routledge, 2011), 97.
128 Dawson, *Australian Aborigines*, 51.
129 Ibid., 57.
130 Ibid., 55.
131 Ibid., 56.
132 Ibid., 52.
133 Gearóid Ó Crualaoich, *The Book of the Cailleach, Stories of the Wise-Woman Healer* (Cork: Cork University Press, 2003), 29.
134 Ó Crualaoich, *The Book of the Cailleach*, 29.
135 Ibid., 50.
136 Emphasis in original. W. E. H. Stanner, *The Dreaming and Other Essays* (Agenda: Melbourne, Black Inc., 2009), 72.
137 Stanner, *The Dreaming and Other Essays*, 60.
138 Dawson, *Australian Aborigines*, 55.
139 A witches' moon dial found near Wayland's Smithy, adjacent to the Uffington White Horse, was made of human bone and had seven sections corresponding to the Seven Hours of Dread. (Grinsell, 'Witchcraft at Some Prehistoric Sites', 72–9, in *The Witch Figure: Folklore Essays by a Group of Scholars in England*, ed. Venetia Newall (London: Routledge, 1973), 76).
140 Cassandra Eason, *Fabulous Creatures, Mythical Monsters, and Animal Power Symbols: A Handbook* (Westport: Greenwood, 2008), 5–6.
141 Alfred Williams, *Villages of the White Horse*, (London: Duckworth, 1913), 238.

142 Richard Jefferies, 'The Valley Spring and the Stream', in *The Jefferies Companion*, ed. S. J. Looker (London: Phoenix House, 1948), 113.
143 Graves, *The White Goddess*, 384.
144 Hajo Duchting, *Kandinsky* (Taschen: Koln, 2007), 42.
145 Ibid.
146 Paul Carter, 'Masters of the Gap: Art, Migration and Eido-Kinesis', in *After the Event, New Perspectives on Art History*, ed. John Potts and Charles Merewether (Manchester: Manchester University Press, 2010), 43–56.
147 De Quincey, *The Collected Writings of Thomas De Quincey*, vol. 1, 287.
148 First formulated in Paul Carter, 'Arcadian Writing: Two Text Into Landscape Proposals', *Studies in the History of Gardens & Designed Landscapes* 21, no. 2 (2001), 137–47.
149 Longman, 'Songs of the Tasmanian Aborigines as recorded by Mrs. Fanny Cochrane Smith', 81.
150 Moyle, 'Tasmanian Music, an Impasse?', 4.
151 García Lorca, 'Theory and Play of The Duende', trans. S. Kline (2007). http://www.poetryintranslation.com/PITBR/Spanish/LorcaDuende.htm

Chapter 6: Recordings

1 George Petrie, *The Ancient Music of Ireland*, cited in Seamus Deane, *A Short History of Irish Literature* (Notre Dame: University of Notre Dame Press, 1994), 79. Similar testimony from Co. Doneghal is found in Kevin Whelan, 'Clachans: landscape and life in Ireland before and after the Famine', in *At The Anvil: Essays in Honour of William J. Smyth*, ed. P. J. Duffy and W. Nolan (Templeogue, Dublin: Geography Publications, 2012), 453–75, 466.
2 Horton, *Six Months in South Australia, with some account of Port Philip (sic) and Portland Bay*.
3 Carter, 'Memory as Desire', 46.
4 Elinor Wrobel, (curator), *Percy Grainger's Paradoxical Quest for 'World Music': Free Music & Free Music Machines*, catalogue, Grainger Museum, University of Melbourne, 1994, 9.
5 Andrey Tarkovsky, *Sculpting in Time*, trans. K. Hunter-Blair (London: Faber & Faber, 1989), 117.
6 Catherine J. Ellis, *Aboriginal Music* (St Lucia, Queensland: University of Queensland Press, 1989), 92.
7 Ibid., 104.
8 Ibid., 94.
9 Catherine Ellis, *Aboriginal Music Making, A study of Central Australian music* (Adelaide: Libraries Board of South Australia, 1964), 8.
10 Ritz, 'The Speech of the Tasmanian Aborigines', 48.

11 Carter, 'Memory as Desire', 30.
12 Jean-Claude Margolin, 'Bachelard and the Refusal of Metaphor', in *The Philosophy and Poetics of Gaston Bachelard*, ed. M. McAllester (Washington DC, Center for Advanced Research in Phenomenology and University Press of America, 1989), 115.
13 Bachelard, *The Poetics of Reverie*, 189.
14 Rosemary Jellis, *Bird Sounds and Their Meaning* (Ithaca and New York: Cornell University Press, 1974), 10.
15 T. A. Coward, *The Birds of the British Isles and Their Eggs* (London: Frederick Warne, 1920).
16 Asturias, 'América, la engañadora', 343.
17 Vicky Unruh, *Latin American Vanguards: The Art of Contentious Encounters* (Berkeley: University of California Press, 1994), 254.
18 Unruh, *Latin American Vanguards: The Art of Contentious Encounters*, 257.
19 Ibid., 258.
20 The script remains unpublished but see Carter, 'Emergency Languages: Echoes of Columbus in Discourses of Precarity'.
21 Michael Silverstein, '"Goodbye Columbus", Language and speech community in Indian-European contact situations', unpublished manuscript, 1972, 33, which the author kindly shared with me. A more formal presentation of the research informing this view is Michael Silverstein, 'Chinook Jargon: Language contact and the problem of multi-level generative systems', parts 1 and 2 in *Language*, vol. 48, nos. 2 and 3, 1972.
22 Silverstein, 'Goodbye Columbus', 46.
23 Carter, *Meeting Place*, 104.
24 Drechsel, 'Metacommunicative Functions of Mobilian Jargon, an American Pidgin of the Lower Mississippi River Region', 434.
25 Kantor, *A Journey through Other Spaces*, 101.
26 Ibid., 100.
27 Ibid., 101.
28 Silverstein, 'Goodbye Columbus', 21.
29 Marianne van Kerkhoven, 'European Dramaturgy in the 21st Century: A Constant Movement', *Performance Research* 14, no. 3 (2009), 7–11, 11.
30 Carter, 'Columbus Echo', unpublished manuscript.
31 Alberto Moreiras, 'Foreword', in *Poetry after the Invention of America*, ed. Andrés Ajens (New York: Palgrave Macmillan, 2011), xix–xxi, xx.
32 Moreiras, 'Foreword', xx.
33 Katrin Sieg, *Ethnic Drag: Performing Race, Nation, Sexuality in West Germany* (Ann Arbor: University of Michigan Press, 2009), 193.
34 Sieg, *Ethnic Drag: Performing Race, Nation, Sexuality in West Germany*, 193, note 7.
35 Ibid.

36 Hubert Fichte, 'Xango', quoted by Klaus Neumann, 'Hubert Fichte as Ethnographer', *Cultural Anthropology*, 263–84, 268.

37 Neumann, 'Hubert Fichte as Ethnographer', 283.

38 Juan Davila and Kate Briggs, *The Moral Meaning of the Wilderness* (Canberra: Australian National University; Drill Hall Gallery, 2010), 24.

39 John Doe, *Speak into the Mirror, A Story of Linguistic Anthropology* (Lanham, MD: University Press of America, 1988), 196, drawing on Mikhail Bakhtin, *The Dialogic Imagination*, ed., M. Holquist, trans. C. Emerson and M. Holquist (Austin: University of Texas Press, 1981), 293–4.

40 Tzvetan Todorov, 'Dialogism and Schizophrenia', in *An Other Tongue, Nation and Ethnicity in the Linguistic Borderlands*, ed. Alfred Arteaga (Durham, NC: Duke University Press, 1994), 214.

41 Major B. Lownsley, *A Glossary of Berkshire Words and Phrases* (Ludgate Hill: Trübner & Co, 1888), 29.

42 Carter, 'What Is Your Name', 52.

43 Ibid., 57.

44 David Kazanjian and Marc Nichanian, 'Between Genocide and Catastrophe', in *Loss, The Politics of Mourning* (Berkeley: University of California Press, 2003), 141.

45 This was not necessarily the end of resistance: there was no question of abandoning one's own name – these were known, they simply weren't shared. Or, more accurately, the people pushed into the mission stations remained enfolded in layers of names, in multiple skins of association and protection. For, it seems, even the 'real' names were screen names: 'we used to go up to listen to the old blokes when we was kids and they used their names to each other. ... They wouldn't be their real tribal name 'cos they never told that – it was a secret' (Phillip Pepper, *You Are What You Make Yourself To Be* (Melbourne: Hyland House, 1980), 20). For the fuller discussion of these issues of 'translation', see Carter, *Ground Truthing*, 246–64.

46 Kazanjian and Nichanian, 'Between Genocide and Catastrophe', 141.

47 Carlo Ginzberg, *Wooden Eyes, Nine Reflections on Distance*, trans. M. Ryle and K. Soper (New York: Columbia University Press, 2001), 140.

48 Ginzberg, *Wooden Eyes, Nine Reflections on Distance*, 142.

49 Carter, 'Memory as Desire', 35.

50 Even reliable sources get confused. Some sources list J. B.'s sons as 'Charles' and 'James', making no mention of a 'Thomas'.

51 W. H. Pyne, *Wine and Walnuts* (London: Longman,1823), 2 vols, vol. 1, 11–12, footnote.

52 Kathleen Kimball, 'Red Handed: An Inquiry Into the Meaning of Prehistoric Red Ochre Handprints', worldhistoryconnected.press.illinois.edu/9.2/forum_kimball.html, viewed 4 February 2016.

53 Tom Cheetham and Green Man, *Earth Angel: The Prophetic Tradition and the Battle for the Soul of the World* (New York: Suny, 2005) 23–4.

54 Jonathan Sterne, *The Audible Past, Cultural Origins of Sound Reproduction* (Durham: Duke University Press, 2006), 23.
55 'Landscape-perception', www.landscape-perception.com/art_and_archaeology/, viewed 30 November 2016.
56 'Landscape-perception', www.landscape-perception.com/archaeoacoustics/, viewed 30 November 2016.
57 R. M. W. Dixon, *Searching for Aboriginal Languages, Memoirs of a Field Worker* (St. Lucia, QLD: University of Queensland Press, 1993), 95.
58 Sterne, *The Audible Past, Cultural Origins of Sound Reproduction*, 40.
59 Ibid., 43.
60 Wagner, *An Anthropology of the Subject, Holographic Worldview in New Guinea and Its Meaning and Significance for the World of Anthropology*, 137.
61 George Grey, *Expeditions into Western Australia, 1837–1839* (London: T. & W. Boone, 1841), 2 vols, vol. 1, 201. See also discussion in Carter, *Living In A New Country*, 51.
62 Jean Gebser, *The Ever-Present Origin*, trans. N. Barstad and A. Mickunas (Athens: Ohio University Press, 1985, orig. pub., 1949), 56–7.
63 Gebser, *The Ever-Present Origin*, 60.
64 Ibid., 61.
65 Ibid., 54–5.
66 Ibid., 64.
67 Ibid., 65.
68 Ibid.
69 Carter, *The Sound In-Between*, 136–7.
70 T. G. H. Strehlow, *Songs of Central Australia* (Sydney: Angus & Robertson, 1970), 448.
71 Strehlow, *Songs of Central Australia*, 351.
72 Ibid., footnote.
73 The epigraph to 'The Native Informant' had quoted Swedenborg's claim of a correspondence between 'the internal human ear' and 'the spiritual world' without perhaps appreciating that Swedenborg had studied the physiology of the ear in detail – and when, later, he heard angelic voices, he was in a position to locate the origin of these voices precisely within the bony labyrinth.
74 Gebser, *The Ever-Present Origin*, 145.
75 F. Joseph Smith, *The Experience of Musical Sound, Prelude to a Phenomenology of Music* (New York: Gordon & Breach, 1978), 154.
76 A. D. Hope, 'Hay Fever', in *'The Scythe Honed Fine', A.D. Hope: A Celebration for his 90th Birthday* (Canberra: National Library of Australia, 1997), vii.
77 A. M. Massee, *The Pests of Fruits and Hops* (London: Crosby Lockwood, 1946), 207.

78 Here lofty mountains lift their azure heads;
    There it's green lap the grassy meadows spreads;
    Enclosures here the sylvan scene divide;
    There plains extended spread their harvests wide;
    Here oaks, their mossy limbs wide stretching meet
    And form impervious thickets at our feet;
    Through aromatic heaps of ripening hay,
    There silver ISIS wins her winding way;
    And many a tower, and many a spire between,
    Shoots from the groves, and cheers the rural scene.

    (Henry James Pye, Faringdon Hill, a poem in two books (Oxford: Daniel Prince, 1778), Book, 1, l.2.)

79 Stephen Duck, 'The Thresher's Labour', *Poems on Several Occasions* (London: printed for the author, 1736), 12.

80 John Berger, 'Uses of Photography', in *About Looking* (New York: Pantheon Books, 1980), 67.

81 Emmanuel Levinas, *Unforeseen History*, trans. N. Poller (Urbana: University of Illinois Press, 2004), 79.

82 Levinas, *Unforeseen History*, 89.

83 Bruno Monsaingeon, *Sviatoslav Richter: Notes and Conversations* (Princeton, NJ: Princeton University Press, 2001), 35.

84 James Clerk Maxwell, 'Essay for the Eranus Club on Psychophysik', 5 February 1878, *The Scientific Letters and Papers*, vol. 3 1874–9 (Cambridge University Press, 2002), 598–607, 599–600.

85 Maxwell, 'Essay for the Eranus Club on Psychophysik', 600–1.

86 Stephanie L. Hawkins, 'William James, Gustav Fechner, and Early Psychophysics', quoting D. C. Lamberth, *Frontiers in Physiology*, published online, 4 October 2011, https://www.ncbi.nlm.nih.gov/pmc/articles/PMC3185290/, viewed 7 August 2016.

87 Hawkins, 'William James, Gustav Fechner, and Early Psychophysics'.

88 Ibid.

89 Ibid.

90 Maxwell, 'Essay for the Eranus Club on Psychophysik', 599.

91 Poul Martin Møller, *The Adventures of a Danish Student* [*En Dansk Students Eventyr*], quoted by Richard Rhodes (*The Making of the Atomic Bomb* (New York: Simon & Schuster, 1986), 58) in his discussion of the intellectual evolution of the formulator of the 'complementarity principle', physicist Niels Bohr.

92 Is this not like Ungaretti's '*re-creative*' process 'which annuls contingent elements … while substituting for them an absolute sphere of emotion peopled by imaginative forms'? (Frederic J. Jones, *Giuseppe Ungaretti, Poet and Critic* (Edinburgh: Edinburgh University Press, 1977), 24.)

93 Susanna Checketts, reproduced by permission. These last lines found their way into 'Remember Me'.

Chapter 7: Voices

1. Church, *Over the Bridge, an Essay in Autobiography*, 191–2, 222.
2. 'He was a boy of high spirits and impatient of rest; but at the age of seven he fell headfirst from the top of a ladder to the floor below, and remained a good five hours without motion or consciousness. ... The surgeon ... observing the broken cranium and considering the long period of unconsciousness, predicted that he would either die or grow up an idiot. However by God's grace neither part of his prediction came true, but as a result of this mischance he grew up with a melancholy and irritable temperament such as belongs to men of ingenuity and depth, who, thanks to one, are quick as lightning in perception, and thanks to the other, take no pleasure in verbal cleverness or falsehood'. *The Autobiography of Giambattista Vico*, trans. M. H. Fisch and T. G. Bergin (New York: Cornell University Press, 1944) 111.
3. Carter, 'The 7448', 185.
4. Bruce Smith, *The Acoustic World of Early Modern England* (Chicago: University of Chicago Press, 1999), 338.
5. Matthew Guerrerí discusses the well-known claim that Beethoven derived his opening notes from the song of the yellowhammer (*The First Four Notes*, New York: Knopf, 2012, 20–2), rendered in English as 'A little bit of bread and noooo – cheese'. But the interval between 'noooo' and 'cheese' is not the same as that between the third and fourth notes of the symphony. Perhaps German yellowhammers used to sing differently.
6. 'Here AP does not refer to the human ability to assign a note name or pitch chroma to a tone, such as "G sharp," but the more general ability to recognize tones on the basis of their AP height' (Micah R. Bregman, Aniruddh D. Patel, Timothy Q. Gentner, 'Songbirds Use Spectral Shape, Not Pitch, for Sound Pattern Recognition', *Proc. Nat. A. of Sci. USA*. 113, no. 6 (2016), 1666–71, https://www.ncbi.nlm.nih.gov/pmc/articles/PMC4760803/, viewed August 14, 2016).
7. Stefan Elmer, et al., 'Bridging the Gap between Perceptual and Cognitive Perspectives on Absolute Pitch', *The Journal of Neuroscience* 35 (7 January 2015), 366–71.
8. Daniel Paul Schreber, *Memories of My Nervous Illness* (New York: New York Review Books, 2000) ('To him all sounds were voices, the universe was full of words: railways, birds and paddle-steamers spoke' Elias Canetti, *Crowds and Power*, trans. C. Stewart (London: Gollancz, 1962), 452).
9. Herbert Mason, *A Legend of Alexander; and, the Merchant and the Parrot* (Notre Dame, IN: University of Notre Dame Press, 1986), 65.
10. Mason, *A Legend of Alexander; and, the Merchant and the Parrot*, 67.
11. Ibid., 70.
12. Ibid., 111.
13. Ibid., 70.
14. Ibid., 111.

15 Carter, 'Memory as Desire', 44.
16 Canetti on Schreber's voices.
17 Carter, 'Cooee Song', 69–93, 82.
18 Ibid., 70.
19 Schreber, *Memories of My Nervous Illness*, 192.
20 Ibid.
21 George Taplin, 'The Narrinyeri, 1873', reprinted in J. D. Woods, *The Native Tribes of South Australia* (Adelaide: E. S. Wigg, 1879), 143.
22 Rafael, 'Gods and Grammar: The Politics of Translation in the Spanish Colonization of the Philippines', 113.
23 In the final act of my association with the Australian Broadcasting Corporation, a version of 'The Letter S' (there were many versions) was produced by Christopher Williams broadcast in the program *Airplay* 15 and 16 September 2006.
24 Canetti, *Crowds and Power*, 15.
25 'Whenever I move my chair, he moves his also, and, in general, imitates all my movements as although he wished to annoy me.' (August Strindberg, *The Inferno*, trans. C. Field (London: William Rider & Son, 1912)), 2's version of this in 'What Is Your Name' is altogether more circumspect. Instead of suspecting persecution, he entertains the thought that he can break down walls, and inaugurate communication: 'When I pushed my chair forward, he pushed his forward; when I drew it back, he drew his back. I thought it was coincidence at first.' ('What Is Your Name', 67.)
26 Carter, 'On the Still Air', 143.
27 Giuseppe Ungaretti, *Vita d'un Uomo, Saggi e Interventi* (Milano: Mondadori, 1975), 237. In flashes, through an intuition of lightning rapidity, the ice is broken and in some sense [an awareness of our true reality] is imparted to us, enabling us to regain in some way our innocence.
28 Carter, 'On the Still Air', 149.
29 Ibid., 164.
30 'The self-centred mimicry of architecture [referring especially to glass-clad buildings] can be likened to Batman's one-sided appropriation of Aboriginal territory: it goes through the motions of dialogue but without allowing any erosion of its original position. In this sense such structures – which are historical and discursive as well as concrete – are the ironic mirror to another kind of communicational structure, one in which both sides mimic each other and from whose free play is produced a new and fluid medium of exchange' (Carter, *The Sound In-Between*, 91).
31 Carter, 'Mirror States', *The Sound In-Between*, 66.
32 Daniel Chalmers, *Semiotics: The Basics* (London: Psychology Press, 2002), 125.
33 Ibid.
34 F. Joseph Smith, *The Experiencing of Musical Sound: Prelude to a Phenomenology of Music* (New York: Gordon & Breach, 1978).

35 Richard L. Lanigen, *Phenomenology of Communication* (Pittsburgh: Duquesne University Press, 1988), 65.
36 Lanigen, *Phenomenology of Communication*, 91.
37 Deborah Eicher-Catt, 'The Authenticity in Ambiguity: Appreciating Maurice Merleau-Ponty's Abductive Logic as Communicative Praxis', *Atlantic Journal of Communication* 13, no. 2 (2005), 113–34, 129.
38 Eicher-Catt, 'The Authenticity in Ambiguity: Appreciating Maurice Merleau-Ponty's Abductive Logic as Communicative Praxis', 129, quoting D. L. Smith.
39 Ibid.
40 *The Tempest*, Act 4, Scene 1, ll.107, 109.
41 Carter, *The Sound In-Between*, 41.
42 Carter, 'The Native Informant', 171.
43 Kantor, *A Journey through Other Spaces*, 101.
44 Ibid.
45 Franz Kafka, 'Third Octavo Notebook', in *Wedding Preparations in the Country and Other Posthumous Writings*, trans. E. Kaiser and E. Wilkins (London: Secker & Warburg, 1954), 73.
46 Cacciari, *L'Angelo Necessario*, 54–5.
47 Ibid., 31.
48 Ibid.
49 Ibid., 23–4.
50 Ibid., 31.
51 Ibid., 157–9.
52 William Skeat, *A Concise Etymological Dictionary* (Oxford: Clarendon Press, 1882), 579.
53 Giuseppe Ungaretti, *Vita d'un Uomo, Saggi e Interventi*, 64. On an ocean/ of tinkling bells/ suddenly/ there floats another morning.' (See Jones, *Giuseppe Ungaretti, Poet and Critic*, 35–6.)
54 Jones, *Giuseppe Ungaretti, Poet and Critic*, 36.
55 Ernesto Livorni, 'Ezra Pound and Giuseppe Ungaretti: Between Haiku and Futurism', in *Ezra Pound and Europe*, ed. R. Taylor and C. Melchior (Amsterdam: Rodopi, 1993), 131–44, 141, note 17.
56 The francophone poet surely puns here on the French *repentir (se)*, to repent of; in Italian one can say *pentirsi* but the combination with the *re-* prefix signifying *again* does not occur.
57 Carl Jung, *Memories, Dreams, Reflections*, trans. R. and C. Winston (London: Pantheon, 1963), 174.
58 Jung, *Memories, Memories, Dreams, Reflections*, 205.
59 Mannoni, *Prospero and Caliban*, 198.
60 James Hillman, *The Dream and the Underworld* (New York: Harper & Row, 1979), 109.
61 John Ball, 'Zermatt in 1845', 121.

62 Greg Louganis, *Breaking the Surface*, with Eric Marcus (New York: Plume, 1996), 205.

63 ... *et je sens tressaillir en moi quelque chose qui se déplace, voudrait s'élever, quelque chose qu'on aurait désancré, à une grande profondeur. Je ne sais ce que c'est, mais cela monte lentement. J'éprouve la résistance et j'entends la rumeur des distances traversées.* (Marcel Proust, *Senses of Consciousness: Swann's Way in Half*, trans. C. K. Scott Moncrieff (San Antonio College, Walnut: MSAC Philosophy Group, 2008), 41. https://fr.wikisource.org/wiki/Page:Proust_-_Du_c%C3%B4t%C3%A9_de_chez_Swann.djvu/63, viewed 14 June 2017.)

64 Ibid., 41.

65 Ibid., 42.

66 Richard Berrong, 'A Significant Source for the Madeleine and Other Major Episodes in Combray: Proust's Intertextual Use of Pierre Loti's *My Brother Yves*', *Studies in 20th & 21st Century Literature* 38, no. 1 (2014), article 3, 1–17, 7.

67 Luzius Keller, *Les Avant-textes de l'épisode de la madeleine dans les cahiers de brouillon de Marcel Proust* (Paris: Jean-Michel Place, 1978), 55.

68 'Proto-madeleine' because in this instance the narrator cannot command the involuntary memory to yield up its riches: 'They offered me an indefinite continuation of the same charm, in an inexhaustible profusion, but without letting me delve into it any more deeply, like those melodies which one can play over a hundred times in succession without coming any nearer to their secret.'

69 Carter, 'Memory as Desire', 40.

70 See Corbin, *The Lure of the Sea*.

71 Carter, 'Underworlds of Jean du Chas', 273.

72 Ibid., 282.

73 Ibid., 284.

74 The rotatory method for tranquillizing mentally disturbed patients was standard nineteenth-century practice in many of the more advanced sanatoria of northern Europe. The threat of repetition – the fear of another terrifying bout of dizziness – was said to be a wonderfully effective sedative. It certainly gives a different gloss to Kierkegaard's advocacy of the 'rotation method' and its principle of periodic repetition – the Danish writer's brilliant reflections on the disease of modernity, boredom, shadow much of the Second Plunge. See Søren Kierkegaard, 'The Rotation Method', *Either/Or*, trans. G. L. Stengren (New York: Harper & Row, 1986), 76–92.

75 Carter, 'Underworlds of Jean du Chas', 299.

76 The original passages are found in Plutarch, 'On the Genius of Socrates', in *Selected Essays of Plutarch*, Vol. II, ed. and trans. A. O. Prickard (Oxford: Oxford University Press, 1918), 6–51, 35–7.

77 Plutarch, 'On the Genius of Socrates', 36–7.

78 Cacciari, *L'Angelo Necessario*, 23.

79 Louise Holland (*Janus and the Bridge*, American Academy in Rome, Papers and Monographs XXI, 1961, 24–5) explains that in this early usage the word 'Janus' refers to a two-faced concept; it comprehends equally the crossing of the river as a bridge over flowing water, and as water crossing over an inaugurated place.

80 Quoted by Juan Eduardo Cirlot, *A Dictionary of Symbols*, trans. J. Sage (New York: Dover Publications, 2002), 356.

81 Ungaretti, *Vita d'un Uomo, Saggi e Interventi*, 369. The most extraordinary and genuine baroque manifest itself in these vases discovered in a twenty-two century old hypogeum.

82 Giuseppe Ungaretti, 'Il porto sepolto', in *Il porto sepolto*, ed. C. Ossola (Milano: il Saggiatore, 1981 [orig. pub. 1916]). 'The poet descends there / then returns to the light with his songs / and scatters them.' 'Of this poetry there stays with me that nullity of a bottomless secret.'

83 Ungaretti, *Il Porto Sepolto*, 19.

84 Ibid., 19–31.

85 Giuseppe Ungaretti, 'In memoria', *Vita d'un Uomo*, 27.

86 Giuseppe Ungaretti, 'Eterno', in *Vita d'un Uomo*, 15.

87 Madsen, 'Written in Air: Experiments in Radio', 169.

88 Dan Lander, 'Radiocastings: Musings on Radio and Art', 19.

89 Ibid.

90 Brecht, 'The Radio as an Apparatus of Communication'.

91 Ibid.

92 Carter, *Lost Subjects*, 169.

93 Ibid.

94 Carter, 'What Is Your Name', 67.

95 Ellison Hawks, *The Romance and Reality of Radio* (London: T.C. & E.K. Jack, 1923), 75.

96 Carter, 'The Letter S', unpublished script. The relapse of the winged word into the 'ocean' of *Finnegans Wake* was intentional: an allegory of the end of our radiophonic project.

97 Christopher Williams, 'Materials and Processes in the *Poiesis* of Radiophonic Art', chapter 4, unpublished PhD draft, 2016, 1–81, 29.

98 Williams, 'Materials and Processes in the *Poiesis* of Radiophonic Art', 25.

99 Ibid., 32.

100 Ibid., 35.

101 Ibid., 37.

102 Ibid.

103 Ibid.

104 Sharon M. Carnicke, *Stanislavsky in Focus* (Amsterdam, 1998), 176.

105 'Vocoder', Wikipedia, viewed 8 April 2017.

106 Mathias S. Oechslin, Martin Meyer and Lutz Jäncke, 'Absolute Pitch – Functional Evidence of Speech-Relevant Auditory Acuity', *Cerebral Cortex* 20, no. 2 (2010 February), 447–55. https://www.ncbi.nlm.nih.gov/pmc/articles/PMC2803739/, viewed 14 September 2017.

107 Ibid.

108 Bregman et al., 'Songbirds use Spectral Shape, not Pitch, for Sound Pattern Recognition', 1666–71. See also R.V. Shannon, 'Is Birdsong More Like Speech or Music?' *Trends Cogn Sci.* 20 April, no. 4 (2016), 245–7, who supports this conclusion.

109 Carter, *Parrot*, 9.

110 Ibid.

111 Rainer Maria Rilke, *Letters of Rainer Maria Rilke*, 1910–1926, trans. J. B. Greene and M. D. Herter (New York: Norton & Co, 1969), 343.

112 Carter, *Parrot*, 154.

113 Rosen, *Metaphysics in Ordinary Language*, 23.

114 Ibid., 24.

115 Ibid., 32.

116 The Neoplatonic theurgist conducts his ritual with the aid of 'signatures' (*sunthemata*). In these the divine energy is mimetically represented or embodied. *Sunthemata* are a kind of divine alphabet. They are illuminated matter, bearing the direct 'imprint of the sun god'. (Gregory Shaw, *Theurgy and the Soul: The Neoplatonism of Iamblichus* (Pennsylvania, 1995), 49).

117 Neilson, *The Collected Verse, A Variorum Edition*, 895.

118 Carter, *Ground Truthing*, 110–13.

119 Carter, 'Mac', unpublished script.

120 Carter, 'A2', notebook in the author' possession.

121 'Another Country: the village as childhood', *Poetry Nation Review* 13, no. 5 (1987), 54–9.

122 See Paul Carter, 'The Letter S' and David Rose', *The Western Desert Code: An Australian Cryptogrammar* (Pacific Linguistics, Australian National University, 2001), 253.

Chapter 8: Callings

1 Nora Post, 'Varèse, Volpe and the Oboe', *PNM*, Fall/Winter (1981), 139–41.

2 Gregory Whitehead, 'Principia Schizophonica: On Noise, Gas, and the Broadcast Disembody', *Art & Text*, no. 37 (1990), 60–2.

3 David Appelbaum, *Voice* (New York: SUNY Press, 1990), 4.

4 Jean-Jacques Cortine, 'Glossolalia: The Meaning of Nonsense', *Art & Text*, no. 37 (September 1990), 54–5.

5 This paragraph draws on Carter, *The Calling to Come*, which prints the script of the installation together with a commentary. It also draws on Paul

Carter, 'Speaking Pantomimes', chapter 6 of *Material Thinking*. The most accessible introduction to the languages that the colonists encountered in and around Sydney Harbour in 1788–95 remains Jakelin Troy, *The Sydney Language*. Troy also gives an excellent introduction to the Dawes material in 'The Sydney Language Notebooks and Responses to Language Contact in Early Colonial NSW', *Australian Journal of Linguistics* 12 (1992), 145–70. The manuscripts drawn on are William Dawes, *Grammatical forms of the language of N.S. Wales in the neighbourhood of Sydney*, c. 1790 and *Vocabulary of the language of N.S. Wales in the neighbourhood of Sydney*, c. 1790–2, both held in the School of Oriental and African Studies, University of London.

6 Appelbaum, *Voice*, 4.
7 Jean-François Cortine, 'Voice of Conscience and Call of Being', in *Who Comes after the Subject*, ed. E. Cadava, P. Connor and J.-L. Nancy (London: Routledge, 1991), 79–93, 88–9.
8 Dawes, *Vocabulary* and Troy, *The Sydney Language*, 73ff.
9 Anon, *Vocabulary of the languages of N.S. Wales in the neighbourhood of Sydney*, c. 1790–92, original held in the School of Oriental and African Studies, University of London. Equally significant is the absence of the interlinear glosses which, in Dawes, indicate an evolving understanding, an ever-present sense of context.
10 The fact that this little bird shared its name with Bennelong's daughter, Dilboong, indicates the poetic integration of calling and naming in this culture; but the insurgent grammarians were deaf to this likelihood.
11 Dawes, *Vocabulary*, 29 under *buna – to tease*. See also Troy, *The Sydney Language*, 73.
12 Jankélévitch, *L'Ironie*, 9.
13 Ernest Goffmann, *Asylums* (London: Penguin, 1968), 277–9.
14 See also discussion in Carter, *Material Thinking*, 163–70.
15 Anna Friz, 'Becoming Radio', in *Re-Inventing Radio: Aspects of Radio as Art*, ed. H. Grundmann et al. (Frankfurt am Main: Revolver, 2008), 87–102, 88–9. Friz quotes Gregory Whitehead and Dan Landers in this regard.
16 Friz, 'Becoming Radio', 92.
17 Ibid., 101.
18 Ibid.
19 Katja Kwastek, 'Art without Time and Space? Radio in the Visual Arts of the Twentieth and Twenty-First Centuries', in *Re-Inventing Radio: Aspects of Radio as Art*, ed. H. Grundmann et al. (Frankfurt am Main: Revolver, 2008), 131–46, 146.
20 Chantal Thomas, 'The War against the Mother Tongue', *Art & Text*, no. 37 (September 1990), 63–5, 64.
21 Maurice Merleau-Ponty, *The Prose of the World*, trans. John O'Neill (Evanston, IL: Northwestern University Press, 1973), 140.
22 Heinrich Schliemann, *Ilios* (New York: Arno Press, 1976 [orig.pub.1881], 9–11).

23 Schliemann, *Ilios*, 11.
24 It was another sound in the ether on my birthday, as Thomas composed his radio drama in those years, offering as one origin his desire to reassert the evidence of beauty in the world in the wake of the Hiroshima bombing. But I did not know these things then, and when I copied Thomas's sonorous tone, I was mainly thinking of my mother's affection for the poet, and the amusement my imitation would have given her.
25 Carter, 'What Is Your Name', 63 and 'The 7448', 209ff.
26 Paul Carter, 'The Empty Space is a Wall, the role of theatrical translation in the public reinscription of the other', *Performance Research* 10, no. 2 (2005), 79–91.
27 See Carter, *Material Thinking*, chapter 5 for a sustained example of this.
28 Carter, 'What Is Your Name', 56.
29 Carter, *Ground Truthing*, 246–64.
30 Klaus Neumann, *Shifting Memories: The Nazi Past in the New Germany* (Ann Arbor: University of Michigan Press, 2000), 5.
31 Neumann, *Shifting Memories: The Nazi Past in the New Germany*, 3.
32 The statement is found at http://www.academia.edu/14287159/The_Radio_of_the_Future_Redux_Rethinking_Transmission_Through_Experiments_in_Radio_Art_2011, viewed 18 April 2017.
33 See the Radio Graffiti website at http://www.radiograffiti.fr, viewed 1 August 2017.
34 Michele Bertomen, *Transmission Towers: On the Long Island Freeway – A Study of the Language of Form* (New York: Princeton Architectural Press, 1991), 9, quoted by Brandon LaBelle, 'Transmission Culture', in *Re-Inventing Radio, Re-Inventing Radio: Aspects of Radio as Art*, eds. H. Grundmann et al., (Frankfurt am Main: Revolver, 2008), 63–86, 70.
35 Geert Lovink, *Dark Fiber: Tracking Critical Internet Culture* (Boston: MIT Press, 2002), 271.
36 Bertomen quoted by LaBelle, 70.
37 Paul Carter, *Places Made After Their Stories, Design and the Art of Choreotopography* (Perth: University of Western Australia Publishing, 2015), 389.
38 Maria Grazia Guido, 'Mediating Linguacultural Asymmetries through ELF in Unequal Immigration Encounters', *Lingue e Linguaggi* 15 (2015), 155–75, 171.
39 Carter, 'The 7448', 260.
40 Sue Wright, *Language Policy and Language Planning: From Nationalism to Globalisation* (London: Palgrave Macmillan, 2016), 214. Wright comments on 'a striking absence of information about language use in refugee camps', and surmises that 'the ability to at least get by linguistically in the new settings must be key to transiting out of the worst situations in which the refugees and economic migrants find themselves'.

INDEX

Adorno, Theodor 74, 76, 162, 211
Akhmatova, Anna 11
Amundesham, John 13
Andreotti, Giulio 202
Appelbaum, David 225, 226
Arabanoo 72
Arch, Joseph 34, 35, 37, 105
Artaud, Antonin 73
Ascham, Roger 51, 52
Asturias, Miguel Ángel 64, 173, 174
Attali, Jacques 123
Auden, W.H. 8, 108

Bachelard, Gaston 10, 30, 142, 150, 162, 172, 173, 205
Bacon, Francis 3
Baggio, R. 70, 109, 136, 212
Bakhtin, Mikail 179
Ball, John 208
Barea, Arturo 103, 104, 105, 106, 107, 117, 122, 123, 130, 134, 190, 235
Barthes, Roland 204
Batman, John 155
Baudelaire, Charles 112
Becket, Thomas à 13
Beckett, Samuel 5, 55, 76, 151, 210
Beethoven, Ludwig van 198
Bennelong 72
Berardi, Franco 125
Berger, John 190
Berio, Luciano 4, 57, 58, 59, 60, 61, 74, 75, 139, 202, 216; 'Outis', 57, 58, 59, 60
Betjeman, John 31
Biasi, Dario di 147
Bidermann, Karl 232, 234, 235
Blackie, John Stuart 66

Blake, Barry 158
Bloch, Ernst 68
Blyton, Enid 101
Bonet, Eugeni 126, 127, 128
Bonwick, James 139
Bordonau, Antonio 106
Bororo (people) 64
Botticelli, Sandro 51
Boulanger, Nadia 108
Brecht, Bertolt 74, 75, 76, 106, 108, 214, 227
Bregman, Albert 43, 219
Britten, Benjamin 98

Cacciari, Massimo 137, 206
Cage, John 122, 123
Calvino, Italo 59
Canetti, Elias 45, 74, 124, 202
Carter, Edmund 148, 149
Carter, F ('Fred') 3, 9, 27, 30, 36
Carter, Margaret 81–102, 150, 151
Carter, Paul
  books: *Baroque Memories* 41, 61, 113, 212; *Ground Truthing* 221, 232; *The Lie of the Land* 50, 143; *Living In A New Country* 185; *Material Thinking* 69; *Meeting Place* 74; *Mythform* 148; *Parrot* 23; *The Road to Botany Bay* 1, 2, 3, 41, 49, 142, 222; *The Sound In-Between* 3, 136, 185, 186, 201
  public art: *Nearamnew* 147, 148; *Relay* 157
  scripts: 'The Calling to Come' 69, 137, 225, 226, 230; 'Columbus Echo' ['Eine Colombisches

Phantasie'] 3, 63; for sources 64, 174, 176, 177, 178, 228; 'Cooee Song' 4, 60, 136, 183, 200; 'Introducing Ulysses' 60; 'Jadi Jadian' 99, 156; 'The Letter S' 4, 202, 213, 215, 216, 217, 223; 'Light' 82, 83, 85, 86, 87, 97, 156; 'Lost Subjects' 4, 69, 138; 'Mac' 221, 222; 'Memory as Desire' 2, 41, 60, 65, 66, 67, 76, 135, 140, 167, 168, 169, 170, 173, 181, 183, 199, 210, 213, 214, 216; 'Mirror States' 3, 54, 149, 150, 152, 154, 155, 156, 202, 203, 204, 205; 'Named in the Margin' 4, 69; 'The Native Informant' 2, 71, 72, 76, 98, 137, 162, 171, 173, 177, 179, 230; 'Nearamnewspeak' 138; 'Old Wives Tales' 156, 157; 'On the Still Air' 2, 52, 112, 202; 'Relay for Radio' 138; 'Remember Me' 2, 25, 26, 44, 76, 190, 194, 195, 196; 'Scarlatti' 2, 5, 76, 77, 106, 107, 114, 116, 117, 119, 120, 121, 129, 133, 134; 'The 7448' 4, 63, 67, 177, 178, 230, 231, 237; 'Siren Sonata' 114, 115, 117, 118, 119, 120, 124, 126, 129, 131; 'Tuned Noises' 42, 54; 'Underworlds of Jean du Chas' 5, 54, 55, 56, 67, 82, 83, 84, 151, 210, 211, 212, 213, 237; 'What Is Your Name' ['Wie ist ihr Name', 'Wie ist dein Name', 'Quel nom toi'] 2, 41, 135, 178, 179, 180, 183, 193, 202, 216, 221, 230, 231, 232, 233, 234, 235, 236, 237, 238
Carter, W.C.M. ('Bill') 27, 30, 36, 56, 133, 187, 188, 189
Cassidy, Frederick 174
Cavalcanti, Guido 50, 52, 53, 73, 112, 212
Cernuda, Luis 130, 131, 132
Chandrabhanu 156, 157
Cheetham, Tom 183
Chesterton, G.K. 20
Churchill, Winston 25
Cixous, Hélène 156
Cobbett, William 17, 19, 33, 47
Cole, G.D.H. 18, 36
Coleby 72
Collins, David 2, 50
Collins, Randell 124
Columbus, Cristofero 62, 63, 64, 65, 136
Coward, T.A. 173
Cox, Christoph 122, 123
Critchett, Jan 157
Curran, Alvin 152

D'Annunzio, Gabriele 50, 51
Dante, Alighieri 50, 52, 53, 112, 210, 212
Davies, Paul 103
Davis, Miles 128
Dawes, William 72, 137, 226, 227
Dawson, Isabella 157, 158
Dawson, James 37, 47, 157, 158, 159, 162, 163
Debussy, Claude 105
Delaunay, Robert 55
Demnig, Gunter 233
Derrida, Jacques 149, 150
Descartes, René 79, 143
Dickinson, Emily 2
Didi-Huberman, Georges 108, 126, 127
Dixon, R.M.W. 184
Djabwurrung (Tjapwurrung) 158
Djadjawurrung (people) 233
Drechsel, Emanuel 71, 175
Duck, Stephen 189
Dvorak, Antonin 198

Eckhart, Meister 206
Eco, Umberto 75, 153
Eliot, George 104
Eliot, T.S. 50, 52, 69, 115

Feld, Steven 73, 124
Fénelon, François 229
Fichte, Hubert 178

INDEX

Finnegans Wake, see Joyce, James
Flagstad, Kirsten   25
Fliess, Wilhelm   51
Flinders, Matthew   143
Foster, Mary LeCron   142
Fox, George   20
Franco, Francisco   77, 104, 108, 115, 116, 117
Freud, Sigmund   51, 75
Friz, Anne   227, 232, 235
Fuertes, Gloria   112

Galeano, Eduardo   62
Galvan, Israel   126, 127
Garfías, Pedro   133
Garvey, Newton   149
Gebser, Jean   184, 185, 186
Gibson, Ian   117
Gilbert, Les   61, 68
Ginzberg, Carlo   181
Gluck, Christoph Willibald   22, 190
Godwin, Joscelyn   53
Goffman, Ernest   227
Gompertz, Terry   173
Grainger, Percy   169
Graves, Robert   94, 95, 112, 121, 164
Grey, Sir George   184, 185
Grock (clown)   58
Guattari, Félix   125
Gunditjmara (people)   158, 161
Gurney, Ivor   112
Gutzmann, Hermann   75, 76, 235

Hagelüken, Andreas   68, 75
Hammond, John and Barbara   47
Handel, Stephen   44
Hardy, Thomas   91, 92, 98
Harrison, Martin   6, 21, 27, 32, 67, 135, 140, 141, 144, 146, 169, 204, 205, 213, 216
Heidegger, Martin   134, 150, 192, 226, 227, 235
Henderson, Gavin (Lord Faringdon)   104, 130, 133
Heraclitus   146
Hernandez, Miguel   144
Herzen, Alexander   47
Hillman, James   109

Holland, Agniewska   26
Hollom. P.A.D.   38
Holtby, Winifred   35, 36
Holton, Kimberley DaCosta   109
Homer   57, 122; *The Odyssey*   57, 84
Horace (Quintus Horatius Flaccus)   122
Hudson, William Henry   31, 40
Hughes, Thomas   40
Husserl, Edmund   75
Huxley, Aldous   19

Jakobson, Roman   43, 144
James, William   192
Jankélévitch, Vladimir   110, 111
Jaynes, Julian   45
Jefferies, Richard   7, 8, 31, 32, 187
Johnson, Samuel   19
Joyce, James   5, 58; *Finnegans Wake*   57
Jung, Carl   207, 208

Kaawirn Kuunnawarn   157
Kafka, Franz   206
Kant, Immanuel   116
Kantor, Tadeusz   69, 70, 72, 175
Keats, John   14, 153, 192
Kefala, Anigone   46, 69
Keil, Charles   124
Kelly, Kevin   125
Kerr, John Hunter   233
Khlebnikov, Velimir   30
Kierkegaard, Søren   211
Kimball, Kathleen   182
Kircher, Athanasius   75
Kirkpatrick, Ralph   76, 106, 107, 109, 120
Kittler, Friedrich   75, 123
Klee, Paul   113
Koch, Ludwig   12
Koekkoek, M.A.   39
Kraus, Karl   134
Krishna, Ananda   128

LaBelle, Brandon   74
Lacoue-Labarthe, Philippe   138
Laennec, René   82, 83
Lamartine   144

Lander, Dan 76, 119, 214
Larkin, Phillip 31
Leaman, Michael 219
Leopardi, Giacomo 144
Levinas, Emanuel 191
Light, William 82, 85, 86, 90, 97
Lispector, Clarice 156
Lope de Vega y Carpio, Félix 104, 105, 121
Lopes, Pedro 121
Lorca, Federico García 104, 105, 107, 108, 112, 115
Loti, Pierre 210
Louganis, Greg 209
Lourandos, Harry 159, 160
Lucretius (Titus Carus) 51

Machado, Antonio 111, 112
Machado, Manuel 111
McLennan, Andrew 67, 213
Madsen, Virginia 4, 73, 74, 214
Mairena, Antonio 107
Mallarmé, Stéphane 112
Malouf, David 49
Mannoni, Octave 63, 66, 208
March, Ausiàs 112, 113, 114, 128
Marion, Jean-Luc 53
Marshall, William 15
Martin, Esau ('Charles') 31, 34
Maxwell, James Clerk 192, 193
Menéndez Pidal, Ramón 112
Merleau-Ponty, Maurice 21, 30, 205, 229
Miles, Bernard 25
Miller, J. Hillis 161
Mitchell, James 158
Møller, Poul Martin 193
Montemayor, Jorge de 114
Morente, Enrique 128
Moyle, Alice 139, 166
Müller, Max 46, 47
Mumford, Lewis 103

Nancy, Jean-Luc 147
Nebrija, Antonio de 62, 201
Neilson, John Shaw 47, 48, 112, 144, 145, 146, 166, 221

Neruda, Pablo 133
Neumann, Klaus 234
Nichanian, Marc 180
Nietzsche, Friedrich 212
Njulkar, Maurice Tjakurl 146
Novalis (Georg Friedrich Philipp Freiherr von Hardenberg) 89, 101

Odyssey, The, see Homer
Ogeil, Jacqueline 108
Ossola, Carlo 213

Pagès, Amadeu 114
Pasolini, Pier Paolo 202
Patyegarang (Eora woman) 137, 227
Peake, Mervyn 8
Pepperberg, Irene 219, 220
Percy, Thomas 112
Pessoa, Fernando 55, 77, 118
Petrie, George 27, 168
Phaedrus, The 74
Phillip, Arthur 227
Pingree, Geoffrey 118
Pitjantjatjara (people) 48, 49, 170
Plutarch (Lucius Mestrius) 211
Poe, Edgar Allan 51
Polo, Marco 65
Porcía, Count Gian Artico di 79
Pound, Ezra 50, 52, 77, 112
Propp, Vladimir 57, 59, 61
Proust, Marcel 207, 209, 212
Purcell, Henry 25, 26, 190
Pye, Henry James 189
Pyne, William Henry 182

Quincey, Thomas de 49, 50, 51, 75, 165

Radhov, V.V. 59
Ravel, Maurice 105, 110, 111
Read, Herbert 223
Richter, Sviatoslav 191
Riley, James 33
Ritz, Hermann B. 46, 49, 63, 139, 142, 171
Robinson, George Augustus 65, 167, 182, 199, 200

Robinson, Rose 167
Rohner, Felix 128
Rolle, Richard 14
Romero, Pedro G. 127
Ronell, Avital 235
Rose, Walter 105
Rosen, Stanley 147, 220
Rūmī, Jalāl ad-Din Muhammad 199, 219

Sachs, Nelly 235
Sanguineti, Edoardo 60
Sapir, Edward 44
Sappho 145, 148
Scarlatti, Alessandro 78
Scarlatti, Domenico 5, 6, 76, 77, 106, 108, 109, 110, 111, 120, 121, 124, 126, 129, 222, 227
Schaefer, Pierre 123
Schafer, Murray R. 42, 43, 72, 73
Schärer, Sabina 128
Schliemann, Heinrich 229, 233, 234
Schöning, Klaus 68
Schopenhauer, Arthur 13
Schreber, Daniel Paul 198, 200, 202
Schuchardt, Hugo 62
Schultz, Andrew 60
Schwartz, Steve 75
Selous, Edmund 33
Shakespeare, William 153, 197, 205
Shaw, George Bernard 36
Shelley, Percy Bysshe 22, 23, 87, 97
Sigar (monk) 13, 14, 16, 24
Silverstein, Michael 62, 176
Sitwell, Sacheverell 109, 110
Skeat, William 207
Smith, Bruce 197
Smith, Fanny Cochrane 135, 139, 165, 171
Smith, F. Joseph 205
Smith, P. Christopher 150
Smyth, Robert Brough 158, 162
Sorabji, Kaikhosru Shapurji 191
Spenser, Edmund 153, 154, 155
Steiner, George 108
Sterne, Jonathan 183
Stewart, Dugald 66

Stokes, Adrian 94
Stokes, John Lort 103, 142
Strehlow, T.G.H. 186
Strindberg, August 202
Sturt, Charles 65, 66, 140, 170
Sturt, Charlotte 167, 182, 200
Sturt, George 31

Taino (people) 64, 65, 136
Taplin, George 161, 201
Tarkovsky, Andrei 169
Tate, Nahum 26
Tench, Watkin 2, 4
Thomas, Caitlin 95
Thomas, Chantal 228
Thomas, Dylan 230
Thomas, Edward 112, 132
Thomas, William 155
Timarchus of Chaeronia, see Plutarch
Tomas, David 73, 136
'The Cuckoo and the Nightingale' 14
Traherne, Thomas 16, 32
Truax, Barry 25, 27
Tsvetayeva, Marina 10, 150

Unamuno, Miguel 134
Ungaretti, Giuseppe 202, 207, 208, 212, 213

Valamanesh, Hossein & Angela 138
Val del Omar, José 126, 127
Valèry, Paul 143
Van Eikels, Kai 124
Varèse, Edgard 8, 225
Verlaine, Paul 143, 210
Vertov, Dziga 127
Vico, Giambattista 4, 58, 77, 78, 79–80, 108, 121, 192, 197
Vygotsky, Lev 149, 153

Wagner, Roy 67, 69, 184, 205
Waller, Steven 182
Watson, Horace 139, 165
Westerkamp, Hildegaard 27
Whalen, D.H. 71
Whewell, William 131
White, Gilbert 15, 38, 94, 164

Whitehead, Gregory   225, 227
Whitman, Walt   151
Williams, Alfred   19, 20, 21, 24, 25, 31, 40, 42, 168
Williams, Christopher   217, 218, 221
Wittgenstein, Ludwig   149
Woiwurrung, *see* Wurundjeri
Wolfson, Louis   228
Wombeet Tuulawarn   158

Worgan, George   3, 227
Wright, Thomas   112
Wurundjeri (people)   54, 150

Yalgear   227
Yaruun Parpur Tarneen   158
Yawuru (people)   145

Zenith, Richard   118

www.ingramcontent.com/pod-product-compliance
Lightning Source LLC
Chambersburg PA
CBHW050337230426
43663CB00010B/1895